Ending the Mendel-Fisher Controversy

In 1859 I obtained a very fertile descendant with large, tasty seeds from a first generation hybrid. Since in the following year, its progeny retained the desirable characteristics and were uniform, the variety was cultivated in our vegetable garden, and many plants were raised every year up to 1865.

> —Gregor Mendel to Carl Nägeli, April 1867, in *The Origin of Genetics: A Mendel Source Book,* ed. C. Stern, C. and E. R. Sherwood, 61–62. San Francisco: W. H. Freeman and Co., 1966.

"Brother Mendel! We grow tired of peas!"

Ending the Mendel-Fisher Controversy

Allan Franklin, A. W. F. Edwards, Daniel J. Fairbanks, Daniel L. Hartl, and Teddy Seidenfeld

University of Pittsburgh Press

Published by the University of Pittsburgh Press, Pittsburgh PA 15260

Copyright © 2008, University of Pittsburgh Press

All rights reserved

Manufactured in the United States of America

Printed on acid-free paper

10 9 8 7 6 5 4 3 2 1

LIBRARY OF CONGRESS CATALOGING-IN-PUBLICATION DATA

Ending the Mendel-Fisher controversy / Allan Franklin . . . [et al.].

    p. ; cm.

    Includes bibliographical references and index.

    ISBN-13: 978-0-8229-4319-8 (cloth : alk. paper)

    ISBN-10: 0-8229-4319-0 (cloth : alk. paper)

    ISBN-13: 978-0-8229-5986-1 (pbk. : alk. paper)

    ISBN-10: 0-8229-5986-0 (pbk. : alk. paper)

    1. Genetics—History.   2. Hybridization, Vegetable.   3. Mendel, Gregor, 1822–1884.   4. Fisher, Ronald Aylmer, Sir, 1890–1962.   I. Franklin, Allan, 1938–

    [DNLM: 1. Mendel, Gregor, 1822–1884.   2. Fisher, Ronald Aylmer, Sir, 1890–1962.   3. Genetics—history.   4. History, 19th Century.   5. Hybridization, Genetic.   6. Scientific Misconduct—history.   QU 11.1 E56 2008]

    QH428.E53 2008

    576.5'2—dc22

                                                2007041848

The History of Science has suffered greatly from the use by teachers of second-hand material, and the consequent obliteration of the circumstances and the intellectual atmosphere in which the great discoveries of the past were made. A first-hand study is always instructive, and often . . . full of surprises.

> —R. A. Fisher, 1955. In *Experiments in Plant Hybridisation by Gregor Mendel, with Commentary and Assessment by Sir Ronald A. Fisher*, ed. J. H. Bennett, 6. Edinburgh: Oliver and Boyd, 1965.

CONTENTS

	Preface	ix
CHAPTER 1	The Mendel-Fisher Controversy: An Overview ALLAN FRANKLIN	1
CHAPTER 2	Experiments in Plant Hybridisation GREGOR MENDEL	78
CHAPTER 3	Has Mendel's Work Been Rediscovered? R. A. FISHER	117
CHAPTER 4	Are Mendel's Results Really Too Close? A. W. F. EDWARDS	141
POSTSCRIPT TO CHAPTER 4	Alternative Hypotheses and Fisher's *The Design of Experiments* A. W. F. EDWARDS	164
CHAPTER 5	Controversies in the Interpretation of Mendel's Discovery VÍTĚZSLAV OREL and DANIEL L. HARTL	167
POSTSCRIPT TO CHAPTER 5	Amiable Wright and Suspicious Fisher on Mendel's "Personal Equation" DANIEL L. HARTL	208
CHAPTER 6	P's in a Pod: Some Recipes for Cooking Mendel's Data TEDDY SEIDENFELD	215
POSTSCRIPT TO CHAPTER 6	A Brief Account of a Trial Conducted at Pillsbury Labs, 2000–2001 TEDDY SEIDENFELD	258

CHAPTER 7	Mendelian Controversies: A Botanical and Historical Review DANIEL J. FAIRBANKS and BRYCE RYTTING	264
POSTSCRIPT TO CHAPTER 7	Mendelian Controversies: An Update DANIEL J. FAIRBANKS	302
APPENDIX	Probability, the Binomial Distribution, and Chi-square Analysis DANIEL J. FAIRBANKS	313
	List of Contributors	325
	Index	327

PREFACE

Gregor Mendel is justly admired as the founder of modern genetics. His experiments on pea plants, reported in 1865, established the principles of segregation and independent assortment, which form the basis of that field of study. Yet Mendel's reputation remains under a cloud. This is due to a 1936 paper by R. A. Fisher, the distinguished British statistician and geneticist, in which he reanalyzed Mendel's data and concluded that they fit his hypotheses too well. They were too good to be true. Fisher concluded that "the data of most, if not all, of the experiments have been falsified so as to agree closely with Mendel's expectations." (To be fair, Fisher attributed the falsification not to Mendel, but to an unnamed assistant.) It was this paper that created the cloud over Mendel's reputation and that, after some time, initiated the Mendel-Fisher controversy.

It is well known that Mendel's work was neglected until its rediscovery in 1900. Similarly, Fisher's paper was overlooked for some 28 years, until just about the centenary of Mendel's work. Since then the Mendel-Fisher controversy has simmered quietly, and sometimes not so quietly, for more than 40 years. It is still ongoing with both attacks on, and defenses of, both Mendel and Fisher.

It is our contention that this controversy should end. The purpose of this book is to present the reader with good reasons for that conclusion. The book includes an overview of the controversy, the original papers of Mendel and Fisher, and four of the most important papers on the controversy (in the judgment of Allan Franklin, who suggested which papers should be included). The book also includes brief updates, by the authors, of the latter four papers. This is not to say that all questions have been answered, but rather that we believe that the most important ones that can be answered have been answered. Barring a miraculous rediscovery of Mendel's notebooks, which are said to have been burned, and which would certainly answer the question of fraud, there seems to be little more of value to be said.

More specifically, we conclude that: (1) Mendel was not guilty of fraud; (2) Fisher's conclusion, based on χ^2 analysis, that Mendel's data fit his expectations extraordinarily well is correct, but may be explained without invoking fraud; (3) Fisher's criticism of Mendel, that in the experiments on the second generation of plants bred from hybrids the ratio of heterozygous to homozygous offspring should not be 2:1 but rather 1.7:1, is incorrect; and, finally, that (4) Fisher had great admiration for Mendel and his work and that he would have been quite unhappy with those who used his work to diminish Mendel's achievements.

It should be emphasized that this is a collaborative work. Since its initial suggestion, we have each benefited from discussions and visits with one another, which have been of great value to all of us. This has not only improved the book, but has also led to a new paper on the controversy by Daniel Hartl and Daniel Fairbanks (2007). Our collaboration has been both enjoyable and rewarding.

The previously published papers that appear in this volume have been reproduced using their original spellings, styles, and reference systems. Stylistic consistency has been imposed only across the new material (the introduction, postscripts, and appendix).

One change has been made to the original articles: this book is intended to be self-contained, so that all page references to papers that are included in this volume refer to those versions contained in this book (even when alternate translations are cited), and these page numbers are noted in italics for ease of use.

Ending the Mendel-Fisher Controversy

CHAPTER 1

■ The Mendel-Fisher Controversy
An Overview

ALLAN FRANKLIN

Gregor Mendel (1822–1884) is regarded as the founder of modern genetics. His experiments on pea plants reported in 1865 established the principles of segregation and of independent assortment. The former states that variation for contrasting traits is associated with a pair of factors that segregate to individual reproductive cells. The latter states that two or more of these factor-pairs assort independently to individual reproductive cells. It is well known that Mendel's work was neglected until its "rediscovery" in 1900 by Hugo de Vries, Carl Erich Correns, and Erich von Tschermak. It is less well known, however, that in 1936, the great British statistician and biologist R. A. Fisher analyzed Mendel's data and found that the fit to Mendel's theoretical expectations was too good (Fisher 1936). Using χ^2 analysis, Fisher found that the probability of obtaining a fit as good as Mendel's was only 7 in 100,000. Fisher also argued that because Mendel used only a limited sample of 10 plants in his experiment to determine the ratio of heterozygous plants (Aa) to homozygous plants (AA) in the F_2 generation produced by the self-pollination of hybrids, there was a 5.63% chance of misidentifying heterozygous plants as being homozygous. Thus, the ratio should be approximately 1.7 to 1, rather than Mendel's expectation of 2 to 1, although Mendel's data agreed more closely with the 2 to 1 ratio. Fisher concluded: "This possibility is supported by independent evidence that the data of most, if not all, of the experiments have been falsified so as to agree closely with Mendel's expectations" (134).[1] Fisher did not believe that Men-

1

del was responsible for the falsification, but attributed it to an unknown assistant.

Fisher's work was overlooked. The first published comments on it appeared in 1964, about the time of the centenary of Mendel's paper, and since then at least 50 papers, letters, and discussions have been published on the controversy as to whether Fisher adequately showed that Mendel's data were falsified. These publications include explanations of Mendel's results and both criticisms and defenses of Fisher.

This chapter will provide an overview of that controversy, including summaries of both Mendel's and Fisher's papers, along with a discussion of most of the papers on the debate. It is not, however, a substitute for reading the original works. Therefore, this book contains the work of both Mendel and Fisher as well as four of the most significant discussions of the controversy, and updates by those four authors. I believe that taken together, these voices argue for an end to the controversy.

Mendel's Experimental Results

Mendel began his experiments on garden peas (*Pisum sativum* L.) in 1856 and continued them until 1863, a period of approximately eight years. His stated purpose was to investigate whether there was a general law for the formation and development of hybrids, something he noted had not yet been formulated:

Those who survey the work done in this department will arrive at the conviction that among all the numerous experiments made, not one has been carried out to such an extent and in such a way as to make it possible to determine the number of different forms under which the offspring of hybrids appear, or to arrange these forms with certainty according to their separate generations, or definitely to ascertain their statistical relations.

It requires indeed some courage to undertake a labour of such far-reaching extent; this appears however, to be the only right way by which we can finally reach the solution of a question the importance of which cannot be overestimated in connection with the history of the evolution of organic forms. (79)[2]

Mendel proposed to remedy the situation and did so. As Fisher remarked, "Mendel's paper is, as has been frequently noted, a model in respect of the order and lucidity with which the successive relevant facts are presented" (Fisher 1936, 121). I will follow Mendel's plan in describing his experiments and will allow Mendel to speak for himself as much as possible.

In order to carry out such experiments successfully, Mendel required: "The experimental plants must necessarily—1. Possess constant differentiating characters. 2. The hybrids of such plants must, during the flowering period, be protected from the influence of all foreign pollen, or be easily capable of such protection. The hybrids and their offspring should suffer no marked disturbance in their fertility in the successive generations" (79). He further noted: "In order to discover the relations in which the hybrid forms stand towards each other and also towards their progenitors it appears to be necessary that all members of the series developed in each successive generation should be, *without exception*, subjected to observation" (79–80).

[F_2] The First Generation [Bred] from the Hybrids

Mendel began with 34 varieties of peas, from which he selected 22 varieties for further experiments. He had confirmed, in two years of experimentation, that these varieties bred true. He reported experiments on seven characters that had two easily distinguishable characteristics. I have listed these below, with the dominant form first:[3]

   1. Seed shape: round or wrinkled
   2. Cotyledon color: yellow or green
   3. Seed-coat color: colored (gray, gray-brown, or leather-brown) or white. Colored seed coats were always associated with violet flower color and reddish markings at the leaf axils. White seed coats were associated with white flowers.
   4. Pod shape: inflated or constricted
   5. Pod color: green or yellow
   6. Flower position: axial (along the stem) or terminal (at the end of the stem)
   7. Stem length: long (six to seven feet) or short (three-quarters of a foot to one and a half feet)

The first two are seed characters because they are observed in seed cotyledons, which consist of embryonic tissue. Each seed is thus a genetically different individual and such characters may differ among the seeds produced on a heterozygous plant. Both yellow and green seeds may be observed on a single heterozygous plant. One may, in fact, observe these characters for the next generation, without the necessity of planting the seeds. The latter five are plant characters. As William Bateson remarked, "It will be observed that the [last] five are *plant-characters*. In order to see the result of crossing, the seeds must be sown and allowed to grow into plants. The [first] two characters belong to the *seeds* themselves. The seeds of course are members of a generation later than that of the plant which bears them" (Bateson 1909, 12). Because of this, Mendel would have had a reasonable expectation of what the results of his plant character experiments would be from his observations of the seed characters, before the plants of the next generation were grown.

Mendel's first experiment was to breed a generation of hybrids from his true breeding plants for each of the seven characters. His results for this generation (F_1) clearly showed dominance. He remarked: "In the case of each of the seven crosses the hybrid-character resembles that of one of the parental forms so closely that the other either escapes observation completely or cannot be detected with certainty" (*84*).

He then allowed these monohybrids to self-fertilize. He found a 3 : 1 ratio for plants that showed the dominant character to those that possessed the recessive character in this generation (F_2).[4] He found that this ratio held for all the characters observed in the experiments and that "*Transitional forms were not observed in any experiment*" (*85*). His results are shown in table 1.1. He concluded, "If now the results of the whole of the experiments be brought together, there is found, as between the number of forms with the dominant character and recessive characters, an average ratio of 2.98 to 1, or 3 to 1" (*87*).

Mendel also noted that the distribution of characters varied in both individual plants and in individual pods. He illustrated this with data from the first ten plants in the seed character experiments (see table 1.2). The variation in both the ratios of the characters and in the number of seeds per plant is considerable. Mendel also presented the extreme variations.

TABLE 1.1 Mendel's results for the F_2 generation of monohybrid experiments (from data in Mendel 1865, 85–87)

Trait	Dominant	Number	Recessive	Number	Ratio
			Expected ratio 3 : 1		
1. Seed shape	Round	5474	Angular	1850	2.96
2. Cotyledon color	Yellow	6022	Green	2001	3.01
3. Seed coat color	Colored	705	White	224	3.15
4. Pod shape	Inflated	882	Constricted	299	2.95
5. Pod color	Green	428	Yellow	152	2.82
6. Flower position	Axial	651	Terminal	207	3.14
7. Stem length	Long	787	Short	277	2.89
Total	Dominant	14,949	Recessive	5010	2.98

TABLE 1.2 Mendel's results for the first 10 plants in the experiments on seed shape and seed color (from data in Mendel 1865, 86)

Plant	Experiment 1: Shape of seeds		Experiment 2: Coloration of albumen	
	Round	Wrinkled	Yellow	Green
1	45	12	25	11
2	27	8	32	7
3	24	7	14	5
4	19	10	70	27
5	32	11	24	13
6	26	6	20	6
7	88	24	32	13
8	22	10	44	9
9	28	6	50	14
10	25	7	44	18
Ratio	3.33 : 1		3.08 : 1	

Note: The fact that the number of seeds in each plant differs for each numbered plant shows clearly that the plants for Experiments 1 and 2 are different plants. Thus, plant 1 in Experiment 1 has 57 seeds, whereas plant 1 in Experiment 2 has 36 seeds.

"As extremes in the distribution of the two seed characters in one plant, there were observed in Expt. 1 an instance of 43 round and only 2 angular, and another of 14 round and 15 angular seeds. In Expt. 2 there was a case of 32 yellow and only 1 green seed, but also one of 20 yellow and 19 green" (86). Mendel was clearly willing to present data that deviated considerably from his expectations.[5]

Mendel also noted: "In well-developed pods which contained on the average six to nine seeds, it often happened that all the seeds were round (Expt. 1) or all yellow (Expt. 2); on the other hand, there were never observed more than 5 wrinkled or five green ones in one pod" (86).[6]

[F_3] The Second Generation [Bred] from the Hybrids

At the end of the section describing the first-generation experiments, Mendel remarked that the dominant character could have a "*double signification.*" It could be either a pure parental (dominant) character or a hybrid character. "In which of the two significations it appears in each separate case can only be determined by the following generation. As a parental character it must pass over unchanged to the whole of the offspring; as a hybrid-character, on the other hand, it must maintain the same behaviour as in the first generation" (87).[7] He further noted that those plants that show the recessive character in the first generation (F_2) do not vary in the second generation (F_3).[8] They breed true. That was not the case for those plants showing the dominant character: "Of these *two*-thirds yield offspring which display the dominant and the recessive characters in the proportion of 3 to 1, and thereby show exactly the same ratio as the hybrid forms, while only *one*-third remains with the dominant character constant"(88). In other words, of those F_2 generation plants showing the dominant character, two-thirds were heterozygous (*Aa*), or hybrid, and one third homozygous (*AA*). For the seed characters Mendel reported the following results: (1) from 565 plants raised from round seeds, 372 produced both round and wrinkled seeds in the proportion of 3 to 1 whereas 193 yielded only round seeds, a ratio of 1.93 to 1; (2) for plants raised from yellow seeds, 353 yielded both yellow and green seeds in the proportion 3 to 1, whereas 166 yielded only yellow seeds, a ratio of 2.13 to 1.

The experiments on plant characters required more effort: "For each separate trial in the following experiments [on plant characters] 100 plants were selected which displayed the dominant character in the first generation [F_2], and in order to ascertain the significance of this, ten [F_3] seeds of each were cultivated" (88).[9] A plant was classified as homozygous if all of the 10 offspring had the dominant character and classified as heterozygous otherwise.[10] Mendel's results for the plant characteristics are shown in table 1.3. Mendel noted that the first two experiments on seed characters were of special importance because of the large number of plants that could be compared. Those experiments yielded a total of 725 hybrid plants and 359 dominant plants that "gave together almost exactly the average ratio of 2 to 1" (89). Experiment 6 also yielded almost the exact ratio expected, whereas for the other experiments, as Mendel noted, "the ratio varies more or less, as was only to be expected in view of the smaller number of 100 trial plants" (89). Mendel was, however concerned about Experiment 5 (the color of unripe pods), in which the result was 60 to 40. He regarded these numbers as deviating too much from the expected 2 to 1 ratio.[11] Mendel repeated the experiment and obtained a ratio of 65 to 35, and was satisfied: "*The average ratio of 2 to 1 appears, therefore, as fixed with certainty*" (89). It is clear that Mendel did not attempt to hide any of his results, especially those that deviated from his expectations, because he presented the results for both the original Experiment 5 as well as its repetition. The sum totals for the six plant characteristic experiments, in-

TABLE 1.3 Mendel's results for the heterozygous-homozygous experiment (the 2 to 1 experiment) (from data in Mendel 1865, 88)

Experiment	Dominant	Hybrid
3. Seed coat color (grey-brown or white)	36	64
4. Pod shape (smooth or constricted)	29	71
5. Pod color (green or yellow)	40	60
6. Flower location (axillary or terminal)	33	67
7. Stem length (long or short)	28	72
8. Repetition of Experiment 5	35	65
Total	201	399
	Ratio (hybrid to dominant) 1.99	

cluding the repetition of Experiment 5, were 399 (hybrid) to 201 (dominant), or 1.99 to 1.

Mendel's conclusion was quite clear:

> The ratio of 3 to 1, in accordance with which the distribution of the dominant and recessive characters results in the first generation, resolves itself into a ratio of 2:1:1 if the dominant character be differentiated according to its significance as a hybrid-character or as a parental one. Since the members of the first generation [F_2] spring directly from the seed of the hybrids [F_1], *it is now clear that the hybrids form seeds having one or the other of the two differentiating characters, and of these one-half develop again the hybrid form, while the other half yield plants which remain constant and receive the dominant or the recessive characters, [respectively], in equal numbers.* (89)

The Subsequent Generations Bred from the Hybrids

Mendel suspected that the results he had obtained from the first and second generations produced from monohybrids were probably valid for all of the subsequent progeny. He continued the experiments on the two seed characters, shape and color, for six generations; the experiments on seed-coat color and stem length for five generations; and the remaining three experiments on pod shape, color of pods, and position of flowers for four generations, "and no departure from the rule has been perceptible. The offspring of the hybrids separated in each generation in the ratio of 2:1:1 into hybrids and constant forms [pure dominant and pure recessive]" (89). He did not, however, present his data for the experiments on the subsequent generations.¹² He went on to state, "If *A* be taken as denoting one of the two constant characters, for instance the dominant, *a*, the recessive, and *Aa* the hybrid form in which both are conjoined, the expression *A* + 2*Aa* + *a* shows the terms in the series for the progeny of the hybrids of two differentiating characters" (89).¹³

The Offspring of Hybrids in which Several Differentiating Characters Are Associated

Mendel's next task, as he put it, was to investigate whether the laws he had found for monohybrid plants also "applied to each pair of differentiating characters when several diverse characters are united in the hybrid by crossing" (90).

He went on to describe the experiments. "Two experiments were made with a considerable number of plants. In the first experiment the parental plants differed in the form of the seed and in the colour of the albumen; in the second in the form of the seed, in the colour of the albumen, and in the colour of the seed-coats. Experiments with seed characters give the result in the simplest and most certain way" (91). He was no doubt referring to the greater number of seeds than plants, which provides data with greater statistical significance, and also to the fact that the shape of the seeds and the color of albumen (cotyledons) could be seen in the second generation, without the need to plant a third generation. Daniel Fairbanks and Bryce Rytting (2001) later remarked with reference to seed-coat color, which, as noted above, was correlated with the presence or absence of axillary pigmentation, could be scored in seedlings, and was also used as the third factor in the trifactorial experiment: "Because this trait can be scored in seedlings, it is an excellent choice for the third trait in the trihybrid experiment because it creates at most a three-week delay between data collection for the first two traits and the third. Garden space is not as critical because many seedlings can be grown in the space occupied by a single mature plant" (276).

In these experiments Mendel distinguished between the differing characters in the seed plant and the pollen plant. *A*, *B*, and *C* represented the dominant characters of the seed plant and *a*, *b*, and *c* the recessive characters of the pollen plant, with hybrids represented as *Aa*, *Bb*, and *Cc*.¹⁴

First Experiment (Bifactorial)

Mendel's first experiment used two seed characters in which the seed plant (*AB*) was *A* (round shape) and *B* (yellow cotyledon), and the pollen plant (*ab*) was *a* (wrinkled shape) and *b* (green albumen). The fertilized

TABLE 1.4   Mendel's results for the bifactorial experiment (from Mendel 1865, 91–92)

	A (round)	Aa (hybrid)	a (angular)
B	AB (round, yellow): 38	AaB (round yellow and angular yellow): 60	aB (angular, yellow): 28
Bb	ABb (round yellow and green): 65	AaBb (round yellow and green and angular yellow and green): 138	aBb (angular yellow and green: green): 68
b	Ab (round green): 35	Aab (round and angular green): 67	ab (angular green): 30

seeds were all round and yellow, as expected. He then raised plants from these seeds and obtained 15 plants with 556 seeds distributed as follows:

   315 round and yellow
   101 wrinkled and yellow
   108 round and green
   32 wrinkled and green[15]

All of these seeds were planted in the following year and Mendel's results are shown in table 1.4.

Mendel separately recorded the results for each set of the 556 seeds (i.e., round and yellow, round and green, wrinkled and yellow, wrinkled and green).[16] He noted that there were nine different forms (we would say genotypes) and classified them this way:

> The whole of the forms may be classed into three essentially different groups. The first includes those with the signs *AB, Ab, aB, ab*: they possess only constant characters and do not vary again in the next generation. Each of these forms is represented on the average thirty-three times. The second group includes the signs *ABb, aBb, AaB, Aab*: these are constant in one character and hybrid in another, and vary in the next generation only as regards the hybrid-character. Each of these appears on an average sixty-five times. The form *AaBb* occurs 138 times: it is hybrid in both characters, and behaves exactly as do the hybrids from which it is derived.
>
> If the numbers in which the forms belonging to these classes appear to be compared, the ratios of 1, 2, 4 are unmistakably evident. The numbers 32, 65, 138 present very fair approximations to the ratio numbers of 33, 66, 132. (92)

Mendel had a very good feel for his data and an ability to see the underlying patterns in his results despite statistical fluctuations. Mendel concluded that these results "indisputably" showed that the results could be explained by the combination of $A + 2Aa + a$ and $B + 2Bb + b$ (i.e., $AB + 2AaB + aB + 2ABb + 4AaBb + 2aBb + Ab + 2Aab + ab$).

Second Experiment (Trifactorial)

In this experiment, Mendel investigated whether the results he had obtained in both the monohybrid and bifactorial experiments held for an experiment in which three different characters were examined, the trifactorial experiment. He remarked, "Among all the experiments it demanded the most time and trouble" (93). The characters investigated for the seed plant (ABC) were: A (round shape), B (yellow albumen), and C (gray-brown seed coat); and for pollen plant (abc): a (wrinkled seed), b (green albumen), and c (white seed coat). The first two were seed characters and could be observed immediately, whereas seed-coat color, a plant character, required plants from the next generation.[17] Mendel obtained 687 seeds from 24 hybrid plants, from which he successfully grew 639 plants and "as further investigations showed,"[18] he obtained the results depicted in table 1.5. He summarized his data as follows:

The whole expression contains 27 terms. Of these 8 are constant in all characters, and each appears on the average 10 times; 12 are constant in two characters, and hybrid in the third; each appears on the average 19 times; 6 are constant in one character and hybrid in the other two; each appears on the average 43 times. One form appears 78 times and is hybrid in all of the characters. The ratios 10, 19, 43, 78 agree so closely with the ratios 10, 20, 40, 80, or 1, 2, 4, 8, that this last undoubtedly represents the true value. (94)[19]

TABLE 1.5 Mendel's results for the trifactorial experiment (Mendel 1865, 93)

8 plants	ABC	22 plants	ABCc	45 plants	ABbCc
14 "	ABc	17 "	AbCc	36 "	aBbCc
9 "	AbC	25 "	aBCc	38 "	AaBCc
11 "	Abc	20 "	abCc	40 "	AabCc
8 "	aBC	15 "	ABbC	49 "	AaBbC
10 "	aBc	18 "	ABbc	48 "	AaBbc
10 "	abC	19 "	aBbC		
7 "	abc	24 "	aBbc		
		14 "	AaBC	78 "	AaBbCc
		18 "	AaBc		
		20 "	AabC		
		16 "	Aabc		

Mendel went on to say that this series resulted from combining $A + 2Aa + a$, $B + 2Bb + b$, and $C + 2Cc + c$. He had a strong feeling about the expected results and was willing to accept conclusions despite limited statistics. As Fisher remarked, "He evidently felt no anxiety lest his counts should be regarded as insufficient to prove his theory" (121).

Mendel remarked that he had conducted several other experiments in which the remaining characters were combined in twos and threes and that these gave approximately equal results, but he presented none of his data for these experiments. He concluded:

There is therefore no doubt that for the whole of the characters involved in the experiments the principle applies that *the offspring of the hybrids in which several essentially different characters are combined exhibit the terms of a series of combinations, in which the developmental series for each pair of differentiating characters are united.* It is demonstrated at the same time that *the relation of each pair of different characters in hybrid union is independent of the other differences in the two original parental stocks.* (94)

In Mendel's opinion, his results justified belief that the same behavior applied to characters that could not be so easily distinguished. He noted, however, the difficulty of such experiments: "An experiment with peduncles of different lengths gave on the whole a fairly satisfactory result, although the differentiation and serial arrangement of the forms could not be effected with that certainty which is indispensable for correct experiment" (95).

The Reproductive Cells of Hybrids

In his bifactorial and trifactorial experiments, Mendel used seed plants with the dominant characters and pollen plants with the recessive characters. The question remained whether his results would remain the same if those parental types were reversed. He stated that in hybrid plants, it was reasonable to assume that there were as many kinds of egg and pollen cells as there were possibilities for constant combination forms. He further noted that this assumption, combined with the idea that the different kinds of egg and pollen cells are produced on average in equal numbers, would explain all of his previous results.

Mendel proposed to investigate these issues explicitly in a series of experiments. He chose true breeding plants as follows: seed plant (AB); where A and B were round shape and yellow albumen, respectively; pollen plant ab, where a and b were wrinkled shape and green albumen, respectively. These were artificially fertilized and the hybrid $AaBb$ obtained. Both the artificially fertilized seeds, together with several seeds from the two parental plants, were sown. He then performed the following fertilizations:

1. The hybrids with the pollen from *AB*
2. The hybrids with the pollen from *ab*
3. *AB* with pollen of the hybrid
4. *ab* with pollen of the hybrid

For each of these experiments, all of the flowers on three plants were fertilized. Mendel stated that if his assumptions were correct, then the hybrids would contain egg and pollen cells of the form *AB*, *Ab*, *aB*, and *ab*. When combined with the egg and pollen cells from the parental plants *AB* and *ab,* the following patterns emerge.

1. *AB, ABb, AaB, AaBb*
2. *AaBb, Aab, aBb, ab*
3. *AB, ABb, AaB, AaBb*
4. *AaBb, Aab, aBb, ab*

These genotypes should occur with equal frequency in each experiment. Experiments 1 and 3, as well as experiments 2 and 4, would demonstrate that the results are independent of which parent is used for pollen and which is used for seed. Mendel also noted that there would be statistical fluctuations in his data.

If, furthermore, the several forms of the egg and pollen cells of the hybrids were produced on an average in equal numbers, then in each experiment the said four combinations should stand in the same ratio to each other. A perfect agreement in the numerical relations was, however, not to be expected, since in each fertilisation, even in normal cases, some egg cells remain undeveloped or subsequently die, and many even of the well-formed seeds fail to germinate when sown. The above assumption is also limited in so far that, while it demands the formation of an equal number of the various sorts of egg and pollen cells, it does not require that this should apply to each separate hybrid with mathematical exactness. (97)

Mendel predicted that in Experiments 1 and 3 all of the seeds produced would be round and yellow, the result of dominance. For Experiments 2 and 4, his expectations were that round yellow seeds, round green seeds, wrinkled yellow seeds, and wrinkled green seeds would be produced in equal proportions. He reported: "The crop fulfilled these expectations perfectly" (98). Experiments 1 and 3 produced 98 and 94 exclusively round and yellow seeds, respectively. Experiment 2 produced 31 round yellow seeds, 26 round green seeds, 27 wrinkled yellow seeds, and 26 wrinkled green seeds. Experiment 4 produced 24 round yellow seeds,

25 round green seeds, 22 wrinkled yellow seeds, and 27 wrinkled green seeds. Mendel noted: "There could scarcely be now any doubt of the success of the experiment; the next generation must afford the final proof" (98).

Mendel sowed all of the seeds obtained in the first experiment, and 90 plants from 98 seeds bore fruit. In the third experiment, 87 plants from 94 seeds bore fruit.[20] Mendel reported on his other results:

> In the second and fourth experiments the round and yellow seeds yielded plants with round and wrinkled yellow and green seeds, *AaBb*.
>
> From the round green seeds plants resulted with round and wrinkled green seeds, *Aab*.
>
> The wrinkled yellow seeds gave plants with wrinkled yellow and green seeds, *aBb*.
>
> From the wrinkled green seeds plants were raised which yielded again only wrinkled green seeds, *ab*. (98)

Mendel's results are also shown in tables 1.6 and 1.7. He concluded, "In all the experiments, therefore, there appeared all the forms which the proposed theory demands, and they came in nearly equal numbers" (99).

TABLE 1.6   Mendel's results from the gametic experiments 1 and 3 (Mendel 1865, 98)

1st Exp.	3rd Exp.		
20	25	round yellow seeds	AB
23	19	round yellow and green seeds	ABb
25	22	round and wrinkled yellow seeds	AaB
22	21	round and wrinkled yellow and green seeds	AaBb

TABLE 1.7   Mendel's results from the gametic experiments 2 and 4 (Mendel 1865, 99)

2nd Exp.	4th Exp.		
31	24	plants of the form	AaBb
26	25	" "	AaB
27	22	" "	aBb
26	27	" "	ab

TABLE 1.8 Mendel's results for the flower color-stem length experiments (Mendel 1865, *100*)

Class	Color of flower	Stem	
1 [*AaBb*]	violet-red	long	47 times
2 [*aBb*]	white	long	40 "
3 [*Aab*]	violet-red	short	38 "
4 [*Ab*]	white	short	41 "

TABLE 1.9 Mendel's subsequent results for the flower color-stem length experiments (from Mendel 1865, *100*)

Trait	Number
violet-red flower color (*Aa*)	85 plants
white flower color (*a*)	81 plants
long stem (*Bb*)	87 plants
short stem (*b*)	79 plants

Mendel conducted a second set of experiments to test his assumptions. For these trials, he made selections so that each character should occur in half the plants if his assumptions were correct. In these experiments, *A* conferred violet-red flowers, *a* conferred white flowers, *B* long stems, and *b* short stems. He fertilized *Ab* (violet-red flowers, short stem) with *ab* (white flowers, short stem) producing hybrid *Aab*. In addition, *aB* (white flowers, long stem) was also fertilized with *ab*, yielding hybrid *aBb*. In the second year, the hybrid *Aab* was used as the seed plant and hybrid *aBb* as pollen plant. This should produce the combinations *AaBb*, *aBb*, *Aab*, and *ab*. In the third year, half the plants would have *Aa* (violet-red flowers), half *a* (white flowers), half *Bb* (long stems), and half *b* (short stems). The results are shown in tables 1.8 and 1.9. Mendel modestly concluded, "The theory adduced is therefore satisfactorily confirmed in this experiment also" (*100*). Mendel also performed other experiments, with fewer plants, on pod shape, pod color, and flower position, and "results obtained in perfect agreement" (*100*). No numerical data were presented.

As a result of this research, Mendel deduced, "Experimentally, therefore, the theory is confirmed that *the pea hybrids form egg and pollen cells which, in their constitution, represent in equal numbers all constant forms which result from the combination of characters united in fertilisation*" (*100*). He also stated, "It was furthermore shown by the whole of the experiments that it is perfectly immaterial whether the dominant character belong to the seed-bearer or to the pollen-parent; the form of the hybrid remains identical in both cases" (*84*).[21]

In discussing his results, Mendel demonstrated that he understood, at least qualitatively, the statistical nature of his data. He stated:

This represents the average results of the self-fertilisation of the hybrids when two differentiating characters are united in them. In individual flowers and in individual plants, however, the ratios in which the forms of the series are produced may suffer not inconsiderable fluctuations. Apart from the fact that the numbers in which both sorts of egg cells occur in the seed vessels can only be regarded as equal on the average, it remains purely a matter of chance which of the two sorts of pollen may fertilise each separate egg cell. For this reason the separate values must necessarily be subject to fluctuations, and there are even extreme cases possible, as were described earlier in connection with the experiments on the form of the seed and the colour of the albumen. The true ratios of the numbers can only be ascertained by an average deduced from the sum of as many single values as possible; the greater the number the more are merely chance effects eliminated. (*102*)

All of Mendel's numerical data from his pea experiments have now been presented, and these are the data on which Fisher based his analysis.

Mendel's Experiments on Other Species

Mendel also reported several experiments on *Phaseolus* (beans). The experiments on *Phaseolus vulgaris* and *Phaseolus nanus* "gave results in perfect agreement" (*103*). Those with *Phaseolus nanus*, L., as the seed plant, and *Phaseolus multiflorus*, W., as the pollen plant, did not. The former had white flowers and small white seeds, whereas the latter had purple-red flowers and seeds with black flecks or splashes on a peach-blood-red background. Mendel reported that the hybrids more closely resembled the pollen plant. He obtained only a few plants but, within limited statistics, he found that for recessive plant characters such as axis length and the form of the pod were the ratio of recessive to dominant was 1:3.

Mendel summarized his work as follows.

Despite the many disturbing factors with which the observations had to contend, it is nevertheless seen by this experiment that the development of the hybrids, with regard to those characters which concern the form of the plants, follows the same laws as in *Pisum*. With regard to the colour characters, it certainly appears difficult to perceive a substantial agreement. Apart from the fact that from the union of a white and a purple-red colouring a whole series of colours results [in F_2], from purple to pale violet and white, the circumstance is a striking one that among thirty-one flowering plants only one received the recessive character of the white colour, while in *Pisum* this occurs on the average in every fourth plant. (*105*)

Thus, Mendel not only reported blending inheritance, but also results that disagreed with his previous experiments.

Mendel also conducted experiments on *Hieracium* (hawkweed) (Mendel 1870). Again, the results did not always agree with those he had obtained previously. He remarked on the difficulty of the experiments and that he had obtained very few hybrids.

If finally we compare the described results, still very uncertain, with those obtained by crosses made between forms of *Pisum*, which I had the honor of communicating in the year 1865, we find a very real distinction. In *Pisum* the hybrids, obtained from the immediate crossing of two forms, all have the same type, but their posterity, on the contrary, are variable and follow a definite law in their variations. In *Hieracium* according to the present experiment the exactly opposite phenomenon seems to be exhibited. (qtd. in Stern and Sherwood 1966, 55)[22]

Summary

There are several points worth noting about Mendel's paper that will be important in the discussion of the Mendel-Fisher controversy. The first is that, as he remarks on several occasions, Mendel did not publish all of his data. The published data, however, also include results that differ considerably from Mendel's expectations. Mendel also knew what results he expected, either from theory or from his early observations. It also seems clear that Mendel had a good understanding of the principles of segregation and of independent assortment that form the basis of modern genetics.

Fisher's Analysis of Mendel's Data

Fisher's Early Thoughts

Although it was not until 1936 that R. A. Fisher published the paper on Mendel that would engender the longstanding controversy, that paper was not his first comment on Mendel's results. In a 1911 talk given to the Cambridge University Eugenics Society, Fisher commented, "It is interesting that Mendel's original results all fall within the limits of probable error;[23] if his experiments were repeated the odds against getting such good results is about 16 to one. It may just have been luck; or it may be that the worthy German abbot, in his ignorance of probable error, unconsciously placed doubtful plants on the side which favoured his hypothesis" (qtd. in Norton and Pearson 1976, 160). Fisher later changed his mind and attributed these results to the work of an assistant.

Fisher, in all probability, based these early comments on the analysis of Mendel's results provided by W. F. R. Weldon (1902). Weldon thought Mendel's work quite interesting and, in a letter to Karl Pearson, wrote, "About pleasanter things I have heard of and read a paper by one, Mendel, on the results of crossing peas, which I think you would like to read" (qtd. in Froggatt and Nevin 1971, 13). In his comments on Mendel, Weldon discussed Mendel's results on the 3:1 ratio in the first generation bred from hybrids. He presented Mendel's data along with the deviation of obser-

TABLE 1.10   Individuals with dominant characters in the second hybrid generation (Weldon 1902, 233)

Characters crossed	Individuals of second hybrid generation	Number of dominant individuals	Dominant individuals on Mendel's theory	Probable error of theory	Deviation of observation from theory
1. (Shape of seeds)	7324	5474	5493	±24.995	−19
2. (Color of cotyledons)	8023	6022	6017.25	±26.160	+4.75
3. (Color of seed coats)	929	705	696.75	±8.902	+8.25
4. (Shape of pod)	1181	882	885.75	±10.037	−3.75
5. Color of pod)	580	428	435	±7.034	−7
6. (Distribution of flowers)	858	651	643.5	±8.555	+7.5
7. (Height of plant)	1064	787	798	±9.527	−11

vation from theory along with a calculation of the probable error (table 1.10). He remarked:

Here are seven determinations of a frequency which is said to obey the law of Chance. Only one determination has a deviation from the hypothetical frequency greater than the probable error of the determination, and one has a deviation sensibly equal to the probable error; so that a discrepancy between the hypothesis and the observations which is greater to or equal to the probable error occurs twice out of seven times, and deviations much greater than the probable error do not occur at all. These results then accord so remarkably with Mendel's summary that if they were repeated a second time, under similar conditions and on a similar scale, the chance that the agreement between observation and hypothesis would be worse than that actually obtained is about 16 to 1. (Weldon 1902, 233)

Weldon also commented on Mendel's experiments on the 2:1 ratio and noted, "Mendel's statement is admirably in accord with his experiment" (Weldon 1902, 234). He then went on to discuss the results of the trifactorial experiment and commented, "Applying the method of Pearson (No. 25)[24] [χ^2 analysis] the chance that a system will exhibit deviations as great or greater than these from the result indicated by Mendel's hypothesis is about 0.95, or if the experiment were repeated a hundred times, we should expect to get a worse result about 95 times, or odds against a result as good as this or better are 20 to 1" (235). This was one of the early uses, perhaps even the first use, of the χ^2 test.

Weldon did not comment further in his paper on the goodness of fit of Mendel's data to his expectations, nor did he give even the slightest hint that he believed that Mendel's results were fraudulent in any way.[25] In a letter to Karl Pearson of November 1901, however, Weldon wrote: "Remembering his shaven crown [an allusion to Mendel's status as a monk] I cannot help wondering if they [Mendel's results] were not too good" (qtd. in Magnello 2004, 23). This line was crossed out and followed by the statement, "I do not see that the results are so good as to be suspicious." This was, in all probability, the first suggestion that Mendel's data were "too good." When Weldon wrote again to Pearson on 28 November 1901, he stated that he was certain that Mendel "cooked his figures, but that he was *substantially* right" (qtd. in Magnello 2004, 23).

In his 1902 paper, Weldon did comment further on both the value of Mendel's work and on some difficulties with Mendel's conclusions:

Mendel's experiments are based upon work extending over eight years. The remarkable results obtained are well worth even the great amount of labour they must have cost, and the question at once arises, how far the laws deduced from them are of general application. It is almost a matter of common knowledge that they do not hold for all characters, even in Peas, and Mendel does not suggest that they do. At the same time I see no escape from the conclusion that they do not hold universally for the characters of Peas which Mendel so carefully describes. In trying to summarise the evidence on which my opinion rests, I have no wish to belittle the importance of Mendel's achievement. I wish simply to call attention to a series of facts which seem to suggest fruitful lines of inquiry. (Weldon 1902, 235)

The rest of Weldon's paper is devoted to a discussion of some of the evidence for his reservations about Mendel's work.[26]

Fisher's Seminal Paper

In 1936, R. A. Fisher published a paper entitled "Has Mendel's Work Been Rediscovered?" (*117*). This is the paper that engendered, albeit after a considerable delay, the so-called Mendel-Fisher controversy. Fisher did not question whether people knew of Mendel's work, but rather whether they really understood what Mendel had written. He noted that the story of Mendel's work and its rediscovery had become traditional in the teaching of biology: "A careful scrutiny can but strengthen the truth in such a tradition, and may serve to free it from such accretions as prejudice or hasty judgment may have woven into the story" (*117*). Fisher proposed to provide such a careful scrutiny and remarked, "When the History of Science is taken seriously the number of enquiries which such a story suggests is somewhat formidable. We want to know first: What did Mendel discover? How did he discover it? And what did he think he discovered? Next, what was the relevance of his discoveries to the science of his time, and what was its reaction to them?" (*118*).

Fisher was concerned that misconceptions about Mendel's work had been propagated by Bateson, particularly claims that Darwinism was responsible for the neglect of Mendel's work and that Mendel was hostile to Darwinism. Fisher presented persuasive arguments against both these views. He was also concerned about Bateson's assertion that Mendel's description of his experiments should not be taken literally. Bateson, in commenting on the monohybrid experiments, stated: "This statement of Mendel's in the light of present knowledge is open to some misconception. Though his work makes evident that such varieties may exist, it is very unlikely that Mendel could have had seven pairs of varieties such that the members of each pair differed from each other in *only* one considerable character (*wesentliches Merkmal*). The point is probably of little theoretical or *practical consequence, but a rather heavy stress is thrown on 'wesentlich'*" (Bateson 1909, 332).[27] Fisher proposed two possible solutions to this problem. Mendel might have arbitrarily chosen one factor for which the particular cross was designated as an experiment and ignored other factors; or he might have scored each plant in all factors and assembled the data for that factor from all of the crosses in which it had been involved and reported the result as a single experiment on a single factor. Fisher noted that the first solution seemed incredibly wasteful of data, but added, "This objection is not so strong as it might seem, since it can be shown that Mendel left uncounted, or at least unpublished, far more material than appears in his paper" (*121*). Fisher believed that the second option was what most modern geneticists would do, but thought it unlikely that Mendel had done so: "[T]he style throughout suggests that he [Mendel] expects to be taken literally; if his facts have suffered much manipulation the style of his report must be judged disingenuous. Consequently, unless real contradictions are encountered in reconstructing his experiments from his paper, regarded as a literal account, this view must be preferred to all alternatives, even though it implies that Mendel had a good understanding of the factorial system, and the frequency ratios which constitute his laws of inheritance, before he carried out the experiments reported in his first and chief paper" (*122*).

Fisher's Reconstruction of Mendel's Data

As far as the subsequent controversy is concerned, the most important section of Fisher's paper is the one entitled "An Attempted Reconstruction." Fisher constructed a chronology of the eight years of Mendel's experiments, including which experiments were done and how many plants were grown in a given year, what Mendel's results were, and in what order those results were obtained.

Fisher inferred that the experiments on seed characters (yellow or green and round or wrinkled) were completed in 1859 and that "Mendel does not test the significance of the deviation, but states the ratios as 2.96 : 1 and 3.01 : 1, without giving any probable error" (*123*). He went on to remark, "The discovery, or demonstration, whichever it may have been, of the 3 : 1 ratio was evidently the critical point in Mendel's researches" (*124*). Fisher believed that Mendel's satisfaction with these approximate ratios was intelligible if "he had convinced himself as to their explanation, and framed the entire Mendelian theory of genetic factors and gametic segregation" (*124*). He further noted:

> In 1930,[28] as a result of a study of the development of Darwin's ideas, I pointed out that the modern genetical system, apart from such special features as dominance and linkages, could have been inferred by any abstract thinker in the middle of the nineteenth century if he were led to postulate that inheritance was particulate, that the germinal material was structural, and that the contributions of the two parents were equivalent. I had at that time no suspicion that Mendel had arrived at his discovery in this way. From an examination of Mendel's work it now appears not improbable that he did so and that his ready assumption of the equivalence of the gametes was a potent factor in leading him to his theory. *In this way his experimental programme becomes intelligible as a carefully planned demonstration of his conclusions.* (*125*, emphasis added)

In other words, Fisher believed that Mendel was, in fact, a Mendelian.[29]

Fisher went on to discuss Mendel's experiments of 1860 in which the 3 : 1 ratio was shown to be 1 : 2 : 1, where 1 is the homozygous dominants or recessives and 2 is the heterozygous hybrid. On several occasions, Fisher commented on the comparison of the observed deviations from the expected results to the standard deviation expected. Thus, in discussing the experiments on plants raised from yellow seeds (which yielded 166 plants with only yellow seeds and 353 plants with both yellow and green seeds) and that on plants grown from round seeds (which yielded 193 plants with only round seeds and 372 plants with both round and wrinkled seeds), Fisher stated: "The ratios in both cases show deviations from the expected 2 : 1 ratio less than their standard errors" (*126*). For the 1861 experiments on plants bred from colored flowers and from tall plants (see table 1.3), Fisher commented, "In neither case does the ratio depart significantly from the 2 : 1 ratio expected, although in the second case the deviation does exceed the standard deviation of random sampling" (*126*). For the experiment on yellow pods (which yielded a 60 : 40 ratio), Fisher remarked on "a relatively large, but not a significant, deviation" (*127*). He further noted, "It is remarkable as the only case in the record in which Mendel was moved to verify a ratio by repeating the trial" (*127*).

TABLE 1.11 Classification of plants grown in the trifactorial experiment (Fisher 1936, table II)

	CC				Cc				cc				Total			
	AA	Aa	aa	Total	AA	Aa	aa	Total	AA	Aa	aa	Total	AA	Aa	aa	Total
BB	8	14	8	30	22	38	25	85	14	18	10	42	44	70	43	157
Bb	15	49	19	83	45	78	36	159	18	48	24	90	78	175	79	332
bb	9	20	10	39	17	40	20	77	11	16	7	34	37	76	37	150
Total	32	83	37	152	84	156	81	321	43	82	41	166	159	321	159	639

Fisher, obviously concerned, went on to critically examine the experiments in which such deviations occurred. It was at this point that he first announced the problem of the 2:1 ratio:

In connection with these tests of homozygosity by examining ten offspring formed by self-fertilization, it is disconcerting to find that the proportion of plants misclassified by this test is not inappreciable. If each offspring has an independent probability, .75, of displaying the dominant character, the probability that all ten will do so is $(.75)^{10}$ or .0563. Consequently, between 5 and 6 per cent of the heterozygous parents will be classified as homozygotes, and the expected ratio of segregating to non-segregating families is not 2:1 but 1.8874:1.1126 or approximately 377.5:222.5 out of 600. Now among the 600 plants tested by Mendel 201 were classified as homozygous and 399 as heterozygous [see table 1.3]. Although these numbers agree extremely closely with his expectations of 200:400, yet, when allowance is made for the limited size of the test progenies, the deviation is one to be taken seriously. It seems extremely improbable that Mendel made any such allowance, or that the numbers he recorded are "corrected" values, rounded off to the nearest integer, obtained by dividing the numbers observed to segregate by .9437. We might suppose that sampling errors in this case caused a deviation in the right direction, and of almost exactly the right magnitude, to compensate for the error in theory. A deviation as fortunate as Mendel's is to be expected once in twenty-nine trials. Unfortunately the same thing occurs again with the trifactorial data [table 1.11] (*127*).

Fisher's further examination of those trifactorial data yielded detailed comments that are also worth examining.

In the case of the 600 plants tested for homozygosity in the first group of experiments Mendel states his practice to have been to sow ten seeds from each self-fertilized [F_2] plant. In the case of the 473 plants with coloured flowers from the trifactorial cross he does not restate his procedure. It was presumably the same as before. As before, however, it leads to the difficulty that between 5 and 6 per cent of heterozygous plants so tested would give only coloured progeny, so that the expected ratio of those showing segregation to those not showing it is really

TABLE 1.12 Comparison of numbers reported with uncorrected and corrected expectations (Fisher 1936, table III)

	Number of plants tested	Number of non-segregating progenies observed	Number expected		Deviation	
			Without correction	Corrected	Without correction	Corrected
1st group of experiments	600	201	200.0	222.5	+1.0	−21.5
Trifactorial experiment	473	152	157.7	175.4	−5.7	−23.4
Total	1073	353	357.7	397.9	−4.7	−44.9

lower than 2:1, while Mendel's reported observations agree with the uncorrected theory.

> The comparisons are shown in Table III [table 1.12]. A total deviation of the magnitude observed, and in the right direction, is only to be expected once in 444 trials; there is therefore here a serious discrepancy. (*130*)

The reliability of Mendel's results had been called into question.

Fisher then offered several possible solutions to the 2:1 ratio problem. He pointed out that if Mendel had backcrossed the 473 trifactorial plants, the probability of misclassification of heterozygotes would be reduced by a factor of 50. (This would have involved a considerable amount of labor.) If, for example, the plants were backcrossed with a recessive plant, then the probability of observing the recessive character in a single plant would be 0.5 for a heterozygote. For 10 plants, the probability of misclassification is then $(0.5)^{10}$, or 0.00098.

A second possibility was that Mendel had used a larger number of progeny in his test, say 15 instead of 10. The probability of misclassification, in this case, is reduced to 0.013, which gives a ratio of 1.974:1.026 = 1.924, much closer to 2. Fisher noted, however, that this would have required a larger number of plants grown in a single year than Mendel had, in fact, ever planted. In addition, it would not apply to the earlier experiments, in which Mendel had explicitly stated that he used 10 progeny.

The third possibility was that the selection of plants for testing favored the heterozygotes. Fisher remarked that in some crosses, it was possible that the heterozygote plants were larger and that "the larger plants might have been unconsciously preferred" (*131*). Fisher presented three arguments against this possible solution: (1) in the trifactorial experiment all plants were counted; (2) it was improbable that the compensating selection would work equally well for all five plant characters; and (3) the total compensation for all plants was unlikely to have given the exact number needed.

Fisher stated, however, that the question of whether the trifactorial data had been manipulated could be tested. He proceeded to use the χ^2 test (discussed further in this book's appendix), and commented:

> The possibility that the data for the trifactorial experiment do not represent objective counts, but are the product of some process of sophistication, is not incapable of being tested. Fictitious data can seldom survive a careful scrutiny, and, since most men underestimate the frequency of large deviations arising by chance, such data may be expected generally to agree more closely with expectation than genuine data would. The twenty-seven classes in the trifactorial experiment supply twenty-six degrees of freedom for the calculation of χ^2. The value obtained is 15.3224, decidedly less than its average value for genuine data, 26, though this value by itself might occur once in twenty genuine trials.[30] (*131*)

Fisher then applied the test to various subdivisions of the trifactorial data, with similar results, and subsequently applied the analysis to all of the experiments performed in 1863. These included the trifactorial experiment, the bifactorial experiment, the experiment on gametic ratios (those involving the question of whether the results depended on which plant produced the pollen and which the egg), and the repetition of the yellow-pod experiment. His results, shown in table 1.13, gave a χ^2 of 15.5464 for 41 degrees of freedom, and prompted him to write: "The discrepancy is strongly significant, and so low a value could scarcely occur by chance once in 2000 trials. There can be no doubt that the data from the later years of the experiment have been biased strongly in the direction of agreement with expectation" (*132*).

Fisher explained that in tests where seeds were deformed or discolored, bias rather than theory might help to explain Mendel's results, but also noted that this would not apply to the tests of gametic ratios or to other experiments based on classification of whole plants.

TABLE 1.13 Measure of deviation expected and observed in 1863 (Fisher 1936, table IV)

	Expectation	χ^2 observed
Trifactorial experiment	17	8.9374
Bifactorial experiment	8	2.8110
Gametic ratios	15	3.6730
Repeated 2 : 1 test	1	0.1250
Total	41	15.5464

TABLE 1.14 Deviations expected and observed in all experiments (Fisher 1936, table V)

		Expectation	χ^2	Probability of exceeding deviations observed
3 : 1 ratios	Seed characters	2	0.2779	
	Plant characters	5	1.8610	
		— 7	— 2.1389	.95
2 : 1 ratios	Seed characters	2	0.5983	
	Plant characters	6	4.5750	
		— 8	— 5.1733	.74
Bifactorial experiment		8	2.8110	.94
Gametic ratios		15	3.6730	.9987
Trifactorial experiment		26	15.3224	.95
Total		64	29.1186	.99987
Illustrations of plant variation		20	12.4870	.90
Total		84	41.6056	.99993

While Fisher did collect the χ^2 values for all the experiments in his Table V (here, table 1.14), the analysis of all of Mendel's results seems to be almost an afterthought. It was the agreement of Mendel's data with what Fisher regarded as the incorrect 2:1 ratio that was most important for Fisher.[31] Fisher's result for all of Mendel's experiments, which included 84 degrees of freedom, was a total χ^2 of 41.6056. The probability of exceeding this χ^2 value was 0.99993, or the probability of getting such a good fit to the expectation was 7 in 100,000. Fisher makes no comment on this extraordinary result except to note that "the bias seems to pervade the whole of the data" (*133*).

Fisher's Conclusions

Fisher's first conclusion (and the one undoubtedly most important to him) was that Mendel's account of his experiments was "to be taken entirely literally." Fisher fully believed that Mendel's experiments "were carried out in just the way and much in the order that they are recounted. The detailed reconstruction of his programme on this assumption leads to no discrepancy whatever" (*134*). Bateson and others who had suggested otherwise were, in Fisher's view, conclusively refuted. However, Fisher went on to state explicitly that he believed that most of Mendel's data has been falsified:

> A serious and almost inexplicable discrepancy has, however, appeared, in that in one series of results the numbers observed agree excellently with the two to one ratio, which Mendel himself expected, but differ significantly from what should have been expected had his theory been corrected to allow for the small size of his test progenies. To suppose that Mendel recognized this theoretical complication, and adjusted the frequencies supposedly observed to allow for it, would be to contravene the weight of the evidence supplied in detail by his paper as a whole. Although no explanation can be expected to be satisfactory, it remains a possibility among others that Mendel was deceived by some assistant who knew too well what was expected. This possibility is supported by independent evidence that the data of most, if not all, of the experiments have been falsified so as to agree closely with Mendel's expectations. (134)

Fisher concluded that Mendel regarded the numerical ratios as demonstrating the truth of his factorial system "and that he was never much concerned to demonstrate either their exactitude or their consistency" (135). Perhaps as a way of rationalizing *why* Mendel might falsify his data, Fisher wrote, "it is clear, from the form his experiments took, that he knew very surely what to expect, and designed them as a demonstration for others rather than for his own enlightenment" (135). Yet Fisher clearly attributes the falsification to someone else, such as an assistant. The last section of Fisher's paper is devoted to examining how Mendel's contemporaries reacted to Mendel's work and why they largely overlooked it. Fisher remarked that the journal in which Mendel published his results was widely distributed and reasonably well known. Moreover, Fisher noted, Mendel's paper was not inaccessible; in fact, "the new ideas are explained most simply, and amply illustrated by the experimental results" (136). Yet Karl Wilhelm von Nägeli, with whom Mendel corresponded, was either unimpressed by Mendel's results or anxious to warn students against paying attention to them. Fisher also cited W. O. Focke, "who, in his *Pflanzenmischlinge* [1881], makes no less than fifteen references to Mendel" (137). As Fisher made clear, however, Focke did not understand Mendel's work and seemed to prefer the more comprehensive contributions of Joseph Gottlieb Kolreuter, Carl Friedrich von Gartner, and others. Fisher remarked that Focke had "overlooked, in his chosen field, experimental researches conclusive in their results, faultlessly lucid in presentation, and vital to the understanding not of one problem of current interest, but of many" (139). It is hard to imagine a more positive opinion of Mendel's work. There is no mention here of any falsification or fraud.

Fisher ended his paper with an exhortation to, and criticism of, his colleagues for failing to carefully examine and understand Mendel's work:

> The peculiar incident in the history of biological thought, which it has been the purpose of this study to elucidate, is not without at least one moral—namely, that there is no substitute for a careful, or even meticulous, examination of all original papers purporting to establish new facts. Mendel's contemporaries may be blamed for failing to recognize his discovery, perhaps through resting too great a confidence on comprehensive compilations. It is equally clear, however, that since 1900, in spite of the immense publicity it has received, his work has not often been examined with sufficient care to prevent its many extraordinary features being overlooked, and the opinions of its author being misrepresented. Each generation, perhaps, found in Mendel's paper only what it expected to find . . . (139)

It is clear that Fisher admired both Mendel and his work.

Fisher's Later Thoughts

Even before Fisher's paper was published, he discussed its contents with E. B. Ford, then departmental demonstrator in the Department of Zoology at Oxford University. In a letter of 2 January 1936, Fisher wrote: "I have had the shocking experience lately of coming to the conclusion that the data given in Mendel's paper must be practically all faked"; he referred to this information as his "abominable discovery" (R. A. Fisher Digital Archive, 2, 4). Just a few days later, on 8 January 1936, Fisher wrote to Douglas McKie, the editor of *Annals of Science*, to submit his paper for publication. In this letter, Fisher stated: "I had not expected to find the strong evidence which has appeared that the data had been cooked. This makes my paper far more sensational than ever I had intended" (R. A. Fisher Digital Archive, 1).

In the letter to Ford, Fisher emphasized, however, that he had concluded that Mendel's experiments were planned and performed exactly as Mendel had recorded and "this is what I was really studying his paper for" (2). He further remarked, "I don't believe that this touches Mendel's own *bona-fides*, or the reality of the experiments he carried out" (3).

Ford was horrified. In his response to Fisher of 5 January 1936, he exclaimed: "I am appalled by your discovery. Your analysis is a remarkable piece of work, but what it reveals is really very shocking. Clearly, as you say, Mendel himself is not to blame" (R. A. Fisher Digital Archive, 3). Ford, like Fisher, refused to believe that Mendel was guilty of fraud: "[I]t is simply incredible that a man of his intelligence could want to fake, *after* he had found out what to look for by honest work" (4). Ford regarded Fisher's paper as extremely important and encouraged its publication. "Too much has been hung on Mendel's results to suppress the matter though one does not want to wash dirty linen in public unnecessarily" (5). In a subsequent letter of 11 January, Ford further noted that Fisher had dealt with "the difficult matter of the faked data in a most tactful way" (R. A. Fisher Digital Archive, 1). He suggested that Fisher might include a summary of the paper because "You see you have naturally had to be so careful in your statements on the faking, that the point seems rather immersed in the paper as a whole. A summary would obviate this" (1–2). Fisher did not act on this suggestion.

It is clear both from Fisher's paper itself and from Ford's comment that the statements concerning Mendel's alleged falsification were not emphasized in the paper. I might suggest that this was more than tact—that it represented Fisher's view that the numerical falsification, if it was that, was relatively less important than Mendel's conclusions. This view seems to have been shared by Ford, who wrote in his letter of 5 January 1936: "When you write, do stress that Mendel's greatness lies not so much in his discoveries as in his deductions—and in planning the work" (4). Fisher clearly agreed. There doesn't seem to be any further Fisher correspondence on Mendel's data. J. Henry Bennett's book (1983), which contains selections from Fisher's letters, does not include any such correspondence.

Fisher did include a discussion of the problems with Mendel's data in his university lectures. Alan Cock reports that in Fisher's lectures in 1942–1943, he presented the following anecdote: "A lay brother assigned the task of weeding Mendel's pea beds, is grumbling to himself about his bad back. 'Brother Gregor and his experiments are all very well, but my back is killing me.' Knowing from experience that if the result were 'bad,' Brother Gregor might insist in repeating the experiment, he 'accidentally' lets his hoe slip to demolish a few chosen plants and ensure a 'good' result" (A. Cock, letter to A. W. F. Edwards, 25 March 1988).

In 1955, at the request of an editor of a proposed series on source papers in science, Fisher wrote both an introduction to, and a commentary with marginal notes on, Mendel's paper. The series never came about, but Fisher's introduction and commentary (Fisher 1965a, 1965b) were included in a book edited by Bennett (1965), containing a translation of Mendel's paper. Fisher's introduction again emphasized the quality and importance of Mendel's work as well as Mendel's contribution to the methodology of research on hybrids. Commenting on the "rediscovery" of Mendel's work in 1900, Fisher wrote, "The facts available in 1900 were at least sufficient to establish Mendel's contribution as one of the greatest experimental advances in the history of biology" (Fisher 1965a, 2).

Fisher then discussed Mendel's method:

If we read his introduction literally we do not find him expressing the purpose of solving a great problem or reporting a resounding discovery. He represents his work rather as a contribution to the *methodology* of research into plant inheritance. He had studied the earlier writers and tells us just in what three respects he thinks their work should be improved upon. If proper care were given, he suggests, to the distinction between generations, to the identification of genotypes, and, to this end, to the frequency ratios exhibited by their progeny, when based on an adequate statistical enumeration, studies in the inheritance of other organisms would yield an understanding of the hereditary process as clear as that which he here exhibits for the varieties of garden pea. There is no hint of a tendency to premature generalization, but an unmistakable emphasis on the question of method. (Fisher 1965a, 3)

Here, using the very same language that he had used in his 1936 paper, Fisher emphasized Mendel's reasoning:

The fact that Mendel was principally concerned to justify a method of investigation, and not primarily to exhibit particular results, is at least a partial explanation of another group of peculiarities of his paper, which flow from the fact that he is reporting a *carefully planned demonstration, rather than the protocol of the first observations which led to the formation of his ideas.* The simplicity of his plan, and the adequacy of the numbers of the first crosses reported, are indications that he knew in advance very much what he intended to do, and what he ought to expect. He constantly omits reference to the confirmation of his first conclusions, which the later generations and other experiments reported must have supplied in abundance. Only once is he led to repeat a test. He seems never to be unsure of the sufficiency of the first evidence reported, even when it is not really so strong as might be wished . . . (Fisher 1965a, 4, emphasis added)

In his marginal notes, Fisher again stated that all of Mendel's data were not reported: "It is remarkable how much of the material, which Mendel must have bred, has not been reported, either in confirmation, or for comparison with what he has given" (Fisher 1965b, 54).

Fisher's introduction made no mention at all either of the problem of the 2:1 ratio or of the overall goodness of fit of Mendel's results. In the marginal notes, only the 2:1 ratio was emphasized. Fisher once again pointed out that if Mendel's method was as reported, the 2:1 ratio should have been 1.8874:1.1126. He noted that the problem appeared in both the monohybrid experiments and in the trifactorial experiment. He did, however, remark that an examination of the general level of agreement between Mendel's expectations and his reported results showed that it is closer than would be expected in the best of several thousand repetitions. "The data have evidently been sophisticated systematically" (Fisher 1965b, 53). Again Fisher placed the blame on a deceiving assistant.

Thus, it is evident that Fisher admired Mendel's achievement and believed that Mendel's most important contribution was his method of experimentation and his deductions from his results and not the numerical results themselves. Fisher also concluded that much of Mendel's data had been falsified. He considered the most important falsification to have been on the 2:1 ratio experiments and believed that the overall "goodness of fit" was less important. It also seems clear that Fisher believed that Mendel's achievements outweighed any possible falsification.

The Mendel-Fisher Controversy

Just as Mendel's work was neglected, so was Fisher's analysis of Mendel's work. This latter neglect lasted for more than 25 years. As discussed earlier, there seems to be no correspondence between Fisher and others on the subject after 1936. Similarly, Vítězslav Orel (1996), a biographer of Mendel, mentions no papers on the controversy published before 1964. Nor do any papers on the controversy published after 1964 make any reference to papers published before that date except, of course, to Fisher's paper.[32] One may speculate that the reason for this neglect was the lack of emphasis on Mendel's possible fraud in Fisher's paper.

The controversy divides itself reasonably into three time periods: (1) the 1960s, about the time of the Mendel centenary; (2) the late 1970s to the mid-1980s; and (3) from 1990 to the present. Many of the papers on the controversy include discussions of other aspects of Mendel's work. I will restrict my discussion to only those sections that deal with Mendel and Fisher. I will also attempt to keep to a chronological account. If there are errors in a paper, I will delay discussion of them until I come to the time in which those criticisms appeared in the published literature. In this way, the reader will get a better feel for the actual history of the controversy.

The 1960s

The year 1965 marked the centenary of Mendel's discovery[33] and, as with other significant scientific discoveries, the occasion was commemorated with conferences, papers, and books. All of these celebrated Mendel's achievement and the evaluations were unanimously and enthusiastically favorable. This attention to Mendel's work also seems to have resulted in attention to Fisher's 1936 analysis paper. This provided the initial impetus for the Mendel-Fisher controversy. In a talk given to the New Jersey Academy of Science on 20 April 1963, published in the *Journal of Heredity,* Conway Zirkle observed that Mendel's results seemed to fit his expectations too well: "Some modern statisticians, who are armed with the mathematical tools of modern statistics, have reported that Mendel's results were significant—in fact, a little too significant. They were a little *too good*, better than we would have a right to expect on the basis of chance. Could the good Father Mendel have fudged his results a little?" (Zirkle 1964, 66, emphasis added). Interestingly, Zirkle made no mention of Fisher's 1936 paper, although he cited several other works by Fisher. This seems to have been the first published mention that Mendel's results were "too good." Zirkle also remarked that if Mendel's results had not been so good "he might never have discovered Mendelism" (66).[34] He illustrated this with a story concerning Darwin's experiments on corn. In examining two types of grain on an ear of corn, Darwin found a ratio of 88:37, which differed only slightly from the 94:31 ratio expected for a 3:1 ratio.[35] Darwin missed its significance. Zirkle concluded, "Mendel, perhaps, was lucky, but was he?" (66). With such statements, Zirkle seemed to acknowledge the possibility that Mendel falsified his data.

In 1964, Gavin de Beer published a paper on Mendel, Darwin, and Fisher in which he also constructed a chronology of Mendel's work, slightly correcting Fisher's chronology. De Beer, like Fisher, was an admirer of Mendel. In discussing Mendel's paper, he stated: "It is not too much to claim that this communication, which was the foundation of the science of genetics and introduced mathematics and the theory of probability into the study of inheritance, represented an advance of knowledge in biological science of an order of magnitude comparable with that of evolution by natural selection" (DeBeer 1964, 192). He noted that it was now possible to appraise the true worth of Mendel's work: "This is largely due to the experiments, demonstrations, and conclusions of Sir Ronald Fisher, F.R.S., who, in 1930, brought out the full significance of Mendel's work" (192). De Beer clearly did not regard Fisher as a detractor of Mendel. He did, however, present a summary of Fisher's 1936 paper including Fisher's conclusion that Mendel's data had been falsified. For De Beer, however, as for Fisher, this conclusion had no effect on his admiration of Mendel: "None of this, of course, detracts in the slightest degree from the genius of Mendel in planning and carrying out his experiments, nor from the validity of the principles of heredity, his Laws, that he discovered" (200). De Beer also suggested that whoever had been responsible for Mendel's results was, in fact, a benefactor to science. He presented an analysis of Darwin's ear of corn experiment similar to Zirkle's analysis, noting that Darwin's observed deviation was "well within the acceptable limits for agreement with a 3:1 ratio" (201).

Another paper commemorating the Mendel centenary was by Leslie Clarence Dunn (1965). Dunn stated that there was a strong indication that Mendel knew what numerical results to expect, and that the agreement of Mendel's results with his expectations could not be accounted for by luck alone. He noted that this had first been pointed out by Fisher and that Fisher had concluded that fraud was involved. Dunn offered several possible explanations of Mendel's results, including the possibility, already mentioned by Fisher, that Mendel had tested more than 10 plants in the 2:1 ratio experiments:

> Dr. Sewall Wright has pointed out to me his view that Mendel, who clearly knew how to compute probabilities, could hardly have been unaware of the likelihood of no recessives would appear in some groups of ten progeny and could have estimated this to be about one in eighteen (0.056). Perhaps he chose the inadequate number ten because of lack of space for growing plants; but perhaps he in fact tested more than ten plants in order to have at least ten left after the inevitable losses. If the average of "at least ten" should be twelve the probability of misclassifying falls from 0.056 to 0.031 and the discrepancy from Mendel's 2:1 expectation is not a serious one.
>
> Those who have experience in tallying such outcomes become aware of the danger that unconscious bias in favor of the expected result will creep in and that the count may be stopped at a point which is favorable to the theory. (Dunn 1965, 194)

Although Dunn acknowledged the correctness of Fisher's analysis, he concluded: "There is no evidence of conscious fraud and he [Mendel] was careful to report wide deviations in some parts of some experiments which he would not have done if bent on fraud" (194). Dunn also believed that Mendel had his theory in mind "when the data *as reported* were tallied" (194). In his final assessment of Mendel, Dunn wrote: "Mendel however stands as a clear example and guide to a new way of studying a biological problem with a sharp, clear experimental design applied to a single question stated with simplicity because it had been reduced to its essentials" (198).

Bennett's aforementioned *Experiments in Plant Hybridisation by Gregor Mendel* (1965) included a translation of Mendel's paper along with Fisher's paper, introduction, and marginal notes. Bennett emphasized Fisher's "remarkable findings" on both the 2:1 ratio experiments and on the overall goodness of fit. This was stated on the first page of the editor's preface along with Fisher's conclusion that fraud had occurred and that this was due to Mendel's assistant.

Fisher and Mendel were also discussed in *A History of Genetics* by Alfred Sturtevant (1965). Sturtevant commented that Fisher had shown that if one examined all of Mendel's experiments, the probability of getting as good a fit as Mendel had was 1 in 14,000. Sturtevant did not seem to be overly impressed by this observation and offered the first technical criticism of Fisher's analysis.

If this were all, one might not be too disturbed, for it is possible to question the logic of the argument that a fit is too close to expectation. If I report that I tossed 1000 coins and got exactly 500 heads and 500 tails, a statistician will raise his eyebrows, though this is the most probable exactly specified result. If I report 480 heads and 520 tails, the statistician will say that is about what one would expect—though this result is less probable than the 500:500 one. He will arrive at this by adding the probabilities for all results between 480:520 and 520:480, whereas for the exact agreement he will consider only the probability of 500:500 itself. If I now report that I tossed 1000 coins ten times, and got 500:500 every time, our statistician will surely conclude that I am lying, though this is the most probable result thus exactly specified. (Sturtevant 1965, 13)

Sturtevant was much more impressed with Fisher's analysis of the 2:1 ratio experiments because "In the present case, however, it appears that in one series of experiments Mendel got an equally close fit to a *wrong* expectation" (13–14). He remarked that Fisher's analysis was correct, but only if exactly 10 seeds were planted. For more than 10 seeds, as we have seen, the correction to the expectation is less and "Fisher's most telling point will be weakened" (14). He went on to note: "The statement by Mendel seems unequivocal, but the possibility remains that he may have used more than 10 seeds in some or many cases" (15).[36]

Sturtevant also examined eight experiments on cotyledon color in peas, including Mendel's experiment and seven others performed between 1900 and 1924, to see if they also reported unexpectedly close agreement with expectation (table 1.15). He noted that half the experiments showed a ratio of the observed deviation to the probable error greater than one, which is what one expects. He concluded, "The over-all impression is that the agreement with expectation is neither too good nor too poor" (16).

Sturtevant also offered a possible botanical explanation for Mendel's results. He stated that in self-pollination an anther will usually break at one point, leading not to a random sample of pollen grains but to one in which all or most of the pollen grains come from one or a few pollen-mother cells. He admitted that this was unlikely to be important in Mendel's experiments, and, "Calculations based on this improbable limiting assumption indicate that Fisher's conclusions would still hold good; but the point remains that in any such analysis one needs to examine the assumptions very carefully, to make sure there may not be some alternative explanation" (15).

TABLE 1.15 F_2 results, pea crosses (from Sturtevant 1965, 15)

Source	Yellow	Green	Total	Deviation from 3 in 4	Prob. error	Dev./P.E.
Mendel 1866	6,022	2,001	8,023	+.0024	±.0130	.18
Correns 1900	1,394	453	1,847	+.0189	±.0272	.70
Tschermak 1900	3,580	1,190	4,770	+.0021	±.0169	.12
Hurst 1904	1,310	445	1,775	−.0142	±.0279	.51
Bateson 1905	11,902	3,903	15,806	+.0123	±.0093	1.23
Lock 1905	1,438	514	1,952	−.0533	±.0264	2.04
Darbishire 1909	109,060	36,186	145,246	+.0035	±.0030	1.16
Winge 1924	19,195	6,553	25,748	−.0180	±.0125	1.44

Despite his criticisms of Fisher, Sturtevant concluded, "In summary, then, Fisher's analysis of Mendel's data must stand essentially as he stated it" (16). He offered three possible explanations for Mendel's data: (1) unconscious bias, (2) omission of aberrant families, or (3) biased action by an assistant. He admitted, "None of these alternatives is wholly satisfactory, since they seem out of character, as judged by the whole tone of the paper" (16). Sturtevant ended his analysis by stating: "Perhaps the best answer—with which I think Fisher would have agreed—is that, after all, Mendel was right!" (16).

Another translation of Mendel's paper appeared in *The Origin of Genetics: A Mendel Source Book*, edited by Curt Stern and Eva Sherwood (1966). The title gives the editors' judgment about the importance of Mendel's work. They stated: "Gregor Mendel's short treatise 'Experiments on Plant Hybrids' is one of the triumphs of the human mind. It does not simply announce the discovery of important facts by new methods of observation and experiment. Rather, in an act of the highest creativity, it presents these facts in a conceptual scheme which gives them general meaning. Mendel's paper is not solely a historical document. It remains alive as a supreme example of scientific experimentation and profound penetration of data" (v).

The book also included Fisher's 1936 analysis paper and a commentary on why Mendel's data were "too good" by Sewall Wright, a leading population geneticist. Wright remarked: "the excessive goodness of fit of Mendel's data is certainly one of the most disconcerting items that a historian of genetics has to deal with" (Wright 1966, 173). He reported that he had repeated Fisher's calculations and obtained substantially the same results and agreed: "There is no question that the data fit the ratios much more closely than can be expected from accidents of sampling" (173). Wright cited Raymond Pearl (1940) on the difficulty of making repeatable counts as evidence for such possible bias. Pearl reported an experiment in which 15 trained observers (two plant pathologists, two professors of agronomy, one professor of philosophy [originally trained as a biologist], four biologists, one computer, one practical corn breeder, and one professor and three assistants in plant physiology) were asked to count the same 532 kernels of corn from a single ear. The traits examined were color (yellow or white) and form of the kernel (starchy or sweet). The Mendelian predictions were nine yellow starchy; three yellow sweet; three white starchy; and one white sweet. Presumably each of these observers knew the results expected. Pearl's results are shown in table 1.16. He remarked, "It must be remembered that each individual handled, sorted, and counted *the same identical kernels of corn.* They were required to discriminate only with reference to the color and the form of each kernel. *Yet no two of the fifteen highly trained and competent observers agreed* as to the distribution of these 532 kernels" (Pearl 1940, 87).

Although Pearl's data show the difficulty of obtaining repeatable results, they do not reveal any subconscious bias on the part of the observers. Even though the observers presumably knew what to expect, the results show fluctuations well beyond what one might expect if these were different sets of kernels that were randomly distributed, and larger fluctuations than he expected for good observers observing the same kernels. These results would seem to argue that Mendel's excessively good fit to expectations *was* rather the result of some sort of "sophistication" or bias, not an unbiased observation.

Wright also considered the agreement in the case of the 2:1 ratio experiments as the most serious evidence of fraud but noted that Mendel had presented wide variations in his data that "would hardly have been reported by one bent on fraud" (174). He also suggested that Mendel's claim that he used only ten plants in such experiments should not be taken literally. He further suggested that it would take only very small changes in Mendel's data, caused by bias, to allow his results to agree so well with expectations. He concluded, "Taking everything into account, I am confident, however, that there was no deliberate effort at falsification" (175).

J. M. Thoday (1966) continued the discussion. He was an admirer of both Mendel and Fisher and remarked, "Fisher in a classic essay, which

TABLE 1.16 Showing the classification of the kernels of ear no. 8 by the different observers (from Pearl 1940, 87)

Observer	Classes of kernels					
	Yellow starchy	Yellow sweet	White starchy	White sweet	Total starchy	Total sweet
Mendelian expectation	299.25	99.75	99.75	33.25	399.00	133.00
I	352	102	52	26	404	128
II	322	49	82	79	404	128
III	298	75	108	51	406	126
IV	332	101	71	28	403	129
V	305	101	86	40	391	141
VI	313	100	90	29	403	129
VII	308	86	95	43	403	129
VIII	311	101	92	28	403	129
IX	327	101	78	26	405	127
X	308	92	95	37	403	129
XI	311	97	92	32	403	129
XII	313	99	91	29	404	128
XIII	308	97	95	32	403	129
XIV	312	104	91	25	403	129
XV	333	97	73	29	406	126
Totals	4753	1402	1291	534	6044	1936
Means	316.87	93.47	86.07	35.60	402.93	129.07

is a model that every would-be historian of science should be required to digest, made a thorough logical and statistical analysis of Mendel's paper" (122). Thoday noted the problems presented by both the overall fit to the data and by the 2:1 ratio experiments. He went on to offer a biological explanation for Mendel's results: "Diffident as I am in taking issue with one of Fisher's standing," Thoday remarked, he believed that there was still a question concerning the explanation of Mendel's data (122). He remarked that the data were too good only if one assumed that the distributions were in accord with binomial expansions.

> In fact, a little reflection shows that they [the distributions] should not, for though egg cells no doubt come at random, pollen grains do not, they come in tetrads, and tetrads give exact ratios. Unless therefore the number of pollen grains is vastly in excess of the number of ovules and the many tetrads are thoroughly randomized, we expect ratios that are better than Fisherian. How much departure from Fisherian ratios we expect will depend on many factors in the biology of the particular organism. No doubt it also will vary with the particular conditions of growth and so on, and we cannot judge these for Mendel's peas in Mendel's garden a century ago. A recent paper by Dr. Ursula Philip has come up with exactly the same hypothesis to explain excessively good data with the sea weed fly (Coleopa frigida). (Thoday 1966, 123)[37]

If there is a reduced expected variance in peas, then Fisher's argument concerning the overall goodness of fit will no longer hold. A reduced expected variance results in a larger χ^2 (see this volume's appendix), thus making Mendel's results more probable on the basis of chance. This explanation does not, however, account for the very good agreement in the gametic ratio experiments in which the plants were artificially fertilized.

George Beadle (1967) made the same suggestion:

> Pollen populations in a given anther or flower are definitely finite and produced in *exactly* a one-to-one ratio for a given gene pair, because each mother cell gives rise to four daughters, the two kinds always distributed in a two-to-two ratio. Furthermore, since pea flowers are self-fertilized before they fully open, the pollen grains that fertilize the flower that produces a given pea pod are very few in number and likely to come from an even smaller number of two-to-two quartets. Thus if one pea seed in a pod carries one allele of a pair from the pollen, the next one is more likely than not to carry the alternate allele. (Beadle 1967, 337)[38]

He further remarked that he and Sturtevant had "explored this possibility to see if it was sufficient to account for the apparent bias. It works in the right direction but is not sufficient, even if alternate forms of pollen grains are assumed to have functioned always in an exact one-to-one ratio" (337). No details of the calculation were presented. Beadle decided that the best explanation for Mendel's results was unconscious bias that resulted from Mendel stopping his count when the results looked good, remarking: "I therefore conclude that Mendel was very human, but not dishonest" (338).

In 1966, Franz Weiling began what would be a twenty-five-year defense of Mendel (Weiling 1965; 1966; 1971). Weiling did not believe that Mendel engaged in fraud and rejected Fisher's views that the falsification was done by an assistant and that Mendel's paper was a demonstration. In fact, Weiling denied that Mendel had an assistant. De Beer, on the other hand, named three such assistants, Alipius Winkelmayer, Josef Lindenthal, and Josef Maresch (De Beer 1964, 200).[39]

TABLE 1.17 Mendel's segregations and their χ values (excerpted from Edwards 1986a, table 2)

Character	Expected	Observed		Total	χ
		F_3, Plant characters			
Color of seed coats	0.63 : 0.37	64	36	100	+0.2252
Form of pods	0.63 : 0.37	71	29	100	+1.6743
Color of unripe pods	0.63 : 0.37	60	40	100	−0.6029
Position of flowers	0.63 : 0.37	67	33	100	+0.8462
Length of stem	0.63 : 0.37	72	28	100	+1/8813
Color of unripe pods	0.63 : 0.37	65	35	100	+0.4322
		F_3, Trifactorial experiment, plant character			
Color of seed coats, among *AaBb*	0.63 : 0.37	78	49	127	−0.3488
Color of seed coats, among *AaBB*	0.63 : 0.37	38	14	52	+1.5174
Color of seed coats, among *AABb*	0.63 : 0.37	45	15	60	+1.9384
Color of seed coats, among *AABB*	0.63 : 0.37	22	8	30	+1.1816
Color of seed coats, among *Aabb*	0.63 : 0.37	40	20	60	+0.6020
Color of seed coats, among *AAbb*	0.63 : 0.37	17	9	26	+0.2610
Color of seed coats, among *aaBb*	0.63 : 0.37	36	19	55	+0.3903
Color of seed coats, among *aaBB*	0.63 : 0.37	25	8	33	+1.5276
Color of seed coats, among *aabb*	0.63 : 0.37	20	10	30	+0.4257
Total	0.63 : 0.37	720	353	1073	+2.8408

Weiling considered other experiments on peas, including several of those discussed by Sturtevant (table 1.17), and concluded that there was no great difference between Mendel's results and those of the later experiments. Weiling further stated that if one thought that Mendel had corrected his data, then the same comment should be made about Correns, von Tschermak, and geneticist Arthur Darbishire. This is in contrast to Sturtevant's conclusion that the data were "neither too good nor too poor."

Weiling also suggested that in the 2:1 ratio experiments, fewer than 10 plants survived. He stated, in contrast to Fisher, Dunn, Sturtevant, and Wright, who suggested that a number larger than 10 was required to improve the fit between Mendel's observed results and his expectations, that this smaller number would achieve that result. Despite Weiling's arguments, the fact is that this smaller number actually worsens the fit.

Weiling also questioned whether χ^2 analysis was appropriate for analyzing Mendel's data because it assumes a binomial distribution. He emphasized that Mendel's pea plants did not, in fact, obey the expected binomial distribution, citing the tetrad model of pollen. He concluded that this would result in a variance that was reduced by a factor c, which was a function of the number of surviving plants, resulting in a higher χ^2. Weiling found that if he assumed that 8 out of 10 plants survived, that c was between 0.7 and 0.8, whereas for 9 out of 10 plants surviving, c was between 0.6 and 0.7. This resulted in a good fit to randomness and a much less improbable value of χ^2. Weiling also raised the issue of whether Mendel's results on the 2:1 ratio experiments should be compared to the 2:1 ratio Mendel expected or to the correct value of 1.8874:1.1126. He calculated the χ^2 using the corrected value and found that the total c^2 increased from 41.6506 to 48.910, which was also highly significant.[40] The probability of a worse fit decreased from 0.99993 to 0.9992. Unfortunately, Weiling seems to have estimated c by asking what value of that parameter would give a reasonable value of χ^2. As Edwards (1986a) subsequently pointed out, this really begs the question.

Theodosius Dobzhansky (1967), in a review of six then-recent books on Mendel and the history of genetics, offered a somewhat different explanation for Mendel's results.

Few experimenters are lucky enough to have no mistakes or accidents happen in any of their experiments, and it is only common sense to have such failures discarded. The evident danger is ascribing to mistakes and expunging from the record perfectly authentic results which do not fit one's expectations. Not having been familiar with chi-squares and other statistical tests, Mendel may have, in perfect conscience, thrown out some crosses which he suspected to involve contaminations with foreign pollen or other accidents. (Dobzhansky 1967, 1588)

Dobzhansky also cited Mendel's publication of his *Hieracium* (hawkweed) data (Mendel 1870), discussed earlier, which disagreed with both his expectations and his previous results, as evidence that Mendel accepted discordant data and had been unlikely to engage in falsification.

A novel suggestion concerning hypothetical experiments was put forth by Bartel Leendert van der Waerden (1968). Van der Waerden examined the effect on the χ^2 if a hypothetical experimenter continued performing an experiment until there was sufficient agreement with his expectations. For calculational purposes he assumed that such an experimenter continued with a sequence of experiments until the χ^2 for that particular experiment was ≤ 1.69.[41] Van der Waerden then showed that such a procedure might account for overall goodness of fit in Mendel's results. There is, however, no evidence that Mendel did this. Van der Waerden went on to discuss the gametic cell experiments and stated, "Yet, I cannot but agree with Fisher's conclusion that the data have probably been biased. By the time the last series of five experiments was performed, Mendel and his gardening assistants knew too well what they had to expect" (van der Waerden 1968, 287). Van der Waerden also discussed the trifactorial experiment, remarking that Mendel might have planted 12 or 15 seeds or used back-crossing as an additional check: "One or two back-crosses of each of his 473 doubtful plants with cc-plants would reduce the systematic error of his counts by a factor of ½ or ¼" (285). He admitted that although the data for the later years were probably biased, the evidence was not quite as strong as Fisher believed, and he thought that the results of the original 3:1 experiments were probably not biased.[42]

A review of the status of the controversy was written by Vítězslav Orel (1968). It included virtually all of the papers that we have discussed in this section. He added both Jaroslav Křiženecký and Robert Olby (1966) to the group of people who believed that Mendel had stopped counting when his results looked good. He also cited the work of Herbert Lamprecht (1968), who thought that either selective pollination or mutant genes that affected pollination might account for the reduced variance in Mendel's results. Orel, however, thought that Lamprecht's omission of the arguments given by others greatly weakened his case. Orel concluded that given all the evidence, "there are no proofs either of the possibility that Mendel was deceived by somebody, or that he had unconsciously or deliberately falsified the data from his experiments" (Orel 1968, 778). He offered the hope that "the story of 'too good' results in Mendel's experiments may be closed by quoting Dobzhansky: 'Far from this, he was a most careful experimenter and a most penetrating analytical mind'" (778).

Contrary to Orel's hope, the controversy continued. His summary was too optimistic. Although a possible biological explanation of Mendel's "too good" results had been offered in the tetrad-pollen model, there was no evidence that it applied to pea plants, or if it did, that it was sufficient to account for the results. None of the plausible explanations of Mendel's results—deception by an assistant, conscious or unconscious bias, or stopping the count when the results agreed with his expectations—had very much supporting evidence, and none of them were universally accepted.

The Middle Period

There were few papers published on the Mendel-Fisher controversy during the 1970s. One interesting mention occurs in A. W. F. Edwards's book, *Likelihood* (1972). In his discussion of the χ^2 test, Edwards presciently remarked, "It would be interesting to rework Fisher's analysis" (190). Edwards also referred the reader to discussion of χ^2 contained in Fisher's *The Design of Experiments* (1935), written only a year before Fisher's analysis of Mendel's data.

Margaret Campbell (1976) summarized the situation regarding the possible explanations of Mendel's results. These included suggestions, previously mentioned in this discussion, that Mendel may have: (1) been deceived by an assistant, (2) faked his results, (3) stopped counting when the results agreed with his expectations, (4) subconsciously favored his expected results, (5) been able to distinguish genotypes from observation of phenotypes, or (6) been lucky. Campbell rejected the first suggestion because she thought that, given the care with which Mendel performed his experiments, it was unlikely that he would have delegated such an important task to an assistant. The second was deemed unlikely because had Mendel wanted to fake his results, he could have withheld results or used selection procedures that would have given better results. She also noted that Mendel had presented data that disagreed with his expectations as an argument against his possible falsification of data. Campbell offered an interesting and novel argument against the view that Mendel had planted seeds from more than 100 plants in the 2:1 ratio experiments. Had he done so, she said, "he could have selected from which 100 lots of 10 he made his observations and there would seem no reason to repeat an experiment [Experiment 5] had this been his procedure" (162). She noted that Experiment 7 on stem length, which gave a large observed deviation from the expected result, was not repeated. If Mendel had selected for hybrid vigor, which this trait showed, then Campbell thought that the results should have been closer to expectation. Campbell also regarded Mendel's comment "that all members of the series developed in each successive generation should be, *without exception*, subjected to observation" (80), as an argument against the view that Mendel stopped counting when the results agreed with his expectations. She believed that Mendel may very well have been able to determine genotypes from phenotypes, in at least some experiments, and conceded that luck might have been a factor in Mendel's results.

Campbell also asked whether an appropriate statistical model had been used. She noted both Beadle's suggestion of the tetrad-pollen model and his failure to account for Mendel's results using this model. She concluded that subconscious bias combined with an ability to detect heterozygosity in the absence of recessives, along with Beadle's tetrad-pollen model, might be sufficient to explain Mendel's results in the 2:1 ratio experiments. No calculations were presented.

Two statisticians proposed alternatives to Fisher's χ^2 analysis. Tom Leonard (1977) suggested a Bayesian approach to multinomial estimates. He applied his technique to Mendel's data and stated, "We have no wish to enter into the general controversy but simply wish to make the point that, in situations like this, with χ^2 smaller than the degrees of freedom, the raw proportions could still give better estimates" (873). He argued that his method gave a lower probability of a worse fit to Mendel's results than did Fisher's calculation. Tim Robertson (1978) proposed an order restriction on multinomial parameters.[43] When he applied these restrictions to an analysis of Mendel's data he, too, found a reduced probability.

In a later review of the controversy, Walter Piegorsch (1983) commented, "Both of these approaches provide somewhat more reasonable interpretations of the fit for this particular subdivision of the data. Their overall impact, however, helps advance the debate only slightly. Still, the various statistical and experimental alternatives proposed do tend to give one the impression that Fisher's original conclusions may have been a bit extreme" (2300). Piegorsch's review also included a detailed summary of Weiling's views, including his argument for a reduced variance because of the tetrad-pollen model.[44] Piegorsch commented that Weiling's view "has not received a great deal of further attention" (2298).[45]

Another review of the controversy was presented by Robert Root-Bernstein (1983). One criticism he raised was that in analyzing the 2:1 ratio experiments, Fisher should have calculated the χ^2 using the corrected values 1.8874:1.1126 rather than 2:1. In that case, Mendel's fits are about what one would expect on the basis of probability. He commented that Fisher used the same "ploy" in analyzing the trifactorial experiment.

Root-Bernstein also added a new explanation of Mendel's "too good" results. He argued that Mendel had attempted to segregate continuously varying quantities into discrete categories. He presented evidence from Pearl's experiments, discussed earlier, and from experiments of his own, on the difficulty of obtaining repeatable results. In one of his experiments, he asked undergraduate students to classify maize kernels into two groups, yellow and purple. He separated the students into two groups. One group was asked to use only the categories "yellow" and "purple," whereas the second group was allowed to add a third category, "indeterminant." The first group obtained a poor fit to the expected results. For the second group, Root-Bernstein first excluded all the "indeterminant" kernels and obtained a good fit to the expectations, "although these were not better than one would expect on statistical grounds.... Finally, students were asked to reassign the 'indeterminant' kernels to the 'ideal' categories so as to achieve the closest possible approximations to the expected Mendelian ratio. This group of students produced results that were as statistically unlikely as those reported by Mendel" (284).[46]

Root-Bernstein argued that this was, in fact, Mendel's method. He also concluded, on the basis of these experiments, that there were a sufficient number of "difficult to classify" (284) individuals to account for Mendel's results. He stated that in Mendel's work on seed-coat color, Mendel originally began with six categories and, when he could not obtain good fits to his expectations, he reduced the number to the three expected. "One could hardly ask for better evidence that Mendel recognized the arbitrary nature of his imposition of discrete categories on nature and was willing to revise them to fit theory" (288).[47]

Root-Bernstein recognized that he had, at least at first glance, offered an argument in favor of fraud on Mendel's part. He concluded, however, "If Mendel was conscious of manipulating his data to achieve the verification of his hypothesis, does this not make him guilty of the charge of fraudulence? Did he not 'doctor' his data as has been alleged? I believe not. One can allege fraud only if one can demonstrate that an objective truth presented itself to Mendel and he ignored it in favor of some preconceptions. In fact, the point of this essay has been to argue that Mendel's peas did not represent an 'objective truth' that Mendel could unambiguously interpret" (289).

I believe that there are several problems with Root-Bernstein's explanation, not the least of which is that it is inconsistent with what Mendel himself stated. Mendel said quite emphatically that he had used traits that could be easily distinguished and that there were others, for which the categorization was not as easy, which were not used. Mendel further commented, "In counting the seeds, also, especially in Expt. 2, some care is requisite, since in some of the seeds of many plants the green colour of the albumen is less developed, and at first may be easily overlooked.... In luxuriant plants this appearance was frequently noted. Seeds which are damaged by insects during their development often vary in colour and form, but, with a little practice in sorting, errors are easily avoided" (86–87). In order to absolve Mendel from fraud, Root-Bernstein would make him a liar. I also question Root-Bernstein's view that an objective truth that was ignored is a requirement for fraud. Mendel clearly did not realize at the time he performed his experiment that he could not accurately count peas, whether this is true or not. In all probability, Mendel also had a good idea of the results he expected. Thus, contrary to Root-Bernstein, if Mendel had manipulated his data to agree with an incorrect, or even impossible, expectation, this would be evidence of fraud.[48]

The defense of Mendel was continued by Ira Pilgrim (1984). Pilgrim stated that his purpose was "to demonstrate that Fisher's reasoning was faulty and to clear the name of an honest man" (501). He faulted Fisher's analysis of Mendel's work this way: "Here is the paradox: the closer his results are to his expectations the less credible they become and the farther they are from his expectations the more credible they become. In other words, if his results are excellent, he is accused of dishonesty, and if his results are poor, they do not support his theory" (501). Pilgrim also maintained that Fisher obtained such a low probability for Mendel's results by successively multiplying the probability of each of Mendel's results. Thus, because the probability of any result is less than one, the probability will decrease with an increasing number of experiments. He further argued that Fisher assumed that getting a royal flush in a poker hand was evidence of cheating because of its low probability. Pilgrim concluded that fictitious data can never be discerned by statistical means, but "only the time honored method of the critical repetition of the work can do that [corroborate results], as it did for Mendel's work" (502). In sum, Pilgrim stated, "There is no evidence that Mendel did anything but report his data with impeccable fidelity. It is to the discredit of science that it did not recognize him during his lifetime. It is a disgrace to slander him now" (502).

Pilgrim was answered by A. W. F. Edwards (1986b). Edwards agreed with Pilgrim that an extremely good fit to one's expectations in a single experiment is not evidence of fraud, but that in Mendel's case it applied to 84 different experiments. "One can applaud the lucky gambler; but when he is lucky again tomorrow, and the next day, and the following day, one is entitled to become a little suspicious" (138). Edwards also corrected Pilgrim's error by asserting that Fisher did not, in fact, perform multiplication of successive probabilities. Fisher added independent χ^2 values, a justified mathematical procedure. Edwards concluded, "[Fisher's] grounds for concluding that Mendel's data were falsified were not that it was exceedingly improbable that they would recur exactly on a repeat of the experiment (which it is) but that it was very improbable than *any* results so close to expectation would recur on a repeat" (138).

Pilgrim (1986) responded by admitting his error on the multiplication of probabilities. He claimed, however, "If data are honest and 'good' adding the chi-squares will increase the P value, making good data seem excessively good" (138).[49] Pilgrim agreed with Edwards about the "lucky gambler" but believed that the evidence was insufficient to support an accusation of cheating. "However, one had better have a good deal more evidence (such as a set of loaded dice or perhaps the information that the man is a known cheat) before accusing someone of cheating, which is what Fisher did to Mendel" (138).

Floyd Monaghan and Alain Corcos (1985) joined the debate on Mendel's behalf. They maintained that there was, in fact, no evidence of bias in Mendel's results. If there were bias, they argued, and Mendel knew what results to expect, then one should expect the later χ^2 values to be smaller than the earlier ones. Monaghan and Corcos found no evidence that this was so. When they examined the gametic ratio experiments they admitted, "Obviously, these chi-square values are small. Since these experiments were done to test his hypothesis of gametic formation, one could think that Mendel in this case could have consciously or unconsciously biased his results" (308). They also calculated and summarized the χ^2 values for all of Mendel's experiments and found a probability of between 0.95 and 0.99 of obtaining a worse fit, which they did not seem to regard as remarkable. Monaghan and Corcos examined the seven early-twentieth-century experiments and noted that the results of Bateson and von Tschermak deviated widely from those of other experimenters. They concluded that although Fisher's statistical procedure "is undoubtedly correct, the conclusion seems to us to be illogical. We have a series of independent experiments, none of which shows evidence of bias and whose chi-square values show no systematic trend. Yet the sum of these individually unbiased experiments is judged as showing bias" (309). Recall, however, Edwards's comment about a gambler being lucky on successive days. Monaghan and Corcos also suggested that the solution to the problem might lie in the biology of egg and pollen formation as suggested by Beadle and Thoday.

Olby (1985), in a revised version of his 1966 book, noted the arguments of Weiling concerning semi-random pollination and van der Waerden's suggestion of sequential experiments. He presented further evidence that Mendel had stopped his counts when the results were close to his expectations by arguing that the number of seeds counted was smaller than one would expect given the number of plants Mendel had examined.

Weiling (1986) continued his defense of Mendel. He added to his argument for reduced variance by citing the work of Evans and Philip (1964), Thoday (1966), and his own previous work. He also examined seven other early-twentieth-century experiments on seed color and, although four of the eight experiments (which included Mendel's experiment) showed χ^2 values greater than that for a probability greater than 0.5 and four less, Weiling believed they provided evidence in support of reduced variance. (The overall χ^2 for the eight experiments was 6.368, whereas the χ^2 for probability equal to 0.5 was 7.34.) Weiling further argued that for biological reasons, the variances for the different experiments performed by Mendel were, in fact, different, unlike the homogeneous variance assumed by Fisher. He also argued that in the 2:1 ratio experiments, one should use a hypergeometric, rather than a binomial, distribution because Mendel's 10 plants were chosen from a finite, not an infinite, number of seeds. Weiling also commented on the fact that Fisher's statement of "too good to be true" data had led popular authors to consider Mendel a "betrayer of the truth" (see Broad and Wade 1982).[50] He expressed the hope that because, as he believed, he had shown that Fisher's two decisive statements were incorrect, that the "defamatory questioning of Mendel's accuracy henceforth will stand corrected" (283).

The next article in the same journal was a correction offered by Corcos and Monaghan (1986) of their 1985 paper. Weiling had pointed out that they had incorrectly stated that the number of degrees of freedom for Mendel's testcross experiments was 9 when it was actually 15. This reduced the χ^2 per degree of freedom considerably and gave a probability of a worse fit of greater than 0.995, which they admitted was evidence of bias. They suggested that because Mendel knew what to expect, he might not have been as careful as he should have been in scoring phenotypes. Weiling had also pointed out that they had made an error in reporting the number of plants observed by von Tschermak. Corcos and Monaghan had found an error in their report of Bateson's results. Both of these errors reduced the χ^2 of these experiments, and they claimed that this supported their view that Mendel and these other experimenters were equally biased. Thus, they stated, "The above errors ... with the possible exception of the testcross data, in no way alter our conclusion" (283).

The middle period closed with A. W. F. Edwards's paper, "Are Mendel's Results Really Too Close?" (1986a; chapter 4 of this volume). In this paper, Edwards provided a survey of previous explanations of Mendel's results, a critical review of the issues raised about Fisher's statistical analysis, a discussion of the relevance and appropriateness of using χ^2 analysis, and a new analysis of Mendel's data. Edwards noted, "Fisher's 1936 conclusion slowly became the received wisdom, but his painstaking analysis and his defence of Mendel's integrity have sometimes been incorrectly reported as having exposed a scientific fraud of major proportions" (*142*).

Edwards began by surveying the explanations offered for Mendel's results. He remarked that while a large number of commentators had accepted the suggestion that Mendel had stopped counting peas when his results agreed with his expectations, "The difficulty is that there is no evidence whatsoever that Mendel did this, and in many of the experiments his description clearly excludes the possibility, so that to accept it as a means of exonerating Mendel from having reported data which had been adjusted in some way is to saddle him with the charge of not having reported his experimental method accurately" (*144*). Edwards also considered the suggestion that peas exhibit a non-binomial variability, the tetrad-pollen model of Beadle and Thoday, and cited Beadle's comment that it was not sufficient to explain Mendel's data. Edwards also presented evidence that peas did exhibit normal variability, a point that will be discussed further later in this introduction. In a detailed consideration of the arguments offered by Weiling, Edwards remarked that Weiling had suggested that one might explain the results of the 2:1 ratio experiment by assuming that only 8 of the 10 seeds sown produced plants. Edwards noted that this would make Mendel's fit to the correct results even worse and cited Sturtevant, Wright, Dunn, and Fisher, all of whom suggested that one could explain the results only if Mendel had examined *more* than 10 plants. "Nothing could illustrate better than these two opposing theories the ingenuity that has been expended on accounting for Mendel's surprisingly good data. Here are two explanations, one postulating that *fewer* than ten plants were scored, and the other that *more* than ten were scored, both with a view to accounting for results judged too close to a 2:1 ratio" (*145*). Edwards further remarked that there was, in fact, no evidence that Mendel had done either.

Edwards devoted considerable attention to previous statistical analyses. He raised the issue of whether Fisher should have computed his χ^2 by comparing Mendel's data to Mendel's expected 2:1 ratio or to the corrected value of 1.8874:1.1126. Edwards noted that this issue was relatively unimportant because Weiling (1966) had adjusted the total χ^2 using the corrected value and found that the value increased from 41.6506 to 48.910 and that this value was also "highly significant" (*147*). (The probability of a worse fit to the data decreased from 0.99993 to 0.9992.) Edwards also considered Weiling's suggestion of a binomial variability reduced by a factor of c. "Of course, once one has estimated c (for which Weiling found the broad limits 0.6–1.0) there is nothing left to test" (*148*). Edwards presented the evidence from the experiments of Bateson and Kilby (1905) and of Darbishire (1908; 1909) for which Weiling had calculated a total χ^2 of 1008.79 for 1062 degrees of freedom. This gave an estimate of c of 1008.79/1062 = 0.95. "In other words these massive data exhibit a standard deviation of about 97.5% of that expected on a binomial model, and far from this lending support to the biological hypothesis that the pea does not segregate randomly, it fills one with admiration for the perfection of the randomizing mechanism!" (*148*). Of Weiling's calculation that had found that 463 one degree of freedom experiments had χ^2 values less than the theoretical median of 0.4529 and 427 above that median, Edwards said the data "lend no real support to the infra-binomial-variance hypothesis" (*148*).

While Edwards offered, "Weiling is to be congratulated for his determined attempts to rescue the Mendelian experiments from the Fisherian conclusion" (*149*), he concluded, "But in the end the attempt fails: there is too much to explain. The diminished-variance hypothesis is not supported by other more extensive data, and, as I show in my own analysis below, simply reducing the variance is in any case not enough to explain the peculiarities of Mendel's data" (*149*).

Edwards found the defense of Mendel offered by Monaghan and Corcos less than admirable. He noted that many of the χ^2 values given were incorrect to the accuracy claimed, that their addition of the χ^2 values was wrong, and that their report of the results of Bateson and of von Tschermak was incorrect. Referring to the data contained in table III of Monaghan and Corcos (1985), Edwards wrote that it "contains the astonishing assertion that $0.95 < P < 0.99$ for $\chi^2_{67} = 32.57$, on which obvious error the authors hang their conclusions" (*149*). (The probability is, in fact, 0.9998.)

Edwards concluded, "My overall impression from reading all of the commentaries since Fisher (1936) is that a good deal of special pleading, not to mention downright advocacy, has failed to make any substantial impact on Fisher's conclusion" (*150*).

The issue of the appropriateness of using χ^2 analysis was also discussed by Edwards. He noted that the total χ^2 was greatly influenced by large individual χ^2 values, for which the probability was low. "The criticism of Mendel's results is that they are too close to expectation in the normal space as judged by being too far from expectation in the χ^2 space" (*151*). He then offered the following amusing illustration. "Suppose the χ^2 test had antedated Mendel, and that in his paper he had reported a value of 84.0000 on 84 d.f. [degrees of freedom] The reaction of a latter-day Fisher might well have been to conclude that Mendel's assistant had known that what Mendel really needed for his paper was not good Mendelian ratios but a good value of χ^2" (*151*). He ended his discussion of χ^2 with this cautionary statement: "To sum up, we must not allow our judgment to be dominated by tests of significance and other calculations of probability which are at best pointers for further thought and at worst misleading" (*152*).

The most significant aspect of Edwards's paper was his new analysis of Mendel's data. Edwards chose to work with χ rather than with χ^2, because it preserved the direction of the departure from expectation. He also chose to compare Mendel's results on the 2:1 ratio experiments to the corrected value of 0.6291:3709 rather than the 2:1 ratio Mendel expected. The results for those 15 experiments are shown in table 1.16. (All of the data are shown in table 2 in chapter 4 of this volume.) He found that of the 15 χ values, 13 were positive and 2 negative, "suggesting something of a bias towards the larger class" (154). The total numbers of plants that Mendel observed in these experiments was 720:353 (heterozygote:homozygote), or 0.6710:0.3290, "with an associated χ value of +2.8408, indicating a very poor fit [to 0.6291:0.3709] indeed" (154). (The probability that such a deviation is due to chance is 0.0045.) Edwards believed this result substantiated "what Sturtevant (1965) called 'Fisher's most telling point' in his argument that the data have been biased in the direction of agreement with what Mendel expected" (154–55).

Edwards then investigated the remaining 69 results obtained by Mendel. He found that the mean values were unexceptionable, but that the sum of χ^2 values for these 69 results was 30.8138, which, he said, is "remarkable on any interpretation of tests of significance" (155). (The probability of a worse fit is 0.9999.) He then went "further than any previous analysis of Mendel's data" (159) and examined the distribution of the 69 χ values, omitting the 15 results from the 2:1 ratio experiments. Edwards's results are shown in figure 1.1. He contrasted these with the first 69 random normal deviates from the tables provided by Lindley and Scott (1984) (figure 1.2). If the data were randomly distributed, these graphs should fit the straight line shown in the figures. This is true for the random deviates (see figure 1.2), but not for Mendel's data (see figure 1.1). "It is immediately obvious that the reduced variance is not characteristic of the whole data, as Weiling's theory would require, but is confined to the tails of the distribution, where the extreme variates are not extreme enough to conform to expectation" (159). The graph of Edwards's results shows no significant deviations in the region between the 25th and 75th percentiles, but does show an excess of observations between the 5th and 25th and the 75th and 95th percentiles, and a lack of observations between 0 and 5% and between 95 and 100%. "The inescapable conclusion," Edwards wrote, "is that some segregations beyond the outer 5-percentiles (approximately) have been systematically biased towards their expectations so as to fall between the 5-percentiles and the 25-percentiles. Further analysis shows that the effect is not confined to particular sample sizes or segregation ratios, but is quite general" (159). Edwards did not speculate on how the data had been adjusted.[51]

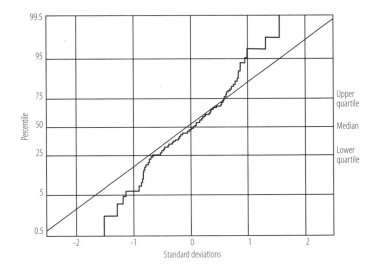

FIG. 1.1 The values of χ for the 69 segregations with undisputed expectations plotted on normal probability paper

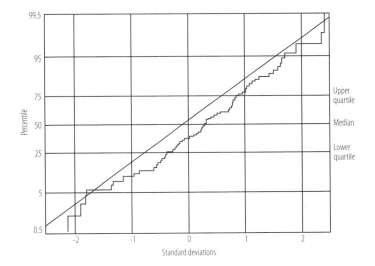

FIG. 1.2 The values of 69 simulated standard normal deviates plotted for comparison with fig. 1.1

Edwards prefaced his concluding remarks by noting that he had originally thought Fisher's analysis might be faulted because of Edwards's own doubts concerning χ^2 analysis, "but a complete review of the whole problem has now persuaded me that his 'abominable discovery' must stand" (*161*). He thought, however, that "any criticism of Mendel himself is quite unwarranted" (*161*). He also suggested, in agreement with Dobzhansky, that Mendel should not be judged by modern standards of data recording; he quoted Dobzhansky's view: "not having been familiar with chi-squares and other statistical tests, Mendel may have, in perfect conscience, thrown out some crosses which he suspected to involve contamination with foreign pollen or other accident" (Dobzhansky 1967, 1589). Edwards's final comment was, "In the words of my title, Mendel's results really are too close" (162).

Thus, by the middle of the 1980s, the Mendel-Fisher controversy was no nearer to a solution than it had been in 1970. Fisher's analysis was still seen as correct by most scholars, there was no generally accepted explanation of why Mendel's results were "too good," and the additional mystery of the distribution of χ values pointed out by Edwards had no apparent explanation.

The Contemporary Period

Weiling (1989) continued his defense of Mendel and argued that the model used by Fisher in calculating the χ^2 values was incorrect. He again argued that, for botanical reasons, the hypergeometric distribution, which gives a smaller variance, and thus a larger χ^2 value and a "less good" fit to Mendel's data, should be preferred to the binomial distribution used by Fisher. Weiling further suggested that the samples contained in Mendel's data were too small to allow the appropriate use of χ^2 analysis. He also claimed that his analysis showed that all of the data were moved to smaller χ^2 values, not just the larger deviations moved inwards as shown by Edwards.[52] With regard to the 2:1 ratio experiments, Weiling criticized Fisher for using Mendel's expected 2:1 ratio rather than the corrected value of 1.8874:1.1126 in calculating the χ^2 values for these experiments. He did not, however, mention his own calculation, discussed earlier, which found that the total χ^2 increased from 41.6506 to 48.910 and that this value was also "highly significant." (The probability of a worse fit to the data decreased from 0.99993 to 0.9992.) Weiling further argued that in considering the 10 seeds planted in these experiments, one should calculate the probability using a "balls taken from an urn without replacement" model. For a plant with an average of 30 seeds, this results in a probability of 23/30 × 22/29 × 21/28 × 20/27 × 19/26 × 18/25 × 17/24 × 16/23 × 15/22 × 14/21 = 0.0381. In that case the probability of misclassification would be 0.0381 rather than the 0.0563 calculated by Fisher.[53]

Fisher's contributions to genetics were reviewed by Piegorsch (1990), who included a discussion of the Mendel-Fisher controversy. He noted that Mendel's data exhibited unusually close agreement with his [Mendel's] expectations and: "The controversy over the data's nature and origin remains unresolved" (921). Piegorsch emphasized that Fisher, despite his concern over the possible fraudulent data, was a great admirer of Mendel: "In spite of the concern over the 'cooked' data, Fisher argued that Mendel's experimental methodology was an advance well ahead of its time" (922). Remarking on Fisher's comment that each generation found in Mendel's work only what it expected to find, he noted:

> Even Fisher's work has fallen prey to some modern writers' inability to read past their own (pre)conceptions. Fisher is often identified among popular writers as Mendel's intellectual assassin and scourge for his discoveries regarding the 'too good' fit of Mendel's data. As noted above, however, Fisher *never* argued that the (possible) falsification of the data was Mendel's doing, and valiantly suggested another alternative: the 'overzealous' assistant. Although this particular solution to the data falsification problem remains in doubt, the fact remains that one of Fisher's basic conclusions espoused support and admiration for Mendel's scientific work. It is inappropriate and misleading to paint Fisher in any other light. (Piegorsch 1990, 922)

Jan Sapp (1990) also supported the view that Fisher was not attempting to discredit Mendel, but rather "meant to celebrate his [Mendel's] power of abstract reasoning" (157). Sapp also discussed several methodological issues raised by the controversy including observer bias, the theory-ladenness of observation, and whether the validity of experimental results could be tested by statistical methods. He discussed a point similar to one made previously by Pilgrim:

> data that looks good to the experimentalist looks bad for the statistician and *vice versa*. There seems to be a methodological incommensurability concerning the nature of statistical and experimental modes of reasoning. This might be called the 'experimentalist-statistician paradox': From a statistical point of view geneticists should not provide 'too much data' and have their results come too close to theoretical expectations, for the closer they come to the 'truth' the less true they will appear to be. This strange paradox is based on faulty reasoning. The reason why data are considered less true the closer they reach theoretical expectations is based on the idea that geneticists *should be* studying a random sample. It assumes that experiments should be carried out independently of the law or theory the observer is using for explanation.... The theory informs the experimenter what kind of experiment to perform, what kind of phenomena to examine how results are to be understood; it also tells the experimenter when the experiment is over. (Sapp 1990, 159)

Sapp regarded this last point as underlying the suggestion that Mendel stopped counting when the results agreed with his expectations.

In 1991, Federico Di Trocchio (1991) proposed a novel way of explaining Mendel's "too good" results. Mendel did not report any linkage between the characters he studied when, in Di Trocchio's view, he should have observed such linkage. Thus, Di Trocchio concluded that Mendel actually performed 22 × 22 = 484 crosses for the 22 varieties of peas he had selected for further experimentation.[54] Di Trocchio agreed with Bateson that Mendel's experiments "are fictitious in the sense that they were made just on paper by disaggregating the data of many polyhybrid experiments" (511). In Di Trocchio's view, Mendel then selected that data that best agreed with his 3 : 1 expectation, and this explains the excessively good fit: "This way of proceeding cannot be considered a form of manipulation or falsification of the data: Mendel simply chose to work with and report the data that showed the closest approximation to the ratio 3 : 1. Since in my opinion Mendel had in his possession a much wider sample than that reported in *Versuche*, he had available to him various tables containing columns of data on the characteristics yellow/green, wrinkled/smooth, tall/dwarf, and so on. There was nothing to prevent him from choosing the table most suitable to his purposes" (Di Trocchio 1991, 514).

As Daniel Hartl and Vítězslav Orel (1992) pointed out, it seems hard to believe that Mendel would have performed such an elaborate procedure "without saying so, since he roundly criticized Gärtner for not describing his experiments in sufficient detail to allow Mendel to repeat them" (246). Hartl also remarked in a personal communication to this author, "Mendel had carried out a trihybrid cross, and complained that of all his experiments it took the most time and effort. Because he knew how much effort went into multifactorial crosses, why would he make these his method of choice, and then conceal it by disaggregation."

Moreover, Fairbanks and Rytting (2001) pointed out on the basis of botanical evidence that "the nature of variation in pea varieties (both old and modern) facilitates, rather than prevents, the construction of monohybrid experiments" and that Di Trocchio's "claim that Mendel hybridized his 22 varieties in all possible combinations runs counter to the experimental design that Mendel described and the logic on which it is based" (283–84). Their analysis fully supports Fisher's conclusion that Mendel's "report is to be taken entirely literally, and that his experiments were carried out in just the way and in much the order that they are recounted" (Fisher 1936, *134*).

Ironically, after having accused Mendel of, at the very least, deception in reporting both his methods and his results and of selectivity and bias in his choice of data, Di Trocchio concluded, "His [Mendel's] research is in no way the fruit of methodological mistakes or forgery, and it remains a landmark in the history of science. . . . We must still consider him the father and founder of genetics" (519).

When Hartl and Orel (1992) discussed and criticized Di Trocchio's view, they did not discuss other aspects of the "goodness of fit" question. They did, however, discuss in detail the questions raised by Fisher in his article on Mendel: What was Mendel trying to discover? What did he discover? What did he think he had discovered? Hartl and Orel also addressed some revisionist views of Mendel's work, including the view that Mendel was not interested in heredity, that he was interested only in hybrids, that he did not perform all the experiments attributed to him, as well as other questions such as the reasons for the neglect of Mendel's work.

In their 1993 text *Gregor Mendel's Experiments on Plant Hybrids: A Guided Study,* Corcos and Monaghan restated their conclusion, citing both their own previous work and that of Weiling, that "Mendel's reputation was indeed restored" (196). They presented a discussion of χ^2 testing and noted that if an experiment had a χ^2 of 0.00015 for one degree of freedom, the probability that that result occurred by chance is less than 1%. They further stated that if a second experiment had the same χ^2, the probability that both experiments would have such a low χ^2 is $0.01 \times 0.01 = 0.0001$. They offered this as part of an argument that there was, in fact, no bias in Mendel's results. Their calculation of probability is correct, but it is not, in fact, what Fisher claimed. His claim was that if all 84 experiments performed by Mendel were repeated, the probability of a worse fit to Mendel's expectations was 0.99993, or the probability of a better fit of 0.00007. This was not calculated by multiplying the probabilities of each of Mendel's results, but by summing their χ^2 values and then computing the probability.[55]

Two new possible explanations of Mendel's results were offered by Moti Nissani (1994). In support of his view that Mendel had not falsified his data, he noted that Mendel had published data that disagreed with his expectations, particularly his results on *Hieracium*, and that Mendel had sent 140 packets of *Pisum* seeds to Nägeli, in the hope that Nägeli would repeat his experiments. After reviewing the history, Nissani stated that he thought that no one had as yet cast any serious doubt on Fisher's analysis, and that the previous explanations of Mendel's results were both out of character and not wholly satisfactory. He then proposed two possible explanations that "involve the contention that Mendel *consciously* presented biased data, but that in doing so he acted honorably and in the best interests of science" (188). Nissani asked the following questions: "What is the proper conduct when a conflict arises between the norm of communicating one's findings and the norm of communicating them as faithfully as possible? What is one to do when the only way to communicate a larger truth is to tamper with some inconsequential details?" (188). Nissani suggested that Mendel, in order to make his contributions, which he believed were of great significance, more acceptable and believable to his intended audience, may have omitted or adjusted his data. Nissani's second explanation posited that Mendel may have omitted data because of length limitations imposed by the editor of the journal.[56]

Nissani quoted Broad and Wade (1982), who argued that "some who commit fraud do so to persuade their refractory colleagues of a theory they know is right. . . . If history has been kind to scientists such as these, it is because the theories turned out to be correct. But for the moralist, no distinction can be made between an Isaac Newton who lied for the truth and was right, and a Cyril Burt who lied for truth and was wrong" (212–13). Nissani concluded, "Sometimes, however, the demand to faithfully report one's data must be sacrificed for the higher value of advancing knowledge. It is, of course, impossible to know whether Mendel (and perhaps also others) faced such a procrustean dilemma. But the information presented in this paper raises the possibility that he may have. In that case, Mendel's choice merits our compassion and thanks, not our disapprobation" (195).

In the year after Nissani's paper was published, Orel and Hartl (1994; chapter 5 of this volume) presented a longer and more detailed discussion of the issues they had raised in their 1992 paper. In particular, they discussed the question of Mendel's 2:1 ratio experiments, which had triggered Fisher's suspicions. They wrote:

Among 600 plants tested, therefore, the true expected ratio is 377:223. However, Mendel reports 399:201, a ratio in much better agreement with 0.67:0.33 (that is, 2:1) than with 0.63:0.37. However, a χ^2 test of the reported ratio against the expected 377:223 yields $\chi^2 = 3.3$, for which $P > 0.05$. The observed result is, therefore, not significantly more deviant from the true expectations than could be expected by chance alone. In other words, this series of progeny tests yields no evidence that the data had been adjusted. (Orel and Hartl 1994, *196*)[57]

The second set of results on the 2:1 ratio, from the trifactorial experiments, was, they noted, more problematic. Assuming 10 seeds were sown from each plant, the probability that the results were due to chance alone was less than 0.01. As Orel and Hartl remarked, Mendel had stated that these experiments had been conducted in a manner quite similar to the preceding one. They also noted that there had been sufficient space in Mendel's garden for more than 5000 plants and that this would have allowed Mendel to have planted more than 10 seeds per plant.

Perhaps what Mendel meant by saying that the method in the second series was "quite similar" to that in the first is that he allocated a certain plot of his garden for the purpose of progeny testing and cultivated as many plants per parent as he could to fill this space. If the space allocation was adequate for 6,000 plants then, in the second series of progeny tests, Mendel could have cultivated an average of 12.7 seeds per parent. He may well have cultivated more than 6,000 plants in the second series because he commented on the amount of work it required, saying that "of all experiments it required the most time and effort." Hence, Fisher's dismissal of the explanation based on more than 10 progeny is too facile, especially in light of Mendel's vague specification of how similar the two experiments actually were and the plausible alternative interpretation of Mendel's text. Fisher's "abominable discovery" is therefore much less damaging than first appears. In short, although Mendel's expectations are certainly wrong, Fisher's expectations may be wrong as well. Thus, the uncertainties in the experiment and the ambiguities in the analysis discredit any inference of deliberate manipulation or falsification of data. (Orel and Hartl 1994, *197*)

Fisher's analysis received support, however, from the work of Charles Novitski (1995). Novitski performed a Monte Carlo simulation of the experiments on gametic ratios, which had enhanced the "too good nature" of Mendel's results (see table 1.6). He generated random events corresponding to Mendel's five experiments on gametic ratios. He considered each set of five experiments as a trial and found that in 100,002 trials only 138, or about 1/725, gave χ^2 values less than the 3.673 calculated by Fisher for these experiments. This agreed quite well with the value 1/728 calculated from the χ^2. Novitski found that Mendel's results were, in fact, even more unusual. He examined the χ^2 values for each of Mendel's experiments and found that in only 47 of the 100,002 trials was the largest χ^2 value as small as the largest value, 1.0843, calculated from Mendel's data. He also computed the variance of Mendel's χ^2 values, which was 0.2183, and found from his simulation that in only 21 of 608 trials with χ^2 between 3.6 and 3.7, the Mendel value, was the variance as low as that of Mendel's data. He estimated that the probability of getting Mendel's results for these experiments was closer to 1/20,000 rather than 1/700. He also concluded that none of the explanations previously offered was sufficient to explain Mendel's results.

An extensive discussion of both Mendel's data and of Fisher's analysis was provided by Teddy Seidenfeld (1998; chapter 6 of this volume). He noted, "Fisher did not write 'Has Mendel's Work Been Rediscovered?' either to question Mendel's integrity or to challenge his rightful place among those at the center of modern genetics" (*216*). He also remarked that Mendel had a good idea of the results he expected. "It hardly needs saying that, therefore, Mendel had well-grounded expectations for his experiments on single (and even double) factor trials involving the 5 plant characteristics, since he had seen the parallel results (at much larger sample sizes) for the two pea characteristics a year earlier" (219-20). Seidenfeld began with a discussion of the misclassification problem in the 2:1 ratio experiments. It was these experiments and the agreement with the "wrong values" that had triggered Fisher's analysis. He remarked that Fisher had argued against the view that Mendel had a selection bias because: "(i) It does not apply to the trifactorial study, where all plants were classified," and "(ii) It is implausible that bias was equally effective for all five characteristics." He continued, "Like Fisher, I find the coincidences of perfectly offsetting selection biases (the second rebuttal point) more difficult to believe even than the alternative that the data were 'cooked'!" (222).

Seidenfeld proposed a new solution to this problem, one he thought compatible with Mendel's own statements. He suggested that when Mendel continued his experiments into subsequent generations, something Mendel stated quite explicitly that he had done, he included an examination of some of the F_4 plants grown from the F_3 plants that showed the dominant trait. If Mendel had tested three such plants, the probability of misclassification was reduced to less than 2×10^{-3}, rather than 0.0563. For only one such plant, the probability is reduced to less than 0.01.[58] Seidenfeld argued, "Thus, one way around Fisher's first objection is to hypothesize that Mendel used an elementary sequential design" (223). Seidenfeld pointed out that a sequential experimental design was quite feasible given the size of Mendel's garden.

Seidenfeld also discussed Weiling's attempt to solve this problem either by suggesting that only 8 of 10 seeds germinate, which, as already noted, worsens the problem, or by invoking a hypergeometric distribution.

> Among several difficulties I have with Weiling's statistics, I do not understand the basis for his use of the hypergeometric distribution. It is true, as he writes, that the process of choosing 10 of 30 particular seeds from a plant (as Mendel is posited to have done to make the 10 F_3 offspring per F_2-parent) follows a hypergeometric distribution, with a smaller variance than the i.i.d. Binomial distribution. However, under Mendelian theory, these 30 seeds follow the i.i.d. Binomial distribution. Hence, the net (marginal) distribution for the 10 seeds, chosen from the 30, is again i.i.d. Binomial, not hypergeometric, contrary to what Weiling asserts. The challenge, taken up below, is to justify the claim that the 30 seeds are not an i.i.d. sample from the Binomial distribution. (Seidenfeld 1998, 251n17)

The question of "too good to be true" data was also discussed by Seidenfeld. He raised several technical objections to Fisher's analysis, particularly the absence of an adequate theory of Fisherian significance testing. Moreover, he discussed the issue of whether Fisher should have used the "corrected" values or Mendel's "incorrect" expectations in computing the χ^2 for the 2:1 ratio experiments. He suggested that it should have been the "corrected" values and calculated the total χ^2 and found, in agreement with Weiling, that the value increased to 48.78 from 41.61, with the probability of a worse fit decreasing from 0.99993 to 0.9975.

Seidenfeld also proposed an alternative model that reduced the variance in Mendel's data, the Correlated Pollen model. This is quite similar to the tetrad-pollen model proposed earlier by Thoday and Beadle.[59] He calculated that this model, under reasonable assumptions, yielded a variance of 74% of that expected for a binomial distribution. This reduced the one-sided probability to approximately 0.96. This was still quite large, but it was not the extraordinary value of 0.99993 obtained by Fisher, and would go some way toward resolving the "too good to be true" aspect of Mendel's data. As noted earlier by Edwards, and discussed in detail below, there are other aspects of Mendel's data in need of explanation.

Seidenfeld remarked that there were two important questions concerning this model. "First, is there evidence to confirm or to refute the speculative genetics that Mendel's peas are not independently distributed within self-fertilizing pods? Second, does it matter to Fisher's analysis if the model of pea genetics is not quite Mendelian but, instead, reflects this alternative distribution of pollen cells? How much of Fisher's .9999 P-value can be explained away with some subtle correlation among the pollen?" (*231*).

With regard to the first question, Seidenfeld noted that Edwards (1986a) had argued that the experiments of Bateson and Kilby (1905) and of Darbishire (1908, 1909) had shown 95% of the variance expected on a binomial model. His own, slightly different, analysis gave 94% of the Mendelian variance. He noted, however, that the number of peas per plant reported by Mendel, 29, was far lower than the 106.5 in the Bateson-Kilby study and the 217.6 reported by Darbishire. A question to be answered, however, is whether peas grown under more severe conditions might exhibit a reduced variance.

Seidenfeld also demonstrated that Mendel's reported result of no more than five recessive seeds in a pod was more likely on the Correlated Pollen model than on a binomial model.[60] He also remarked that the model had no effect on the excellent fit in the gametic experiments, in which artificial fertilization was used.

One important discussion included in Seidenfeld's paper was on the possible model of cheating. In agreement with Edwards, he found that the data "were adjusted (rather than censored) to avoid extreme segregations in the record" (242). He presented the results of Edwards's analysis in deciles of probability values (figure 1.3) and noted that there was a deficit in the lowest deciles, as one would expect if extreme values were omitted, but there is no excess at high probability values. There is, rather, a bulge near the median values. This is also true for Seidenfeld's own analysis of Mendel's experiments (figure 1.4).

What model of cheating, then, can the reader propose that replaces extremely discrepant outcomes with ones clustered about the median of χ^2s? I challenge the reader to try to adjust binomial data from sample sizes in Mendel's experiments, so that the following three features appear in the resulting distribution of P-values from the (1df) [degree of freedom] χ^2s.

1. There is a significant reduction in the left-tail of the Ps.
2. There is no significant departure from uniformity in the right tail of the Ps.
3. There is a significant concentration of the Ps about their median, i.e., about .50.

To be fair to Mendel, this exercise should be attempted without the aid of χ^2 Tables, which distribution, the reader recalls, K. Pearson discovered only in 1900! (Seidenfeld 1998, 242–43)

Seidenfeld concluded that his suggestion of sequential experiments by Mendel removed the problem of Mendel's fit to the "wrong" values in the 2:1 ratio experiments. "Regarding the '3:1' and '2:1' laws of segregation for self-fertilizing hybrids, I propose the Correlated Pollen model, an alternative to the usual Mendelian (i.i.d.) distribution of peas in a pod. The C-P model has the same first moment as the Mendelian model with about 75% of the Mendelian variance. This speculative model is enough to recover the P-values in Mendel's data for each of the two, main Mendelian laws" (245). He admitted, however, that there was no support for the model in the large-scale studies of Bateson and Kilby and of Darbishire. On the other hand, one of Darbishire's experiments, which had a smaller yield of peas per plant, closer to that of Mendel, also showed an anomalous small sum of χ^2s. Seidenfeld offered no explanation for the excellent fit in the gametic experiments: "Where do we stand, more than sixty years after Fisher's shocking allegations against the authenticity of data in Mendel's paper? The allegation of misclassification (of hybrids) admits such a straightforward reply that I no longer find merit in that aspect of Fisher's criticism. But, unless some alternative model with reduced variance, like the C-P model, can be justified, I see little hope of explaining away the Ps that are 'too good to be true'" (244–45).

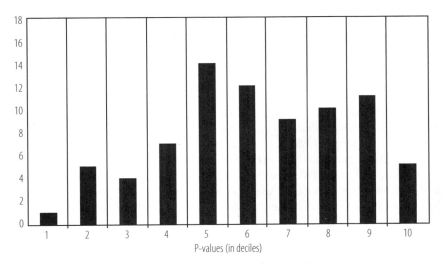

FIG. 1.3   P-values, by deciles, in Edwards's 1-df partition of Mendel's data (N = 78)

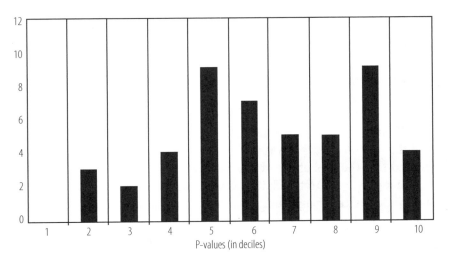

FIG. 1.4   Mendel's data (by experiment) (N = 48)

Seidenfeld also remarked that the C-P model could be subject to field trials. "Regardless the outcome, no matter how peas self-fertilize, I urge the reader to study Mendel's classic paper and Fisher's provocative article. Mendel's work is a standard of clarity and a delight for its intelligent, sequential designs. Fisher, as always, is a brilliant statistician and imposing geneticist. As with many of his other writings, coming to an understanding of how he argues is the key, regardless what the reader thinks, in the end, of his conclusion" (246).

In 2001, Daniel Fairbanks and Bryce Rytting published an extensive review of the Mendel-Fisher controversy. In discussing Mendel's work, they noted:

> There is substantial disagreement about his objectives, the accuracy of his presentation, the statistical validity of his data, and the relationship of his work to evolutionary theories of his day. In the following pages we address five of the most contentiously debated issues by looking at the historical record through the lens of current botanical science: (1) Are Mendel's data too good to be true? (2) Is Mendel's description of his experiments fictitious? (3) Did Mendel articulate the laws of inheritance attributed to him? (4) Did Mendel detect but not mention linkage? (5) Did Mendel support or oppose Darwin? (Fairbanks and Rytting 2001, 265)

For the issue I want to address, the most important section of this paper is "Are Mendel's Data Too Good to Be True?" As Fairbanks and Rytting noted, Fisher believed that the fit of Mendel's data to his expectations was so good that it must be questioned. They further commented that Fisher's analysis was based on consistently low χ^2 values, and that his most telling point was Mendel's close fit to an incorrect value in the 2:1 ratio experiments. For those experiments, Fairbanks and Rytting continued, Mendel had reported a total of 399 heterozygous plants to 201 homozygous plants, an excellent fit to the 2:1 ratio expected. As Fisher had pointed out, the correct expectations were 377.5:222.5. These authors asserted that the observation was not, in fact, a statistically significant deviation and that the probability that it was due to chance was 0.0692. They stated that one needs to take half this value (0.0346) to obtain the chance of a deviation of this magnitude, and in the direction toward Mendel's expectation, which agreed with Fisher's estimate of 1 in 29.

Fairbanks and Rytting also addressed Weiling's claim that one could not use Fisher's independent model, but that one should use a model of balls selected from an urn without replacement. They, and Seidenfeld, argued that Weiling was wrong.

Weiling argued that Mendel sampled ten seeds per plant without replacement in the F_3 progeny tests, and that the sampling, therefore, was not independent. He assumed that the average pea plant in Mendel's experiments had 30 seeds per plant, 23 of which had the dominant phenotype ($0.75 \times 30 = 22.5$, rounded to 23). Based on this assumption, Weiling determined the average probability of misclassification was $23/30 \times 22/29 \times 21/28 \times 20/27 \times 19/26 \times 18/25 \times 17/24 \times 16/23 \times 15/22 \times 14/21 = 0.0381$, instead of 0.0563 as determined by Fisher.

However, although Weiling's estimate is correct for a plant with 30 seeds, 23 of which have the dominant phenotype, it cannot be used to estimate the average probability of misclassification for a population of plants. For any particular number of seeds per plant, the average probability of misclassification must be determined as the sum of the probabilities of misclassification for all possible combinations weighted by the expected frequencies of those combinations according to the binomial distribution. When this is done, the average probability of misclassification is consistently 0.0563. In other words, if Mendel's data are from random seed samples collected from a binomially distributed population, Fisher's estimate of 0.0563 as the probability of misclassification is correct, even when the effect of sampling seeds without replacement is taken into account. (Fairbanks and Rytting 2001, 274)

Fairbanks and Rytting further discussed the possibility of an experimental test of the urn model in peas: "There is currently no empirical evidence to support the urn model in *Pisum*, but it is one that can be empirically tested because it should produce a significant deviation from a binomial distribution for phenotypes of individual seeds from the same pod (or plants grown from those seeds). We have initiated the necessary experiments but do not yet have the results" (279).[61]

The question of whether Mendel had scored exactly 10 plants was also discussed: "The proximity of Mendel's F_3 progeny data to an incorrect expectation is not as questionable as it might seem when viewed in a botanical context" (274). Fairbanks and Rytting suggested that Mendel had probably sown more than 10 seeds to guard against losses due to germination failure. Had Mendel done so, he could have scored seedlings for two of the plant characters: stem length and seed-coat color. The latter was perfectly correlated with axial pigmentation. This would have mitigated the "too good" effect because these two experiments had provided almost half the total deviation of that set of experiments. They pointed out that the same misclassification problem also applied to the trifactorial experiment. They cited Orel and Hartl (1994), who had also presented arguments for Mendel having cultivated more than 10 seeds per plant and that there was sufficient room in Mendel's garden to allow this. Fairbanks and Rytting posited: "When these statistical and botanical aspects of Mendel's F_3 progeny tests are considered, there is no reason for us to question his results from these experiments. *However, we must still account for the bias that is evident when the data for all of the experiments that he reported are compared as a whole*" (276, emphasis added).

They discussed the usual explanations for the goodness-of-fit bias, including the theory that Mendel stopped counting, and found all these arguments insufficient. The most likely explanation of the "goodness of fit,"

Fairbanks and Rytting suggested, was that Mendel had selected his best data for presentation in a public lecture and its subsequent publication: "We believe that the most likely explanation of the bias in Mendel's data is also the simplest. If Mendel selected for his presentation a subset of his experiments that best represented his theories, χ^2 analyses of these experiments should display a bias. His paper contains multiple references to experiments for which he did not report numerical data, particularly di- and trihybrid experiments" (*279–80*). They also cited as supporting evidence a letter from Mendel to Nägeli: "In Mendel's second letter to Nägeli, he referred to his paper as 'the unchanged reprint of the draft of the lecture mentioned: thus the brevity of the exposition, as is essential for a public lecture'" (qtd. in Stern and Sherwood 1966, 61). They admitted the bias in Mendel's data but stated, "there are reasonable statistical and botanical explanations for the bias, and insufficient evidence to indicate that Mendel or anyone else falsified the data" (*281*).

This plausible explanation would help to lessen the overall "too good to be true" character of Mendel's data.[62] It does not, however, address the demonstration, by both Edwards and Seidenfeld, that Mendel's data had not merely been truncated, but adjusted.

Kenneth Weiss (2002) accepted the views of Fairbanks and Rytting on the question of Mendel's research integrity. In discussing the 2:1 ratio experiments, he accepted their calculation that the results for the monohybrid experiments (399:201) are only slightly unlikely and that this exonerated Mendel from the most serious charge of falsification. He did not, however, discuss the question of the overall goodness of fit. Instead, he commented, "I think the whole issue has been greatly overblown in the first place. The reasons to lighten up on the poor monk have to do with his context, with statistical considerations, and with what science is all about" (42). He suggested that because modern statistical tests were unavailable to Mendel that this would excuse the falsification of data. He also agreed that Mendel's statements that he reported only part of his data were evidence against fraud. "And even if he or his assistant selectively pitched (or nibbled) some peas, well, before I put blueberries on my cereal I discard the green ones, but I still say that blueberries are blue" (42). Weiss found no distinction between fraudulent data in scientific experiments and an unwillingness to eat unripe blueberries. In his view, the correctness of Mendel's results provides justification for such behavior. "And he was right: 2:1 makes theoretical sense; 1.88:1.11 makes none" (43).

Weiss's views were criticized by DeGusta (2003a; 2003b). He agreed that Mendel's data were clearly biased, citing Fisher, Fairbanks and Rytting, and Piegorsch. DeGusta noted, "While Mendel was a monk, he was no saint. At least some of the data he published are clearly biased beyond what would be expected by chance. Furthermore, this bias is uniformly in the direction of Mendel's expectations, which has long raised suspicion that its source was something other than innocent error" (2003b, 1). Contrary to Weiss, he argued that the lack of available statistical tests did not provide an excuse for fraudulent data: "The beauty of statistical methods (probably their only beauty) is that the results are independent of any particular context. . . . Of course, as Weiss notes, statistical tests of significance (i.e., the chi-squared test) were not known to Mendel. But such tests cannot *prevent* bias, they can only *detect* it after the fact. So the lack of such tests in Mendel's time does not exculpate any 19th century data manipulators" (2003a, 5). He further remarked that Fairbanks and Rytting had accepted that there was bias and he agreed that their explanation of Mendel's goodness of fit—the selection of that data that best represented his theories—was the best available.

In *The Great Betrayal: Fraud in Science* (2004), Horace Judson discussed the Mendel-Fisher controversy. Citing the work of Corcos and Monaghan, he concluded that Mendel had been vindicated.

Fisher's analysis of the 2:1 ratio experiments was questioned by Edward Novitski (2004). Novitski suggested that Mendel would have accepted as evidence for heterozygosity plants with fewer than 10 progeny, if one of those progeny showed the recessive trait. On the other hand, he would have excluded plants with fewer than 10 progeny if all of them were dominant. "If Mendel demanded 10 progeny for an adequate test only when the 9 or fewer existing progeny failed to reveal the heterozygosity of the parent, we can reasonably assume that when fewer than 10 progeny matured, he would be eliminating primarily *AA* (homozygous) individuals" (1134). This would tend to compensate for the loss of heterozygous plants according to Fisher's correction. Suppose that Mendel excluded all plants with 9 dominant progeny. One-third of the time they would have been homozygous parents. Two-thirds of the time the parent would have been heterozygous. The probability of obtaining 9 dominant progeny from a heterozygous parent is $(3/4)^9 = 0.075$. Thus, the chance of excluding homozygous parents compared with excluding heterozygous parents is $(1/3)/[(2/3)(0.075)] = 6.67$. Homozygous plants are preferentially excluded if one excluded sets with 9 progeny. Similar results occur with sets with fewer progeny. This would mitigate the undercounting of heterozygous plants pointed out by Fisher. Novitski assumed that Mendel would have cultivated other sets of 10 progeny to make up for those lost to germination failure. Including those sets would increase the number of heterozygotes counted. In fact, if the failure rate were high enough, then the expected ratio of heterozygous to homozygous plants would be greater than 2:1: "The selective elimination of *AA* [homozygous] individuals would shift the ratio calculated by Fisher from 1.8874 *Aa* to 1.1126 *AA* toward the ideal of 2. In fact, if the failure rate were high enough the ratio might well exceed 2:1" (1134). Novitski presented a numerical example in which one looked at 100 plants with 10 seeds each. For a 2% failure rate, he reported a calculation by C. E. Novitski, who found that 22.5 heterozygous plants would have been excluded by Fisher's correction and 23 added because of the procedure to deal with failures. "The similarity of the loss (22.5) and the gain (23) is purely fortuitous and not to be taken too literally. The essential point is that the two values are of similar magnitude and of opposite effect. They do show, however, how these two complicating factors, working in opposite directions, could give Mendel a final ratio closer to the 2:1 theoretically expected, Fisher's analysis notwithstanding" (1134).

There are, however, two serious problems with this scenario. First, Mendel did not report any such procedure. Second, Mendel's failure rate was, as discussed earlier, 5.6%, which would raise the corrected ratio to larger than 2:1. Recent work by Daniel Hartl and Daniel Fairbanks (Hartl and Fairbanks 2007; see also postscripts to chapters 5 and 7 in this volume) presented a detailed calculation supporting this point.

With regard to the trifactorial experiment, Novitski remarked, "It can be argued, however, that Mendel was almost certainly using the correct expectation, and it is Fisher who was using the incorrect one" (1135). Novitski argued that Fisher had misread Mendel's paper and that the third characteristic used in this experiment was seed-coat color rather than flower color. Novitski remarked that the hybrid seed coats are spotted even when that trait was absent in the parent and that "it is reasonable to assume that the maternal plant might be classified unambiguously as either a homozygote or a heterozygote" (1135). In other words, for this characteristic, the genotype could often be inferred from the phenotype. (Hartl and Fairbanks [2007] suggest that this is probably incorrect.)

Novitski noted, however, that his analysis did "not alter Fisher's conclusions that, overall, Mendel's results are closer to theory than expected on a chance basis" (1136). He cited the results of both Edwards (1986a) and C. E. Novitski (1995) as support for this conclusion. He further suggested that Mendel might have selected his data, or even altered it, to make his results more understandable to his audience and noted that Mendel expected and welcomed repetitions of his experiments. He concluded: "Fisher's criticism of Mendel's data—that Mendel was obtaining data too close to false expectations in the two sets of experiments involving the determination of segregation ratios [2:1]—is undoubtedly unfounded" (1136).

In an adjoining paper, Charles Novitski (2004) presented the details of his calculations. Using the binomial distribution, Fisher's correction, and a failure rate of 2%, he found that R, the ratio of those counted as heterozygotes to those counted as homozygotes was 2.068, which was in reasonable agreement with Mendel's expectation of 2. By minimizing the χ^2 and allowing the failure rate to float, he found a failure rate of 1.54% and R = 1.975. I note, however, that for the more accurate estimate of Mendel's failure rate, 5.6%, the problem is exacerbated. R would then be considerably greater than 2, and would, in fact, be closer to 3.

In an article on the use of statistical techniques to detect fraud, Michael O'Kelly (2004) cited Fisher's study of Mendel as the pioneering use of such techniques to investigate potential fraud. Similarly, in "A Beginner's Guide to Scientific Misconduct," Bob Montgomerie and Tim Birkhead (2005) stated, "Gregor Mendel might well be called the father of scientific misconduct, not because he was necessarily a wrongdoer, but because his published work sparked more than a century of controversy about the validity of his data" (17). After a very brief discussion of the controversy, they cited Fairbanks and Rytting (2001) and concluded, "More recent analyses appear to have exonerated Mendel in any wrongdoing, though the details are complex and not entirely convincing" (17).

Finally, after describing Mendel's achievement as "one of the most brilliant in the entire history of science," Yongsheng Liu (2005, 314) went on to state, "It is well known that truth is the essence of science. Mendel's gravest mischief was that he cooked his experimental data" (315). After briefly discussing the controversy, Liu summed up, "The controversy over the 'too good to be true' data remains unresolved and has been the subject of passionate scientific debate" (315).

Recently, Fairbanks has reported experimental results on peas (see postscript to chapter 7 in this volume). In a letter to this author, he explained: "All analyses (chi-square analysis and linear regression) show no evidence of a bias toward expectation for distributions of phenotypes within pods, and thus no evidence that the tetrad-pollen model (sampling without replacement) is valid. In fact, the pattern that emerged conforms to the one expected with completely random sampling of gametes with replacement. In other words, the pea plant is, as Edwards put it, an excellent randomizer and there is no reason to suspect that gamete sampling is responsible for the bias in Mendel's experiments." Seidenfeld also reported that an experiment to investigate whether pea plants grown under severe conditions exhibit reduced variability has, unfortunately, not yielded usable results (see the tables in the appendix to chapter 6 in this volume). It is interesting to note that one of the preliminary experiments yielded an excellent fit to the expected 9:3:3:1 ratios for that experiment. The data were 156:54:54:18. The probability of a worse fit is 0.9919. Likewise, in the summed data of his test of the tetrad-pollen model, Fairbanks examined a total of 5,204 F_2 plants for flower/axillary pigmentation and scored 3,899 as pigmented and 1305 as not pigmented. The probability of a worse fit to a 3:1 ratio is 0.8981. Extraordinarily good results do occur.

Hartl and Fairbanks (2007 and postscripts to chapters 5 and 7 in this volume) recently argued, quite persuasively, that Fisher's analysis of the 2:1 ratio experiments is incorrect. They suggested that there is no factual basis for suspecting any tampering with the data from the first series of experiments on the 2:1 ratio. They showed that examination of Mendel's data reveals no significant deviation from Fisher's expectations, and, that for the one experiment repeated by Mendel (Experiment 5 on pod color), the pooled results of both experiments fit Fisher's expectation almost exactly. They also noted that Mendel's contemporaries agreed that he was a superb gardener who would certainly have known that in some of the experiments fewer than 10 seeds would germinate. They suggested that Mendel planted 2–3 seeds per hill and thinned them sometime after germination or that he planted more than 10 seeds in the greenhouse and then from the resulting seedlings selected 10 to transplant to his garden. In this way, Mendel could guarantee 10 progeny per parent plant.

They also offered two plausible explanations of the trifactorial experiment, which also involves the 2:1 ratio. They suggested that Mendel used axillary pigmentation rather than flower color, as stated by Fisher. Although these characteristics are perfectly correlated, axillary pigmentation can be easily identified in seedlings, whereas flower color cannot. Thus, the entire experiment could be completed within a single growing season. They further suggested that Mendel had planted more than 10 seeds in order to guarantee at least 10 seedlings and that he might have scored all of the seedlings for axillary pigmentation. They calculated that if Mendel had scored as few as 11 seedlings for 70% of the progeny plants and only 10 seedlings for the remaining 30%, then any significant discrepancy from Fisher's expectations disappears. Hartl and Fairbanks noted that this was consistent with what Mendel had written. He did not state that the experiment had been done in the same way as the monohybrid experiments, but referred only to "further investigations." The second explanation was that Mendel might have been able to identify genotypes by observing the phenotypes. This had been suggested by both Wright (1966) and by E. Novitski (2004). Novitski had suggested that the observed trait was seed-coat color and that this trait could be used to distinguish heterozygotes from homozygotes in plants with the dominant phenotype, which Hartl and Fairbanks argued, on botanical grounds, is probably not correct. Nevertheless, they showed that even a 2% observation rate of such a character would eliminate any significant discrepancy between Mendel's data and the correct expectations.

Hartl and Fairbanks concluded, "Let us hope against all experience that Fisher's allegation of deliberate falsification can finally be put to rest, because on closer analysis it has proved to be unsupported by convincing evidence" (2007, 13).

Conclusions

It has now been 140 years since the publication of Mendel's seminal work, 70 years since Fisher's analysis, and more than 40 years since the beginning of the Mendel-Fisher controversy. Is the issue still unresolved? Perhaps more importantly, should it be? I think the answer is no.

There are, I believe, several conclusions that are supported by both this reexamination of the controversy, as well as the articles included in this volume. They are: (1) Mendel was not guilty of deliberate fraud in the presentation of his experimental results; (2) the problem of the 2:1 ratio experiments has been solved; (3) Fisher's analysis of the "too good to be true" data is still correct; and (4) Fisher would have been quite unhappy with those who used his work to diminish Mendel's achievement.

Let us consider first the question of whether Mendel committed deliberate fraud. I believe the evidence is overwhelmingly against such a conclusion. Mendel published data that disagreed with his expectations in both his original paper on plant hybridization and in his paper on *Hieracium*. He also made it clear that he had not included significant amounts of data. In addition, he sent 140 packets of seeds to Nägeli in the hopes that his experiments would be replicated. These are not the actions of someone guilty of fraud. Perhaps, more importantly, Seidenfeld's challenge to provide a method of cheating that would reproduce the oddities in Mendel's data shown by Edwards and himself is still unanswered. I also believe that these oddities argue against other possible explanations of Mendel's results such as unconscious bias, stopping the count when results agreed with expectations, or fraud by another.

With regard to Fisher's analysis of the 2:1 ratio experiments, I believe that Seidenfeld's explanation that Mendel used sequential experiments in subsequent generations provides an adequate solution. This is consistent with Mendel's statement that he had continued the experiments for several generations. In addition, as Seidenfeld showed, it would take examination of only very few such plants to reduce the misclassification problem to negligible proportions. I also believe that the calculations of Hartl and Fairbanks for both the monohybrid and the trifactorial experiments argue strongly against Fisher's analysis of these experiments. Both explanations may, in fact, be correct. Either argues persuasively against Fisher's analysis of the 2:1 ratio experiments. It is also possible, as suggested by many authors including Orel and Hartl (chapter 5) and Fairbanks and Rytting (chapter 7), that Mendel sowed more than 10 plants in these experiments. That explanation is inconsistent with what Mendel explicitly stated in his report of the early monohybrid experiments, but it is a plausible, although I believe a less probable, explanation of the 2:1 ratios in the trifactorial experiment. Thus, both sets of 2:1 ratio experiments, the experiments that had initially triggered Fisher's suspicions, can be explained without any fraud.

The issue of the "too good to be true" aspect of Mendel's data found by Fisher still stands, however. No one has yet raised any valid criticism of Fisher's analysis, or of the later analyses by Edwards and Seidenfeld. There is also no empirical evidence to support a botanical explanation of reduced variability such as the tetrad or urn model. In fact, good evidence has been provided by both Edwards and Fairbanks that peas are good randomizers. The analyses by Seidenfeld and by Hartl and Fairbanks argue that Fisher's use of 2:1 as the ratio in the monohybrid and trifactorial experiments is correct. In addition, as Weiling showed, there is little difference to the goodness-of-fit problem whether one uses 2:1 or 1.7:1 for the expected ratio in the monohybrid and trifactorial experiments.

Finally, it seems clear that Fisher would have been quite unhappy with those who have used his work to diminish Mendel's achievement. As we have seen, Fisher was unstinting in his praise of both Mendel's methods and his conclusions. As Fisher himself said in describing Mendel's experiments, they are "experimental researches conclusive in their results, faultlessly lucid in presentation, and vital to the understanding not of one problem of current interest, but of many" (Fisher 1936, *139*).

It is time to end the controversy.

NOTES

I want to thank my collaborators Anthony Edwards, Daniel Fairbanks, Daniel Hartl, and Teddy Seidenfeld for valuable discussions, for their comments and always constructive and gentle criticism, and most importantly, for their work, which gave me a sufficient understanding of the issues involved in the controversy so that I could write this essay. I am grateful to Anthony Edwards for both his wonderful hospitality during my research trip to Cambridge and for sharing his files on the Mendel-Fisher controversy with me. Our discussions were invaluable.

1. All page references to papers that are included in this volume refer to those versions contained in this book, and are noted in italics for ease of use. For all quoted material, all italics are from the original source unless otherwise noted.

2. The translation of Mendel's 1865 paper reprinted in this volume is contained in Bateson 1909, which is the same translation used by Fisher.

3. Not all of the experiments described in the same section of Mendel's paper necessarily occurred at the same time.

4. In modern notation, self-fertilization of a heterozygous plant is genetically equivalent to the cross $Aa \times Aa \to AA + 2Aa + aa$. Both the AA and Aa plants will display the dominant character, whereas an aa plant will display the recessive character. Mendel used A, Aa, and a, respectively, to denote genotypes we currently symbolize as AA, Aa, and aa. For a detailed discussion of this see chapters 5 and 7. I will use Mendel's notation of A and a for dominant and recessive characters, respectively, rather than the modern AA and aa.

5. Mendel noted that care must be taken in these experiments. He stated, "These two experiments are important for the determination of the average ratios, because with a smaller number of experimental plants they show that very considerable fluctuations may occur. In counting the seeds, also, especially in Expt. 2, some care is requisite, since in some of the seeds of many plants the green colour of the albumen is less developed, and at first may be easily overlooked. . . . In luxuriant plants this appearance was frequently noted. Seeds which are damaged by insects during their development often vary in colour and form, but, with a little practice in sorting, errors are easily avoided" (*87*).

6. This rather unexpected result is discussed in some detail in Seidenfeld (1998), which is reproduced in this volume as chapter 6.

7. This is true for the plant characteristics. The seed characteristics would appear in the same generation.

8. The F_3 generation experiments are the first experiments on the 2 to 1 ratio that would be of concern to Fisher. A second instance is the trifactorial hybrid experiment, discussed below.

9. There is considerable discussion in the ensuing controversy over whether the German phrase *von jeder 10 Samen angebot* should be translated as "10 seeds were sown" or "10 seeds were cultivated." In their translation, Stern and Sherwood (1966) used "sown." The choice of 10 seeds is also of significance. Fisher would later argue, as discussed below, that because only 10 seeds were planted, that there is a 5.6% probability that heterozygous plants would be classified as homozygous and thus be undercounted. Because of this, he argued that the ratio should be 1.8874 to 1.1126, or about 1.7 to 1, rather than 2 to 1.

10. This is a reasonable reading of what Mendel wrote, and it was Fisher's interpretation. There is considerable discussion in the later literature about whether Fisher was correct. This is discussed in some detail below and in the other chapters in this volume.

11. A modern statistician would not regard this as a significant deviation.

12. Mendel did not present all of his data from several of his other experiments. In ad-

dition, as discussed below and in chapter 6, this continuation of the experiments is an important point in the discussion of the 2:1 ratio experiments.

13. The remainder of this section of Mendel's paper discusses the reversion to parental forms, which is not important for our story.

14. In other experiments, discussed below, Mendel would investigate whether there was any difference in results depending on which characters were associated with the seed and pollen plants, respectively. He would conclude that there was no difference.

15. These results are a good fit to the expected 9:3:3:1 ratio.

16. Although he did not take any notice of this, Mendel obtained results for the failure rates for growing plants from seeds, a quantity which will be of some importance in our later discussion. From 315 round and yellow seeds a total of 301 plants were obtained, a failure rate of 4.4%. For wrinkled and yellow seeds, round and green seeds, and wrinkled and green seeds, the results were 96 plants from 101 seeds (5.0%); 102 plants from 108 seeds (5.6%); and 30 plants from 32 seeds (6.2%), respectively.

17. These experiments are other instances of the 2:1 ratio experiments.

18. I interpret this as indicating that Mendel used the same procedure that he had used previously to investigate plant characters, i.e., that he grew 10 plants in the next generation for each plant that showed the dominant gray-brown seed-coat color. The failure rate for growing plants from seeds was 7.0% for this experiment. Hartl and Orel (1992) suggest that Mendel may have meant that he allotted the same amount of space for the experiments and that there was sufficient room for Mendel to have sown more than 10 plants.

19. Notice again the excellent feel that Mendel has for his data. He sees the significant pattern and neglects the small deviations from that pattern. This is true of all of his analyses.

20. The failure rates were 8.2% and 7.4%, respectively.

21. Mendel remarked that this phenomenon had also been emphasized by C. F. Gärtner but gave no reference.

22. Hawkweed was not a good choice for Mendel, although he was not aware of this, at least when he began his hawkweed experiments. It sometimes produces seeds by apomixis, in which seeds are produced from unfertilized ova. In this process, genes are inherited only from the female parent and would not exhibit Mendelian patterns.

23. This was an exaggeration.

24. This was a reference to Pearson (1900), in which the χ^2 test was introduced.

25. In a letter to Karl Pearson, Weldon remarked, "If only one could know whether the whole thing is not a damned lie!" (Karl Pearson Papers, University College London, file 625). Weldon was discussing Mendel's entire scheme, however. He does not suggest that Mendel's results were fraudulent.

26. Weldon's reservations were too much for Bateson, the arch Mendelian. Within a month of the publication of Weldon's article, Bateson published *Mendel's Principles of Heredity: A Defence* (1902) in which he devoted more than 100 pages to the attempted refutation of Weldon. Citing Weldon's statement that he wished to "suggest fruitful lines of inquiry," Bateson concluded: "In this purpose I venture to assist him, for I am disposed to think that unaided he is—to borrow Horace Walpole's phrase—about as likely to light a fire with a wet dish-clout as to kindle interest in Mendel's discoveries by his tempered appreciation" (Bateson 1902, 208).

This was just the latest salvo in the bitter and nasty battle between the biometicians, headed by Pearson and Weldon, who used statistical methods on large populations and supported Darwinism and blending inheritance, and the Mendelians, led by Bateson, proponents of applying Mendel's laws to small populations. (For details of this controversy see

Provine [1971], Froggatt and Nevin [1971a, b], Farrall [1975], Kevles [1980], and Morrison [2002]). Fisher, himself, remarked that this battle was "one of the most needless controversies in the history of science" (1924, 192) and David Hull (1985) referred to it as "an inexplicable embarrassment."

Fisher himself became a casualty of this battle. In 1916 he wrote a paper entitled, "On the Correlation Between Relatives on the Supposition of Mendelian Inheritance," which showed that the two opposing views could be reconciled. This became one of the founding papers of population genetics and was referred to by Kempthorne as "a work of genius" (qtd. in Norton and Pearson 1976, 151). Fisher originally submitted the paper to the Royal Society of London and it was sent to Karl Pearson, a biometrician, and to Reginald Crundall Punnett, a Mendelian, for refereeing. Both had reservations about the paper. Pearson thought that Fisher's assumptions were not supported by either observational or experimental evidence, and that his assumption was only one of many that might lead to similar results. He suggested that the paper's "publication should depend on whether Mendelians consider its hypotheses of value as actually representing observational facts" (qtd. in Norton and Pearson 1976, 153–54). Punnett, on the other hand, admitted that he could not follow the mathematics of Fisher's paper but did not question its correctness and remarked,

> I do not in any way wish to suggest that the mathematics were not all they should be as I have not the least doubt that the author is perfectly competent on this head. And as a contribution to biometry it may have real value—but I am not qualified to judge it from that point of view. However, whatever its value from the standpoint of statistics & population I do not feel that this kind of work affects us biologists much at present. It is too much of the order of problem that deals with weightless elephants upon frictionless surfaces, where at the same time we are largely ignorant of the other properties of the said elephants and surfaces. (qtd. in Norton and Pearson 1976, 155)

Although the referees' reports did not explicitly recommend rejection, Fisher saw that he was caught between Scylla and Charybdis. He withdrew the paper and it was later published as Fisher (1918).

27. The same charge was made later by DiTrocchio (1991). This has been a point of contention. Fisher, and others, have concluded that Mendel did the experiments as described.

28. Here, Fisher is referencing his previous work, *The Genetical Theory of Natural Selection* (Fisher 1930).

29. Some authors have questioned this. See, for example Olby (1979).

30. For randomly distributed data, the average χ^2 per degree of freedom is 1, giving an expected χ^2 of 26 for 26 degrees of freedom, rather than the 15.3224 calculated.

31. It is interesting to note that Fisher gives the probability of a worse fit to the eight experiments on the 2:1 ratio for monohybrids as 0.74, the lowest for any set of Mendel's experiments. See table 1.14.

32. There is an earlier reference to Fisher's paper in the 1963 biographical memoir written by Frank Yates and Kenneth Mather (1963). In discussing Fisher's paper they remark, "His study of Mendel's experiments (1936) was a delightful example of statistical analysis applied to the better understanding of an important chapter in the history of science." There is no mention of any controversy. Nor is any earlier controversy mentioned in the biography of Fisher written by his daughter Joan Fisher Box (1978).

33. One might also argue for 1966 as the centenary. Mendel's talks were given in 1865 and the journal issue in which his paper was published is dated 1865, but the journal did not appear until 1866.

34. This assumes that Mendel did not already have a good idea of his system before he began his experiments. This is a controversial issue. Fisher, and others, disagree, and

believe that Mendel did have a theory either before he began his experiments, or shortly thereafter. Other scholars agree with Zirkle.

35. The result is only slightly more than one standard deviation from the expected result, a not unlikely result.

36. This is true only for the monohybrid experiments, not for the trifactorial experiment.

37. Thoday is referring to Evans and Philip (1964). Although not explicitly applicable to peas, the experiments by Evans and Philip raised the possibility of such an explanation of reduced variance.

38. Teddy Seidenfeld would later independently suggest the same model (see chapter 6, this volume).

39. Campbell (1976) named the same three assistants.

40. Scholars disagree as to whether Mendel's results should be compared to Fisher's corrected predictions or to Mendel's expectations. As we see, the conclusion doesn't change significantly.

41. This was the value van der Waerden calculated for Mendel's Experiment 5.

42. Van der Waerden also calculated that the probability of Mendel obtaining seven pairs varieties of plants which differed in only one character was 0.97: "The difficulty raised by Bateson does not exist. A simple calculation of probabilities shows that it is not at all unlikely that Mendel had, from the beginning, seven pairs of varieties, each pair differing in only one essential character" (van der Waerden 1968, 277).

43. In the case of Mendel's 3:1 ratio experiments, this would require that $P_1 > P_2$, where P_1 and P_2 are the probabilities observing the dominant trait or the recessive trait, respectively.

44. Piegorsch had translated Weiling (1966). It appeared as Paper BU-718-M, Biometrics Unit, Cornell University, 1980.

45. Piegorsch also noted that "only three experiments in 100,000 attempts would show ratios as close or closer to agreement with Mendel's ratios" (2291). He commented that the probability of getting a worse fit was 0.99997, in contrast to Fisher's result of 0.99993. He attributed this to a lack of precision in either Fisher's algorithm or in his calculating machine. This makes Mendel's results even more unlikely.

46. In one experiment, the students obtained a perfect fit to the expected Mendelian ratio. With a sufficient number of "indeterminants" this is not unexpected.

47. This was based on a fragment of Mendel's writings called the Notizblatt. Weiling (1991) dated this fragment as later than 1874. It also does not specify what type of plants were used in the experiments.

48. My arguments against Root-Bernstein's explanation of Mendel's results should not, however, be taken to imply that I believe Mendel was guilty of fraud. They are merely comments on the inadequacy of his analysis.

49. There is no mathematical justification for this statement.

50. Opinion was not unanimous. Gardner (1981) stated, "Mendel's figures are suspect for just this reason. They are too good to be true. Did the priest consciously fudge his data? Let us be charitable. Perhaps he was guilty only of 'wishful seeing' when he classified and counted his talls and dwarfs" (124). Kohn (1986), on the other hand, citing the arguments given by Pearl and Root-Bernstein concluded, "It seems to me that all insinuations about Mendel's possible unethical behavior should be discounted" (43).

51. Using modern computational techniques, Matthew Stephens (1994) confirmed Edwards's conclusions concerning both Mendel's data and excellent randomizing by peas shown in the data of Darbishire and by Bateson and Kilby. "There seems to be good statis-

tical grounds on which to argue that either Mendel was mistaken in his statement or that the model is incorrect" (20). He further remarked, "It is clear that Mendel's results do not fit the binomial model very well; nor do they agree well with the other two data sets considered, which suggests that it is more than just the model which is wrong" (53). Stephens also calculated that the probability of never having more than five recessives in a pod, under a binomial model is approximately 1 in 2750. This is also discussed by Seidenfeld, as detailed later in this introduction.

52. This claim is not only in disagreement with Edwards's analysis, but also with that of Seidenfeld, as discussed later in this introduction and in chapter 6.

53. This point is discussed in detail by both Seidenfeld and by Fairbanks and Rytting, chapters 6 and 7, respectively. Both disagree with Weiling's analysis. Weiling does not calculate the effect that this would have on the χ^2 obtained. In addition, as discussed later in this introduction, the recent experimental results obtained by Fairbanks argue against an urn model.

54. The question of whether Mendel should have observed linkage is also discussed in Piegorsch (1986), van der Waerden (1968), and in the references cited in those papers. Piegorsch concluded that it was quite plausible that Mendel did not observe linkage.

55. Corcos and Monaghan had repeated the error that Pilgrim had made. In fact, the probability that a χ^2 of 0.00030 for two degrees of freedom is due to chance is 0.0002.

56. Nissani cites no evidence to support this point.

57. This refers to the total χ^2 for all six experiments, not to those for the separate experiments. Recall that Fisher had reported a χ^2 probability of 0.74 for the entire set of eight experiments (see table 1.14). This is a good fit, but certainly nothing extraordinary.

58. Seidenfeld is being quite conservative here. The actual probability is 3×10^{-3}. For a probability of 0.01, only six plants out of 600 would have been misclassified, which is negligible.

59. Seidenfeld's model was proposed independently. When he told me of this model in a private discussion, I informed him of the earlier work and gave him the references.

60. Recall that Stephens (1994, 53n52) had calculated that such a result had a probability of 1/2750 on a binomial model.

61. Recent results, discussed briefly below and in chapter 7, indicate no support for the urn model for peas and that the pea is an excellent randomizer.

62. It is amusing to speculate about how Mendel's exclusion of data might have affected the goodness of fit. It is generally agreed that Mendel excluded at least as much data as he presented. Let us assume that the amount of excluded data was equal to the amount of data that Mendel presented and that it was normally distributed, i.e., that it had a total χ^2 of 84 for 84 degrees of freedom. Adding this set of data to Mendel's published data, which had a χ^2 of 41.6056 (Fisher's value), we get a total χ^2 of 125.6056 for 168 degrees of freedom, which gives a probability of a worse fit of 0.9938. This is still extremely good, and it does nothing to solve the problem of the bulge in the probability distribution. If one assumes, quite plausibly, that the excluded data had a somewhat worse than normal distribution, then the probability goes down further. Suppose we wish to add an equal amount of excluded data such that the total χ^2 would be 168 for 168 degrees of freedom, a reasonable result. This would mean that the excluded data had a χ^2 of 126.39 for 84 degrees of freedom, which has a probability of 0.0018 of arising by chance, a very unlikely result.

REFERENCES

Bateson, W. 1902. *Mendel's Principles of Heredity: A Defence.* Cambridge: Cambridge University Press.

———. 1909. *Mendel's Principles of Heredity*. Cambridge: Cambridge University Press.
Bateson, W., and H. Kilby. 1905. "Experimental Studies in the Physiology of Heredity: Peas." *Royal Society Reports to the Evolution Committee* 2:55–80.
Beadle, G. W. 1967. "Mendelism, 1965." In *Heritage from Mendel*, ed. R. A. Brink, 335–50. Madison: University of Wisconsin Press.
Bennett, J. H., ed. 1965. *Experiments in Plant Hybridisation by Gregor Mendel, with Commentary and Assessment by Sir Ronald A. Fisher*. Edinburgh: Oliver and Boyd.
———, ed. 1983. *Natural Selection, Heredity, and Eugenics: Including Selected Correspondence of R. A. Fisher with Leonard Darwin and Others*. Oxford: Clarendon Press.
Box, Joan Fisher. 1978. *R. A. Fisher: The Life of a Scientist*. Wiley Series in Probability and Mathematical Statistics. New York: John Wiley and Sons.
Broad, W., and N. Wade. 1982. *Betrayers of the Truth*. New York: Simon and Schuster.
Campbell, M. 1976. "Explanations of Mendel's Results." *Centaurus* 20:159–74.
Corcos, A., and F. Monaghan. 1986. "Correction: Chi-Square and Mendel's Experiments." *Journal of Heredity* 77:283.
———. 1993. *Gregor Mendel's Experiments on Plant Hybrids: A Guided Study*. New Brunswick: Rutgers University Press.
Darbishire, A. D. 1908. "On the Results of Crossing Round with Wrinkled Peas, with Especial Reference to Their Starch-grains." *Journal of the Royal Society of London B, Proceedings* 80:122–35.
———. 1909. "An Experimental Estimation of the Theory of Ancestral Contributions in Heredity." *Journal of the Royal Society of London B, Proceedings* 81:61–79.
De Beer, G. 1964. "Mendel, Darwin, and Fisher." *Notes and Records of the Royal Society* 19:192–225.
DeGusta, D. 2003a. "More Digging in Mendel's Garden." Crotchety Comments, Kenneth M. Weiss's Lab on the Web, Department of Anthropology, College of Liberal Art, Penn State University, http://www.anthro.psu.edu/weiss_lab/DeGusta_FULL.doc/, 1–14.
———. 2003b. "More Digging in Mendel's Garden." *Evolutionary Anthropology* 12:1.
DiTrocchio, F. 1991. "Mendel's Experiments: A Reinterpretation." *Journal of the History of Biology* 24:485–519.
Dobzhansky, T. 1967. "Looking Back at Mendel's Discovery." *Science* 156:1588–89.
Dunn, L. C. 1965. "Mendel, His Work and His Place in History." *Proceedings of the American Philosophical Society* 109:189–98.
Edwards, A. W. F. 1972. *Likelihood*. Cambridge: Cambridge University Press.
———. 1986a. "Are Mendel's Results Really Too Close?" *Biological Reviews* 61:295–312.
———. 1986b. "More on the Too-Good-to-Be-True Paradox and Gregor Mendel." *Journal of Heredity* 77:138.
Evans, D. A., and U. Philip. 1964. "On the Distribution of Mendelian Ratios." *Biometrics* 20:794–817.
Fairbanks, D. J., and B. Rytting. 2001. "Mendelian Controversies: A Botanical and Historical Review." *American Journal of Botany* 88:737–52.
Farrall, L. A. 1975. "Controversy and Conflict in Science: A Case Study—The English Biometric School and Mendel's Laws." *Social Studies of Science* 5:269–301.
Fisher, R. A. 1918. "On the Correlation between Relatives on the Supposition of Mendelian Inheritance." *Transactions of the Royal Society of Edinburgh* 52:399–433.
———. 1924. "The Biometrical Study of Heredity." *Eugenics Review* 16:189–210.
———. 1930. *The Genetical Theory of Natural Selection*. Oxford: Clarendon Press.
———. 1935. *The Design of Experiments*. Edinburgh: Oliver and Boyd.
———. 1936. "Has Mendel's Work Been Rediscovered?" *Annals of Science* 1:115–37.

———. 1965a. "Introductory Notes on Mendel's Paper." In *Experiments in Plant Hybridisation by Gregor Mendel, with Commentary and Assessment by Sir Ronald Fisher*, ed. J. H. Bennett, 1–6. Edinburgh: Oliver and Boyd.

———. 1965b. "Marginal Comments on Mendel's Paper." In *Experiments in Plant Hybridisation by Gregor Mendel, with Commentary and Assessment by Sir Ronald Fisher*, ed. J. H. Bennett, 52–58. Edinburgh: Oliver and Boyd.

Focke, W. O. 1881. *Die Pflanzen-mischlinge ein Beitrag zur Biologie der Gewächse*. Berlin.

Froggatt, P., and N. C. Nevin. 1971a. "Galton's 'Law of Ancestral Heredity': Its Influence on the Early Development of Human Genetics." *History of Science* 10:1–27.

———. 1971b. "The 'Law of Ancestral Heredity' and the Mendelian Ancestrian Controversy in England, 1899–1906." *Journal of Medical Genetics* 8:1–36.

Gardner, M. 1981. *Science: Good, Bad, and Bogus*. Buffalo: Prometheus Books.

Hartl, D. L., and D. J. Fairbanks. 2007. "Mud Sticks: On the Alleged Falsification of Mendel's Data." *Genetics* 175:975–79.

Hartl, D. L., and V. Orel. 1992. "What Did Gregor Mendel Think He Discovered?" *Genetics* 131:245–53.

Judson, H. F. 2004. *The Great Betrayal: Fraud in Science*. New York: Harcourt.

Kevles, D. J. 1980. "Genetics in the United States and Great Britain, 1890–1930: A Review with Speculations." *Isis* 71:441–55.

Kohn, A. 1986. *False Prophets*. Oxford: Basil Blackwell.

Lamprecht, H. 1968. *Die Grundlagen der Mendelschen Gesetze*. Berlin: Paul Parey.

Leonard, T. 1977. "A Bayesian Approach to Some Multinomial Estimation and Pretesting." *Journal of the American Statistical Association* 72:869–74.

Lindley, D. V., and W. F. Scott. 1984. *New Cambridge Elementary Statistical Tables*. Cambridge: Cambridge University Press.

Liu, Y. 2005. "Darwin and Mendel: Who Was the Pioneer of Genetics?" *Rivista di Biologia/Biology Forum* 98:305–22.

Magnello, E. 2004. "The Reception of Mendelism by the Biometricians and the Early Mendelians (1899–1909)." In *A Century of Mendelism in Human Genetics*, ed. M. Keynes, A. W. F. Edwards, and R. Peel, 19–32. Boca Raton: CRC Press.

Mendel, G. 1870. "Uber einige aus kunstlicher Befruchtung gewonnen Hieracium-Bastarde." *Verhandlungen des naturforschenden Vereines in Brunn* 8:26–31.

Monaghan, F., and A. Corcos. 1985. "Chi-Square and Mendel's Experiments: Where's the Bias?" *Journal of Heredity* 76:307–9.

Montgomerie, B., and T. Birkhead. 2005. "A Beginner's Guide to Scientific Misconduct." *International Society for Behavioral Ecology* 17:16–24.

Morrison, M. 2002. "Modelling Populations: Pearson and Fisher on Mendelism and Biometry." *British Journal for the Philosophy of Science* 53:39–68.

Nissani, M. 1994. "Psychological, Historical, and Ethical Reflections on the Mendelian Paradox." *Perspectives in Biology and Medicine* 37:182–96.

Norton, B., and E. S. Pearson. 1976. "A Note on the Background to, and Refereeing of, R. A. Fisher's 1918 Paper, 'On the Correlation between Relatives on the Superposition of Mendelian Inheritance.'" *Notes and Records of the Royal Society* 31:151–62.

Novitski, C. E. 1995. "Another Look at Some of Mendel's Results." *Journal of Heredity* 86:62–66.

———. 2004. "Revision of Fisher's Analysis of Mendel's Garden Pea Experiments." *Genetics* 166:1139–40.

Novitski, E. 2004. "On Fisher's Criticism of Mendel's Results With the Garden Pea." *Genetics* 166:1133–36.

O'Kelly, M. 2004. "Using Statistical Techniques to Detect Fraud: A Test Case." *Pharmaceutical Statistics* 3: 237–46.

Olby, R. C. 1966. *Origins of Mendelism*. New York: Schocken Books.

———. 1979. "Mendel No Mendelian?" *History of Science* 17:55–72.

———. 1985. *Origins of Mendelism*. Chicago: University of Chicago Press.

Orel, V. 1968. "Will the Story on 'Too Good' Results of Mendel's Data Continue?" *BioScience* 18:776–78.

———. 1996. *Gregor Mendel: The First Geneticist*. Oxford: Oxford University Press.

Orel, V., and D. L. Hartl. 1994. "Controversies in the Interpretation of Mendel's Discovery." *History and Philosophy of the Life Sciences* 16:423–64.

Pearl, R. 1940. *Introduction to Medical Biometry and Statistics*. Philadelphia: W. B. Saunders.

Pearson, K. 1900. "On the Criterion that a Given System of Deviations from the Probable in the Case of a Correlated System of Variables Is Such that It Can Be Reasonably Supposed to Have Arisen from Random Sampling." *Philosophical Magazine* 50:157–75.

Piegorsch, W. 1983. "The Question of Fit in the Gregor Mendel Controversy." *Communications in Statistics: Theory and Methods* 12:2289–304.

———. 1986. "The Gregor Mendel Controversy: Early Issues of Goodness-of-Fit and Recent Issues of Genetic Linkage." *History of Science* 24:173–82.

———. 1990. "Fisher's Contributions to Genetics and Heredity, with Special Emphasis on the Gregor Mendel Controversy." *Biometrics* 46:915–24.

Pilgrim, I. 1984. "The Too-Good-to-Be-True Paradox and Gregor Mendel." *Journal of Heredity* 75:501–2.

———. 1986. "Rebuttal." *Journal of Heredity* 77:138.

Provine, W. 1971. *The Origins of Theoretical Population Genetics*. Chicago: University of Chicago Press.

R. A. Fisher Digital Archive. The University of Adelaide Digital Library. University of Adelaide, North Terrace, Adelaide, Australia. http://digital.library.adelaide.edu.au/coll/special/fisher/.

Robertson, T. 1978. "Testing For and Against an Order Restriction on Multinomial Parameters." *Journal of the American Statistical Association* 73:197–202.

Root-Bernstein, R. S. 1983. "Mendel and Methodology." *History of Science* 21:275–95.

Sapp, J. 1990. "The Nine Lives of Gregor Mendel." In *Experimental Inquiries*, ed. H. E. LeGrand, 137–66. Dordrecht: Kluwer Academic Publishers.

Seidenfeld, T. 1998. "P's in a Pod: Some Recipes for Cooking Mendel's Data." PhilSci Archive, Department of History and Philosophy of Science and Department of Philosophy, University of Pittsburgh, http://philsci-archive.pitt.edu/view/subjects/confirmation-induction.html.

Stephens, M. "The Results of Gregor Mendel: An Analysis and Comparison with the Results of Other Researchers." Diploma in Mathematical Statistics Thesis, University of Cambridge.

Stern, C., and E. R. Sherwood, eds. 1966. *The Origin of Genetics: A Mendel Source Book*. San Francisco: W. H. Freeman and Co.

Sturtevant, A. H. 1965. *A History of Genetics*. New York: Harper and Row.

Thoday, J. M. 1966. "Mendel's Work as an Introduction to Genetics." *Advancement of Science* 23:120–24.

Van der Waerden, B. L. 1968. "Mendel's Experiments." *Centaurus* 12:275–88.

Weiling, F. 1965. "Die Mendelschen Ebversuche in biometrischer Sicht/Zum 100/Jahrestag des ersten Mendelschen Vortrages vor dem Naturforschenden Verein in Brunn am 8.2.1865." *Biometrische Zeitschrift* 7:230–62.

———. 1966. "Hat J. G. Mendel bei seinen Versuchen 'zu genau' gearbeitet? Der Chi-2 Test und seien Bedeutung fur die Beurteilung genetischer Spaltungsverhaltnisse." *Der Zuchter* 36:359–65.

———. 1971. "Mendel's 'Too Good' Data in Pisum-Experiments." *Folia Mendeliana* 6:75–77.

———. 1986. "What about R. A. Fisher's Statement of the 'Too Good' Data of J. G. Mendel's Pisum Paper?" *Journal of Heredity* 77:281–83.

———. 1989. "Which Points Are Incorrect in R. A. Fisher's Statistical Conclusion? Mendel's Experimental Data Agree Too Closely with His Expectations." *Angewandte Botanik* 63:129–43.

———. 1991. "Historical Study: Johann Gregor Mendel 1822–1884." *American Journal of Medical Genetics* 40:1–25.

Weiss, K. 2002. "Goings on in Mendel's Garden." *Evolutionary Anthropology* 11:40–44.

Weldon, W. F. R. 1902. "Mendel's Laws of Alternative Inheritance in Peas." *Biometrika* 1:228–54.

Wright, S. 1966. "Mendel's Ratios." In Stern and Sherwood 1966, 173–75.

Yates, Frank, and Kenneth Mather. 1963. "Ronald Aylmer Fisher." *Biographical Memoirs of Fellows of the Royal Society of London* 9:91–120.

Zirkle, C. 1964. "Some Oddities in the Delayed Discovery of Mendelism." *Journal of Heredity* 55 (1964): 65–72.

CHAPTER 2

■ Experiments in Plant Hybridisation[1]

GREGOR MENDEL

(*Read at the Meetings of the 8th February and 8th March, 1865.*)

Introductory Remarks

Experience of artificial fertilisation, such as is effected with ornamental plants in order to obtain new variations in colour, has led to the experiments which will here be discussed. The striking regularity with which the same hybrid forms always reappeared whenever fertilisation took place between the same species induced further experiments to be undertaken, the object of which was to follow up the developments of the hybrids in their progeny.

To this object numerous careful observers, such as Kölreuter, Gärtner, Herbert, Lecoq, Wichura and others, have devoted a part of their lives with inexhaustible perseverance. Gärtner especially, in his work "Die Bastarderzeugung im Pflanzenreiche"(The Production of Hybrids in the Vegetable Kingdom), has recorded very valuable observations; and quite recently Wichura published the results of some profound investigations into the hybrids of the Willow. That, so far, no generally applicable law governing the formation and development of hybrids has been successfully formulated can hardly be wondered at by anyone who is acquainted with the extent of the task, and can appreciate the difficulties with which experiments of this class have to contend. A final decision can only be arrived at when we shall have before us the results of detailed experiments made on plants belonging to the most diverse orders.

Those who survey the work done in this department will arrive at the conviction that among all the numerous experiments made, not one has been carried out to such an extent and in such a way as to make it possible to determine the number of different forms under which the offspring of hybrids appear, or to arrange these forms with certainty according to their separate generations, or definitely to ascertain their statistical relations.[2]

It requires indeed some courage to undertake a labour of such far-reaching extent; this appears, however, to be the only right way by which we can finally reach the solution of a question the importance of which cannot be overestimated in connection with the history of the evolution of organic forms.

The paper now presented records the results of such a detailed experiment. This experiment was practically confined to a small plant group, and is now, after eight years' pursuit, concluded in all essentials. Whether the plan upon which the separate experiments were conducted and carried out was the best suited to attain the desired end is left to the friendly decision of the reader.

Selection of the Experimental Plants

The value and utility of any experiment are determined by the fitness of the material to the purpose for which it is used, and thus in the case before us it cannot be immaterial what plants are subjected to experiment and in what manner such experiments are conducted.

The selection of the plant group which shall serve for experiments of this kind must be made with all possible care if it be desired to avoid from the outset every risk of questionable results.

The experimental plants must necessarily—

1. Possess constant differentiating characters.
2. The hybrids of such plants must, during the flowering period, be protected from the influence of all foreign pollen, or be easily capable of such protection.

The hybrids and their offspring should suffer no marked disturbance in their fertility in the successive generations.

Accidental impregnation by foreign pollen, if it occurred during the experiments and were not recognized, would lead to entirely erroneous conclusions. Reduced fertility or entire sterility of certain forms, such as occurs in the offspring of many hybrids, would render the experiments very difficult or entirely frustrate them. In order to discover the relations in which the hybrid forms stand towards each other and also towards their progenitors it appears to be necessary that all members of the series developed in each successive generation should be, *without exception,* subjected to observation.

At the very outset special attention was devoted to the *Leguminosae* on account of their peculiar floral structure. Experiments which were made with several members of this family led to the result that the genus *Pisum* was found to possess the necessary qualifications.

Some thoroughly distinct forms of this genus possess characters which are constant, and easily and certainly recognizable, and when their hybrids are mutually crossed they yield perfectly fertile progeny. Furthermore, a disturbance through foreign pollen cannot easily occur, since the fertilising organs are closely packed inside the keel and the anther bursts within the bud, so that the stigma becomes covered with pollen even before the flower opens. This circumstance is of especial importance. As additional advantages worth mentioning, there may be cited the easy culture of these plants in the open ground and in pots, and also their relatively short period of growth. Artificial fertilisation is certainly a somewhat elaborate process, but nearly always succeeds. For this purpose the bud is opened before it is perfectly developed, the keel is removed, and each stamen carefully extracted by means of forceps, after which the stigma can at once be dusted over with the foreign pollen.

In all, thirty-four more or less distinct varieties of Peas were obtained from several seedsmen and subjected to a two years' trial. In the case of one variety there were noticed, among a larger number of plants all alike, a few forms which were markedly different. These, however, did not vary in the following year, and agreed entirely with another variety obtained from the same seedsman; the seeds were therefore doubtless merely accidentally mixed. All the other varieties yielded perfectly constant and similar offspring; at any rate, no essential difference was observed during two trial years. For fertilisation twenty-two of these were selected and cultivated during the whole period of the experiments. They remained constant without any exception.

Their systematic classification is difficult and uncertain. If we adopt the strictest definition of a species, according to which only those individuals belong to a species which under precisely the same circumstances display precisely similar characters, no two of these varieties could be referred to one species. According to the opinion of experts, however, the majority belong to the species *Pisum sativum;* while the rest are regarded and classed, some as sub-species of *P. sativum,* and some as independent species, such as *P. quadratum, P. saccharatum,* and *P. umbellatum.* The positions, however, which may be assigned to them in a classificatory system are quite immaterial for the purposes of the experiments in question. It has so far been found to be just as impossible to draw a sharp line between the hybrids of species and varieties as between species and varieties themselves.

Division and Arrangement of the Experiments

If two plants which differ constantly in one or several characters be crossed, numerous experiments have demonstrated that the common characters are transmitted unchanged to the hybrids and their progeny; but each pair of differentiating characters, on the other hand, unite in the hybrid to form a new character, which in the progeny of the hybrid is usually variable. The object of the experiment was to observe these variations in the case of each pair of differentiating characters, and to deduce the law according to which they appear in the successive generations. The experiment resolves itself therefore into just as many separate experiments as there are constantly differentiating characters presented in the experimental plants.

The various forms of Peas selected for crossing showed differences in the length and colour of the stem; in the size and form of the leaves; in the position, colour, and size of the flowers; in the length of the flower stalk; in the colour, form, and size of the pods; in the form and size of the seeds; and in the colour of the seed-coats and of the albumen [cotyledons]. Some of the characters noted do not permit of a sharp and certain separation, since the difference is of a "more or less" nature, which is often difficult to define. Such characters could not be utilised for the separate experiments; these could only be applied to characters which stand out clearly and definitely in the plants. Lastly, the result must show whether they, in their entirety, observe a regular behaviour in their hybrid unions, and whether from these facts any conclusion can be come to regarding those characters which possess a subordinate significance in the type.

The characters which were selected for experiment relate:

1. To the *difference in the form of the ripe seeds*. These are either round or roundish, the depressions, if any, occur on the surface, being always only shallow; or they are irregularly angular and deeply wrinkled (*P. quadratum*).

2. To the *difference in the colour of the seed albumen* (endosperm).[3] The albumen of the ripe seeds is either pale yellow, bright yellow and orange coloured, or it possesses a more or less intense green tint. This difference

of colour is easily seen in the seeds as [= if] their coats are transparent.

3. To the *difference in the colour of the seed-coat*. This is either white, with which character white flowers are constantly correlated; or it is grey, grey-brown, leather-brown, with or without violet spotting, in which case the colour of the standards is violet, that of the wings purple, and the stein in the axils of the leaves is of a reddish tint. The grey seed-coats become dark brown in boiling water.

4. To the *difference in the form of the ripe pods*. These are either simply inflated, not contracted in places; or they are deeply constricted between the seeds and more or less wrinkled *(P. saccharatum)*.

5. To the *difference in the colour of the unripe pods*. They are either light to dark green, or vividly yellow, in which colouring the stalks, leaf-veins, and calyx participate.[4]

6. To the *difference in the position of the flowers*. They are either axial, that is, distributed along the main stem; or they are terminal, that is, bunched at the top of the stem and arranged almost in a false umbel; in this case the upper part of the stem is more or less widened in section *(P. umbellatum)*.[5]

7. To the *difference in the length of the stem*. The length of the stem[6] is very various in some forms; it is, however, a constant character for each, in so far that healthy plants, grown in the same soil, are only subject to unimportant variations in this character.

In experiments with this character, in order to be able to discriminate with certainty, the long axis of 6 to 7 ft. was always crossed with the short one of ¾ ft. to 1½ ft.

Each two of the differentiating characters enumerated above were united by cross-fertilisation. There were made for the

1st trial	60	fertilizations on	15 plants	
2nd "	58	"	"	10 "
3rd "	35	"	"	10 "
4th "	40	"	"	10 "
5th "	23	"	"	5 "
6th "	34	"	"	10 "
7th "	37	"	"	10 "

From a larger number of plants of the same variety only the most vigorous were chosen for fertilisation. Weakly plants always afford uncertain results, because even in the first generation of hybrids, and still more so in the subsequent ones, many of the offspring either entirely fail to flower or only form a few and inferior seeds.

Furthermore, in all the experiments reciprocal crossings were effected in such a way that each of the two varieties which in one set of fertilisations served as seed-bearer in the other set was used as the pollen plant.

The plants were grown in garden beds, a few also in pots, and were maintained in their naturally upright position by means of sticks, branches of trees, and strings stretched between. For each experiment a number of pot plants were placed during the blooming period in a greenhouse, to serve as control plants for the main experiment in the open as regards possible disturbance by insects. Among the insects[7] which visit Peas the beetle *Bruchus pisi* might be detrimental to the experiments should it appear in numbers. The female of this species is known to lay the eggs in the flower, and in so doing opens the keel; upon the tarsi of one specimen, which was caught in a flower, some pollen grains could clearly be seen under a lens. Mention must also be made of a circumstance which possibly might lead to the introduction of foreign pollen. It occurs, for instance, in some rare cases that certain parts of an otherwise quite normally developed flower wither, resulting in a partial exposure of the fertilising organs. A defective development of the keel has also been observed, owing to which the stigma and anthers remained partially uncovered.[8] It also sometimes happens that the pollen does not reach full perfection. In this event there occurs a gradual lengthening of the pistil during the blooming period, until the stigmatic tip protrudes at the point of the keel. This remarkable appearance has also been observed in hybrids of *Phaseolus* and *Lathyrus*.

The risk of false impregnation by foreign pollen is, however, a very slight one with *Pisum,* and is quite incapable of disturbing the general result. Among more than 10,000 plants which were carefully examined there were only a very few cases where an indubitable false impregnation had occurred. Since in the greenhouse such a case was never remarked, it may well be supposed that *Bruchus pisi,* and possibly also the described abnormalities in the floral structure, were to blame.

[F_1] The Forms of the Hybrids[9]

Experiments which in previous years were made with ornamental plants have already afforded evidence that the hybrids, as a rule, are not exactly intermediate between the parental species. With some of the more striking characters, those, for instance, which relate to the form and size of the leaves, the pubescence of the several parts, &c., the intermediate, indeed, is nearly always to be seen; in other cases, however, one of the two parental characters is so preponderant that it is difficult, or quite impossible, to detect the other in the hybrid.

This is, precisely the case with the Pea hybrids. In the case of each of the seven crosses the hybrid-character resembles[10] that of one of the parental forms so closely that the other either escapes observation completely or cannot be detected with certainty. This circumstance is of great importance in the determination and classification of the forms under which the offspring of the hybrids appear. Henceforth in this paper those characters which are transmitted entire, or almost unchanged in the hybridisation, and therefore in themselves constitute the characters of the hybrid, are termed the *dominant,* and those which become latent in the process *recessive.* The expression "recessive" has been chosen because the characters thereby designated withdraw or entirely disappear in the hybrids, but nevertheless reappear unchanged in their progeny, as will be demonstrated later on.

It was furthermore shown by the whole of the experiments that it is perfectly immaterial whether the dominant character belong to the seed-bearer or to the pollen-parent; the form of the hybrid remains identical in both cases. This interesting fact was also emphasised by Gärtner, with the remark that even the most practised expert is not in a position to determine in a hybrid which of the two parental species was the seed or the pollen plant.[11]

Of the differentiating characters which were used in the experiments the following are dominant:

1. The round or roundish form of the seed with or without shallow depressions.

2. The yellow colouring of the seed albumen [cotyledons].

3. The grey, grey-brown, or leather-brown colour of the seed-coat, in association with violet-red blossoms and reddish spots in the leaf axils.

4. The simply inflated form of the pod.

5. The green colouring of the unripe pod in association with the same colour in the stems, the leaf-veins and the calyx.

6. The distribution of the flowers along the stem.

7. The greater length of stem.

With regard to this last character it must be stated that the longer of the two parental stems is usually exceeded by the hybrid, a fact which is possibly only attributable to the greater luxuriance which appears in all parts of plants when stems of very different length are crossed. Thus, for instance, in repeated experiments, stems of 1 ft. and 6 ft. in length yielded without exception hybrids which varied in length between 6 ft. and 7 ½ ft.

The hybrid seeds in the experiments with seed-coat are often more spotted, and the spots sometimes coalesce into small bluish-violet patches. The spotting also frequently appears even when it is absent as a parental character.[12]

The hybrid forms of the seed-shape and of the albumen [colour] are developed immediately after the artificial fertilisation by the mere influence of the foreign pollen. They can, therefore, be observed even in the first year of experiment, whilst all the other characters naturally only appear in the following year in such plants as have been raised from the crossed seed.

[F_2] The First Generation [Bred] from the Hybrids

In this generation there reappear, together with the dominant characters, also the recessive ones with their peculiarities fully developed, and this occurs in the definitely expressed average proportion of three to one, so that among each four plants of this generation three display the dominant character and one the recessive. This relates without exception to all the characters which were investigated in the experiments. The angular wrinkled form of the seed, the green colour of the albumen, the white colour of the seed-coats and the flowers, the constrictions of the pods, the yellow colour of the unripe pod, of the stalk, of the calyx, and of the leaf venation, the umbel-like form of the inflorescence, and the dwarfed stem, all reappear in the numerical proportion given, without any essential alteration. *Transitional forms were not observed in any experiment.*

Since the hybrids resulting from reciprocal crosses are formed alike and present no appreciable difference in their subsequent development, consequently the results [of the reciprocal crosses] can be reckoned together in each experiment. The relative numbers which were obtained for each pair of differentiating characters are as follows:

Expt. 1. Form of seed.—From 253 hybrids 7,324 seeds were obtained in the second trial year. Among them were 5,474 round or roundish ones and 1,850 angular wrinkled ones. Therefrom the ratio 2.96 to 1 is deduced.

Expt. 2. Colour of albumen.—258 plants yielded 8,023 seeds, 6,022 yellow, and 2,001 green; their ratio, therefore, is as 3.01 to 1.

In these two experiments each pod yielded usually both kinds of seed. In well-developed pods which contained on the average six to nine seeds, it often happened that all the seeds were round (Expt. 1) or all yellow (Expt. 2); on the other hand there were never observed more than five wrinkled or five green ones in one pod. It appears to make no difference whether the pods are developed early or later in the hybrid or whether they spring from the main axis or from a lateral one. In some few plants only a few seeds developed in the first formed pods, and these possessed exclusively one of the two characters, but in the subsequently developed pods the normal proportions were maintained nevertheless.

As in separate pods, so did the distribution of the characters vary in separate plants. By way of illustration the first ten individuals from both series of experiments may serve.

	Experiment 1		Experiment 2	
	Form of seed		Color of albumen	
Plants	Round	Angular	Yellow	Green
1	45	12	25	11
2	27	8	32	7
3	24	7	14	5
4	19	10	70	27
5	32	11	24	13
6	26	6	20	6
7	88	24	32	13
8	22	10	44	9
9	28	6	50	14
10	25	7	44	18

As extremes in the distribution of the two seed characters in one plant, there were observed in Expt. 1 an instance of 43 round and only 2 angular, and another of 14 round and 15 angular seeds. In Expt. 2 there was a case of 32 yellow and only 1 green seed, but also one of 20 yellow and 19 green.

These two experiments are important for the determination of the average ratios, because with a smaller number of experimental plants they show that very considerable fluctuations may occur. In counting the seeds, also, especially in Expt. 2, some care is requisite, since in some of the seeds of many plants the green colour of the albumen is less developed, and at first may be easily overlooked. The cause of this partial disappearance of the green colouring has no connection with the hybrid-character of the plants, as it likewise occurs in the parental variety. This peculiarity [bleaching] is also confined to the individual and is not inherited by the offspring. In luxuriant plants this appearance was frequently noted. Seeds which are damaged by insects during their development often vary in colour and form, but, with a little practice in sorting, errors are easily avoided. It is almost superfluous to mention that the pods must remain on the plants until they are thoroughly ripened and have become dried, since it is only then that the shape and colour of the seed are fully developed.

> **Expt. 3.** Colour of the seed-coats.—Among 929 plants 705 bore violet-red flowers and grey-brown seed-coats; 224 had white flowers and white seed-coats, giving the proportion 3.15 to 1.
>
> **Expt. 4.** Form of pods.—Of 1,181 plants 882 had them simply inflated, and in 299 they were constricted. Resulting ratio, 2.95 to 1.
>
> **Expt. 5.** Colour of the unripe pods.—The number of trial plants was 580, of which 428 had green pods and 152 yellow ones. Consequently these stand in the ratio 2.82 to 1.
>
> **Expt. 6.** Position of flowers.—Among 858 cases 651 had inflorescences axial and 207 terminal. Ratio, 3.14 to 1.
>
> **Expt. 7.** Length of stem.—Out of 1,064 plants, in 787 cases the stem was long, and in 277 short. Hence a mutual ratio of 2.84 to 1. In this experiment the dwarfed plants were carefully lifted and transferred to a special bed. This precaution was necessary, as otherwise they would have perished through being overgrown by their tall relatives. Even in their quite young state they can be easily picked out by their compact growth and thick dark-green foliage.[13]

If now the results of the whole of the experiments be brought together, there is found, as between the number of forms with the dominant and recessive characters, an average ratio of 2.98 to 1, or 3 to 1.

The dominant character can have here a *double signification*—viz. that of a parental character, or a hybrid-character.[14] In which of the two significations it appears in each separate case can only be determined by the following generation. As a parental character it must pass over unchanged to the whole of the offspring; as a hybrid-character, on the other hand, it must maintain the same behaviour as in the first generation [F_2].

[F_3] The Second Generation [Bred] from the Hybrids

Those forms which in the first generation [F_2] exhibit the recessive character do not further vary in the second generation [F_3] as regards this character; they remain constant in their offspring.

It is otherwise with those which possess the dominant character in the first generation [bred from the hybrids]. Of these *two*-thirds yield offspring which display the dominant and recessive characters in the proportion of 3 to 1, and thereby show exactly the same ratio as the hybrid forms, while only *one*-third remains with the dominant character constant.

The separate experiments yielded the following results:

Expt. 1. Among 565 plants which were raised from round seeds of the first generation, 193 yielded round seeds only, and remained therefore constant in this character; 372, however, gave both round and wrinkled seeds, in the proportion of 3 to 1. The number of the hybrids, therefore, as compared with the constants is 1.93 to 1.

Expt. 2. Of 519 plants which were raised from seeds whose albumen was of yellow colour in the first generation, 166 yielded exclusively yellow, while 353 yielded yellow and green seeds in the proportion of 3 to 1. There resulted, therefore, a division into hybrid and constant forms in the proportion of 2.13 to 1.

For each separate trial in the following experiments 100 plants were selected which displayed the dominant character in the first generation, and in order to ascertain the significance of this, ten seeds of each were cultivated.

Expt. 3. The offspring of 36 plants yielded exclusively grey-brown seed-coats, while of the offspring of 64 plants some had grey-brown and some had white.

Expt. 4. The offspring of 29 plants had only simply inflated pods; of the offspring of 71, on the other hand, some had inflated and some constricted.

Expt. 5. The offspring of 40 plants had only green pods; of the offspring of 60 plants some had green, some yellow ones.

Expt. 6. The offspring of 33 plants had only axial flowers; of the offspring of 67, on the other hand, some had axial and some terminal flowers.

Expt. 7. The offspring of 28 plants inherited the long axis, and those of 72 plants some the long and some the short axis.

In each of these experiments a certain number of the plants came constant with the dominant character. For the determination of the proportion in which the separation of the forms with the constantly persistent character results, the two first experiments are of especial importance, since in these a larger number of plants can be compared. The ratios 1.93 to 1 and 2.13 to 1 gave together almost exactly the average ratio of 2 to 1. The sixth experiment gave a quite concordant result; in the others the ratio varies more or less, as was only to be expected in view of the smaller number of 100 trial plants. Experiment 5, which shows the greatest departure, was repeated, and then, in lieu of the ratio of 60 and 40, that of 65 and 35 resulted. *The average ratio of 2 to 1 appears, therefore, as fixed with certainty.* It is therefore demonstrated that, of those forms which possess the dominant character in the first generation, two-thirds have the hybrid-character, while one-third remains constant with the dominant character.

The ratio of 3 to 1, in accordance with which the distribution of the dominant and recessive characters results in the first generation, resolves itself therefore in all experiments into the ratio of 2:1:1 if the dominant character be differentiated according to its significance as a hybrid-character or as a parental one. Since the members of the first generation [F_2] spring directly from the seed of the hybrids [F_1], *it is now clear that the hybrids form seeds having one or other of the two differentiating characters, and of these one-half develop again the hybrid form, while the other half yield plants which remain constant and receive the dominant or the recessive characters [respectively] in equal numbers.*

The Subsequent Generations [Bred] from the Hybrids

The proportions in which the descendants of the hybrids develop and split up in the first and second generations presumably hold good for all subsequent progeny. Experiments 1 and 2 have already been carried through six generations, 3 and 7 through five, and 4, 5, and 6 through four, these experiments being continued from the third generation with a small number of plants, and no departure from the rule has been perceptible. The offspring of the hybrids separated in each generation in the ratio of 2:1:1 into hybrids and constant forms.

If A be taken as denoting one of the two constant characters, for instance the dominant, a, the recessive, and Aa the hybrid form in which both are conjoined, the expression

$A + 2Aa + a$

shows the terms in the series for the progeny of the hybrids of two differentiating characters.

The observation made by Gärtner, Kölreuter, and others, that hybrids are inclined to revert to the parental forms, is also confirmed by the experiments described. It is seen that the number of the hybrids which arise from one fertilisation, as compared with the number of forms which become constant, and their progeny from generation to generation, is continually diminishing, but that nevertheless they could not entirely disappear. If an average equality of fertility in all plants in all generations be assumed, and if, furthermore, each hybrid forms seed of which one-half yields hybrids again, while the other half is constant to both characters in equal proportions, the ratio of numbers for the offspring in each generation is seen by the following summary, in which A and a denote again the two parental characters, and Aa the hybrid forms. For brevity's sake it may be assumed that each plant in each generation furnishes only 4 seeds.

Generation	A	Aa	a	Ratios				
				A	:	Aa	:	a
1	1	2	1	1	:	2	:	1
2	6	4	6	3	:	2	:	3
3	28	8	28	7	:	2	:	7
4	120	16	120	15	:	2	:	15
5	496	32	496	31	:	2	:	31
n				2^n-1	:	2	:	2^n-1

In the tenth generation, for instance, $2^n-1=1023$. There result, therefore, in each 2,048 plants which arise in this generation 1,023 with the constant dominant character, 1,023 with the recessive character, and only two hybrids.

The Offspring of Hybrids in which Several Differentiating Characters Are Associated

In the experiments above described plants were used which differed only in one essential character.[15] The next task consisted in ascertaining whether the law of development discovered in these applied to each pair of differentiating characters when several diverse characters are united in the hybrid by crossing. As regards the form of the hybrids in these cases, the experiments showed throughout that this invariably more nearly approaches to that one of the two parental plants which possesses the greater number of dominant characters. If, for instance, the seed plant has a short stem, terminal white flowers, and simply inflated pods; the pollen plant, on the other hand, a long stem, violet-red flowers distributed along the stem, and constricted pods; the hybrid resembles the seed parent only in the form of the pod; in the other characters it agrees with the pollen parent. Should one of the two parental types possess only dominant characters, then the hybrid is scarcely or not at all distinguishable from it.

Two experiments were made with a considerable number of plants. In the first experiment the parental plants differed in the form of the seed and in the colour of the albumen; in the second in the form of the seed, in the colour of the albumen, and in the colour of the seed-coats. Experiments with seed characters give the result in the simplest and most certain way.

In order to facilitate study of the data in these experiments, the different characters of the seed plant will be indicated by A, B, C, those of the pollen plant by a, b, c, and the hybrid forms of the characters by Aa, Bb, and Cc.

Expt. 1.—AB, seed parents; ab, pollen parents;
           A, form round; a, form wrinkled;
           B, albumen yellow. b, albumen green.

The fertilised seeds appeared round and yellow like those of the seed parents. The plants raised therefrom yielded seeds of four sorts, which frequently presented themselves in one pod. In all, 556 seeds were yielded by 15 plants, and of these there were:

    315 round and yellow,
    101 wrinkled and yellow,
    108 round and green,
     32 wrinkled and green.

All were sown the following year. Eleven of the round yellow seeds did not yield plants, and three plants did not form seeds. Among the rest:

    38   had round yellow seeds AB
    65   round yellow and green seeds ABb
    60   round yellow and wrinkled yellow seeds AaB
  138   round yellow and green, wrinkled yellow
           and green seeds $AaBb$.

From the wrinkled yellow seeds 96 resulting plants bore seed, of which:

    28   had only wrinkled yellow seeds aB
    68   wrinkled yellow and green seeds aBb.

From 108 round green seeds 102 resulting plants fruited, of which:

35 had only round green seeds   *Ab*
67 round and wrinkled green seeds   *Aab.*

The wrinkled green seeds yielded 30 plants which bore seeds all of like character; they remained constant *ab*.

The offspring of the hybrids appeared therefore under nine different forms, some of them in very unequal numbers. When these are collected and co-ordinated we find:

38	plants with the sign	*AB*
35	" " " "	*Ab*
28	" " " "	*aB*
30	" " " "	*ab*
65	" " " "	*ABb*
68	" " " "	*aBb*
60	" " " "	*AaB*
67	" " " "	*Aab*
138	" " " "	*AaBb*

The whole of the forms may be classed into three essentially different groups. The first includes those with the signs *AB, Ab, aB,* and *ab*: they possess only constant characters and do not vary again in the next generation. Each of these forms is represented on the average thirty-three times. The second group includes the signs *ABb, aBb, AaB, Aab*: these are constant in one character and hybrid in another, and vary in the next generation only as regards the hybrid-character. Each of these appears on an average sixty-five times. The form *AaBb* occurs 138 times: it is hybrid in both characters, and behaves exactly as do the hybrids from which it is derived.

If the numbers in which the forms belonging to these classes appear be compared, the ratios of 1, 2, 4 are unmistakably evident. The numbers 32, 65, 138 present very fair approximations to the ratio numbers of 33, 66, 132.

The developmental series consists, therefore, of nine classes, of which four appear therein always once and are constant in both characters; the forms *AB, ab,* resemble the parental forms, the two others present combinations between the conjoined characters *A, a, B, b,* which combinations are likewise possibly constant. Four classes appear always twice, and are constant in one character and hybrid in the other. One class appears four times, and is hybrid in both characters. Consequently the offspring of the hybrids, if two kinds of differentiating characters are combined therein, are represented by the expression

$AB + Ab + aB + ab + 2ABb + 2aBb + 2AaB + 2Aab + 4AaBb.$

This expression is indisputably a combination series in which the two expressions for the characters A and a, B and b are combined. We arrive at the full number of the classes of the series by the combination of the expressions:

$A + 2Aa + a$
$B + 2Bb + b.$

Expt. 2.

ABC, seed parents; abc, pollen parents;
   A, form round; a, form wrinkled;
   B, albumen yellow; b, albumen green;
   C, seed-coat grey-brown. c, seed-coat white.

This experiment was made in precisely the same way as the previous one. Among all the experiments it demanded the most time and trouble. From 24 hybrids 687 seeds were obtained in all: these were all either spotted, grey-brown or grey-green, round or wrinkled.[16] From these in the following year 639 plants fruited, and, as further investigation showed, there were among them:

8 plants	ABC	22 plants	$ABCc$	45 plants	$ABbCc$
14 "	ABc	17 "	$AbCc$	36 "	$aBbCc$
9 "	AbC	25 "	$aBCc$	38 "	$AaBCc$
11 "	Abc	20 "	$abCc$	40 "	$AabCc$
8 "	aBC	15 "	$ABbC$	49 "	$AaBbC$
10 "	aBc	18 "	$ABbc$	48 "	$AaBbc$
10 "	abC	19 "	$aBbC$		
7 "	abc	24 "	$aBbc$		
		14 "	$AaBC$	78 "	$AaBbCc$
		18 "	$AaBc$		
		20 "	$AabC$		
		16 "	$Aabc$		

The whole expression contains 27 terms. Of these 8 are constant in all characters, and each appears on the average 10 times; 12 are constant in two characters, and hybrid in the third; each appears on the average 19 times; 6 are constant in one character and hybrid in the other two; each appears on the average 43 times. One form appears 78 times and is hybrid in all of the characters. The ratios 10, 19, 43, 78 agree so closely with the ratios 10, 20, 40, 80, or 1, 2, 4, 8, that this last undoubtedly represents the true value.

The development of the hybrids when the original parents differ in three characters results therefore according to the following expression:

$ABC + ABc + AbC + Abc + aBC + aBc + abC + abc +$
$2\ ABCc + 2\ AbCc + 2\ aBCc + 2\ abCc + 2\ ABbC + 2\ ABbc +$
$2\ aBbC + 2\ aBbc + 2\ AaBC + 2\ AaBc + 2\ AabC + 2\ Aabc +$
$4\ ABbCc + 4\ aBbCc + 4\ AaBCc + 4\ AabCc + 4\ AaBbC +$
$4\ AaBbc + 8\ AaBbCc.$

Here also is involved a combination series in which the expressions for the characters A and a, B and b, C and c, are united. The expressions

$A + 2Aa + a$
$B + 2Bb + b$
$C + 2Cc + c$

give all the classes of the series. The constant combinations which occur therein agree with all combinations which are possible between the characters A, B, C, a, b, c; two thereof, ABC and abc, resemble the two original parental stocks.

In addition, further experiments were made with a smaller number of experimental plants in which the remaining characters by twos and threes were united as hybrids: all yielded approximately the same results. There is therefore no doubt that for the whole of the characters involved in the experiments the principle applies that *the offspring of the hybrids in which several essentially different characters are combined exhibit the terms of a series of combinations, in which the developmental series for each pair of differentiating characters are united.* It is demonstrated at the same time that *the relation of each pair of different characters in hybrid union is independent of the other differences in the two original parental stocks.*

If n represent the number of the differentiating characters in the two original stocks, 3^n gives the number of terms of the combination series, 4^n the number of individuals which belong to the series, and 2^n the number of unions which remain constant. The series therefore contains, if the original stocks differ in four characters, $3^4 = 81$ classes, $4^4 = 256$ individuals, and $2^4 = 16$ constant forms; or, which is the same, among each 256 offspring of the hybrids there are 81 different combinations, 16 of which are constant.

All constant combinations which in Peas are possible by the combination of the said seven differentiating characters were actually obtained by repeated crossing. Their number is given by $2^7 = 128$. Thereby is simultaneously given the practical proof *that the constant characters which appear in the several varieties of a group of plants may be obtained in all the associations which are possible according to the [mathematical] laws of combination, by means of repeated artificial fertilisation.*

As regards the flowering time of the hybrids, the experiments are not yet concluded. It can, however, already be stated that the time stands almost exactly between those of the seed and pollen parents, and that the constitution of the hybrids with respect to this character probably follows the rule ascertained in the case of the other characters. The forms which are selected for experiments of this class must have a difference of at least twenty days from the middle flowering period of one to that of the other; furthermore, the seeds when sown must all be placed at the same depth in the earth, so that they may germinate simultaneously. Also, during the whole flowering period, the more important variations in temperature must be taken into account, and the partial hastening or delaying of the flowering which may result therefrom. It is clear that this experiment presents many difficulties to be overcome and necessitates great attention.

If we endeavour to collate in a brief form the results arrived at, we find that those differentiating characters, which admit of easy and certain recognition in the experimental plants, all behave exactly alike in their hybrid associations. The offspring of the hybrids of each pair of differentiating characters are, one-half, hybrid again, while the other half are constant in equal proportions having the characters of the seed and pollen parents respectively. If several differentiating characters are combined by cross-fertilisation in a hybrid, the resulting offspring form the terms of a combination series in which the combination series for each pair of differentiating characters are united.

The uniformity of behaviour shown by the whole of the characters submitted to experiment permits, and fully justifies, the acceptance of the principle that a similar relation exists in the other characters which appear less sharply defined in plants, and therefore could not be included in the separate experiments. An experiment with peduncles of different lengths gave on the whole a fairly satisfactory result, although the differentiation and serial arrangement of the forms could not be effected with that certainty which is indispensable for correct experiment.

The Reproductive Cells of the Hybrids

The results of the previously described experiments led to further experiments, the results of which appear fitted to afford some conclusions as regards the composition of the egg and pollen cells of hybrids. An important clue is afforded in *Pisum* by the circumstance that among the progeny of the hybrids constant forms appear, and that this occurs, too, in respect of all combinations of the associated characters. So far as experience goes, we find it in every case confirmed that constant progeny can only be formed when the egg cells and the fertilising pollen are of like character, so that both are provided with the material for creating quite similar individuals, as is the case with the normal fertilisation of pure species. We must therefore regard it as certain that exactly similar factors must be at work also in the production of the constant forms in the hybrid plants. Since the various constant forms are produced in *one* plant, or even in *one* flower of a plant, the conclusion appears logical that in the ovaries of the hybrids there are formed as many sorts of egg cells, and in the anthers as many sorts of pollen cells, as there are possible constant combination forms, and that these egg and pollen cells agree in their internal composition with those of the separate forms.

In point of fact it is possible to demonstrate theoretically that this hypothesis would fully suffice to account for the development of the hybrids in the separate generations, if we might at the same time assume that the various kinds of egg and pollen cells were formed in the hybrids on the average in equal numbers.[17]

In order to bring these assumptions to an experimental proof, the following experiments were designed. Two forms which were constantly different in the form of the seed and the colour of the albumen were united by fertilisation.

If the differentiating characters are again indicated as *A, B, a, b,* we have:

AB, seed parent;   *ab,* pollen parent;
A, form round;   *a,* form wrinkled;
B, albumen yellow.   *b,* albumen green.

The artificially fertilised seeds were sown together with several seeds of both original stocks, and the most vigorous examples were chosen for the reciprocal crossing. There were, fertilised:

1. The hybrids with the pollen of *AB.*
2. The hybrids " " *ab.*
3. *AB* " " " the hybrids.
4. *ab* " " " the hybrids.

For each of these four experiments the whole of the flowers on three plants were fertilised. If the above theory be correct, there must be developed on the hybrids egg and pollen cells of the forms *AB, Ab, aB, ab,* and there would be combined:

1. The egg cells *AB, Ab, aB, ab* with the pollen cells *AB*.
2. The egg cells *AB, Ab, aB, ab* with the pollen cells *ab*.
3. The egg cells *AB* with the pollen cells *AB, Ab, aB, ab*.
4. The egg cells *ab* with the pollen cells *AB, Ab, aB, ab*.

From each of these experiments there could then result only the following forms:

1. *AB, ABb, AaB, AaBb.*
2. *AaBb, Aab, aBb, ab.*
3. *AB, ABb, AaB, AaBb.*
4. *AaBb, Aab, aBb, ab.*

If, furthermore, the several forms of the egg and pollen cells of the hybrids were produced on an average in equal numbers, then in each experiment the said four combinations should stand in the same ratio to each other. A perfect agreement in the numerical relations was, however, not to be expected, since in each fertilisation, even in normal cases, some egg cells remain undeveloped or subsequently die, and many even of the well-formed seeds fail to germinate when sown. The above assumption is also limited in so far that, while it demands the formation of an equal number of the various sorts of egg and pollen cells, it does not require that this should apply to each separate hybrid with mathematical exactness.

The first and second experiments had primarily the object of proving the composition of the hybrid egg cells, while the third and fourth experiments were to decide that of the pollen cells.[18] As is shown by the above demonstration the first, and third experiments and the second and fourth experiments should produce precisely the same combinations, and even in the second year the result should be partially visible in the form and colour of the artificially fertilised seed. In the first and third experiments the dominant characters of form and colour, *A* and *B*, appear in each union, and are also partly constant and partly in hybrid union with the recessive characters *a* and *b*, for which reason they must impress their peculiarity upon the whole of the seeds. All seeds should therefore appear round and yellow, if the theory be justified. In the second and fourth experiments, on the other hand, one union is hybrid in form and in colour, and consequently the seeds are round and yellow; another is hybrid in form, but constant in the recessive character of colour, whence the seeds are round and green; the third is constant in the recessive character of form but hybrid in colour, consequently the seeds are wrinkled and yellow; the fourth is constant in both recessive characters, so that the seeds are wrinkled and green. In both these experiments there were consequently four sorts of seed to be expected—viz. round and yellow, round and green, wrinkled and yellow, wrinkled and green.

The crop fulfilled these expectations perfectly. There were obtained in the

1st Experiment, 98 exclusively round yellow seeds;
3rd   "   94   "   "   "   "

In the 2nd Experiment, 31 round and yellow, 26 round and green, 27 wrinkled and yellow, 26 wrinkled and green seeds.

In the 4th Experiment, 24 round and yellow, 25 round and green, 22 wrinkled and yellow, 27 wrinkled and green seeds.

There could scarcely be now any doubt of the success of the experiment; the next generation must afford the final proof. From the seed sown there resulted for the first experiment 90 plants, and for the third 87 plants which fruited: these yielded for the

1st Exp.	3rd Exp.		
20	25	round yellow seeds	AB
23	19	round yellow and green seeds	ABb
25	22	round and wrinkled yellow seeds	AaB
22	21	round and wrinkled green and yellow seeds	AaBb

In the second and fourth experiments the round and yellow seeds yielded plants with round and wrinkled yellow and green seeds, *AaBb*.

From the round green seeds plants resulted with round and wrinkled green seeds, *Aab*.

The wrinkled yellow seeds gave plants with wrinkled yellow and green seeds, *aBb*.

From the wrinkled green seeds plants were raised which yielded again only wrinkled and green seeds, *ab*.

Although in these two experiments likewise some seeds did not germinate, the figures arrived at already in the previous year were not affected thereby, since each kind of seed gave plants which, as regards their seed, were like each other and different from the others. There resulted therefore from the

2nd Exp.	4th Exp.			
31	24	plants of the form		AaBb
26	25	"	"	AaB
27	22	"	"	aBb
26	27	"	"	ab

In all the experiments, therefore, there appeared all the forms which the proposed theory demands, and they came in nearly equal numbers.

In a further experiment the characters of flower-colour and length of stem were experimented upon, and selection was so made that in the third year of the experiment each character ought to appear in half of all the plants if the above theory were correct. *A, B, a, b* serve again as indicating the various characters.

A, violet-red flowers.   *a*, white flowers.
B, axis long.   *b*, axis short.

The form *Ab* was fertilised with *ab*, which produced the hybrid *Aab*. Furthermore, *aB* was also fertilised with *ab*, whence the hybrid *aBb*. In the second year, for further fertilisation, the hybrid *Aab* was used as seed parent, and hybrid *aBb* as pollen parent.

Seed parent, *Aab*.   Pollen parent, *aBb*.
Possible egg cells, *Abab*.   Pollen cells, *aBab*.

From the fertilisation between the possible egg and pollen cells four combinations should result, viz.:

AaBb + aBb + Aab + ab.

From this it is perceived that, according to the above theory, in the third year of the experiment out of all the plants

Half should have violet-red flowers *(Aa)*, Classes 1, 3
 "   "   "   white flowers   *(a)*   "   2, 4
 "   "   "   a long axis   *(Bb)*   "   1, 2
 "   "   "   a short axis   *(b)*   "   3, 4

From 45 fertilisations of the second year 187 seeds resulted, of which only 166 reached the flowering stage in the third year. Among these the separate classes appeared in the numbers following:

Class	Color of flower	Stem	
1	violet-red	long	47 times
2	white	long	40 "
3	violet-red	short	38 "
4	white	short	41 "

There subsequently appeared

The violet-red flower-colour (*Aa*) in 85 plants.
" white   "   "   (*a*) in 81   "
" long stem   (*Bb*) in 87   "
" short   "   (*b*) in 79   "

The theory adduced is therefore satisfactorily confirmed in this experiment also.

For the characters of form of pod, colour of pod, and position of flowers experiments were also made on a small scale, and results obtained in perfect agreement. All combinations which were possible through the union of the differentiating characters duly appeared, and in nearly equal numbers.

Experimentally, therefore, the theory is confirmed that *the pea hybrids form egg and pollen cells which, in their constitution, represent in equal numbers all constant forms which result from the combination of the characters united in fertilisation.*

The difference of the forms among the progeny of the hybrids, as well as the respective ratios of the numbers in which they are observed, find a sufficient explanation in the principle above deduced. The simplest case is afforded by the developmental series of each pair of differentiating characters. This series is represented by the expression $A + 2Aa + a$, in which A and a signify the forms with constant differentiating characters, and Aa the hybrid form of both. It includes in three different classes four individuals. In the formation of these, pollen and egg cells of the form A and a take part on the average equally in the fertilisation; hence each form [occurs] twice, since four individuals are formed. There participate consequently in the fertilisation

The pollen cells $A + A + a + a$
The egg cells $A + A + a + a$.

It remains, therefore, purely a matter of chance which of the two sorts of pollen will become united with each separate egg cell. According, however, to the law of probability, it will always happen, on the average of many cases, that each pollen form A and a will unite equally often with each egg cell form A and a, consequently one of the two pollen cells A in the fertilisation will meet with the egg cell A and the other with an egg cell a, and so likewise one pollen cell a will unite with an egg cell A, and the other with egg cell a.

The result of the fertilisation may be made clear by putting the signs for the conjoined egg and pollen cells in the form of fractions, those for the pollen cells above and those for the egg cells below the line. We then have

$$\frac{A}{A} + \frac{A}{a} + \frac{a}{A} + \frac{a}{a}.$$

In the first and fourth term the egg and pollen cells are of like kind, consequently the product of their union must be constant, viz. A and a; in the second and third, on the other hand, there again results a union of the two differentiating characters of the stocks, consequently the forms resulting from these fertilisations are identical with those of the hybrid from which they sprang. *There occurs accordingly a repeated hybridisation.* This explains the striking fact that the hybrids are able to produce, besides the two parental forms, offspring which are like themselves; A / a and a / A both give the same union Aa, since, as already remarked above, it makes no difference in the result of fertilisation to which of the two characters the pollen or egg cells belong. We may write then

$$\frac{A}{A} + \frac{A}{a} + \frac{a}{A} + \frac{a}{a} = A + 2Aa + a$$

This represents the average result of the self-fertilisation of the hybrids when two differentiating characters are united in them. In individual flowers and in individual plants, however, the ratios in which the forms of the series are produced may suffer not inconsiderable fluctuations.[19] Apart from the fact that the numbers in which both sorts of egg cells occur in the seed vessels can only be regarded as equal on the average, it remains purely a matter of chance which of the two sorts of pollen may fertilise each separate egg cell. For this reason the separate values must necessarily be subject to fluctuations, and there are even extreme cases possible, as were described earlier in connection with the experiments on the form of the seed and the colour of the albumen. The true ratios of the numbers can only be ascertained by an average deduced from the sum of as many single values as possible; the greater the number the more are merely chance effects eliminated.

The developmental series for hybrids in which two kinds of differentiating characters are united contains among sixteen individuals nine different forms, viz.:

$AB + Ab + aB + ab + 2ABb + 2aBb + 2AaB + 2Aab + 4AaBb.$

Between the differentiating characters of the original stocks Aa and Bb four constant combinations are possible, and consequently the hybrids produce the corresponding four forms of egg and pollen cells AB, Ab, aB, ab, and each of these will on the average figure four times in the fertilisation, since sixteen individuals are included in the series. Therefore the participators in the fertilisation are

Pollen cells $\quad AB + AB + AB + AB + Ab + Ab + Ab + Ab + aB + aB + aB + aB + ab + ab + ab + ab.$

Egg cells $\quad AB + AB + AB + AB + Ab + Ab + Ab + Ab + aB + aB + aB + aB + ab + ab + ab + ab.$

In the process of fertilisation each pollen form unites on an average equally often with each egg cell form, so that each of the four pollen cells AB unites once with one of the forms of egg cell AB, Ab, aB, ab. In precisely the same way the rest of the pollen cells of the forms Ab, aB, ab unite with all the other egg cells. We obtain therefore

$$\frac{AB}{AB} + \frac{AB}{Ab} + \frac{AB}{aB} + \frac{AB}{ab} + \frac{Ab}{AB} + \frac{Ab}{Ab} + \frac{Ab}{aB} + \frac{Ab}{ab} +$$

$$\frac{aB}{AB} + \frac{aB}{Ab} + \frac{aB}{aB} + \frac{aB}{ab} + \frac{ab}{AB} + \frac{ab}{Ab} + \frac{ab}{aB} + \frac{ab}{ab},$$

or

$AB + ABb + AaB + AaBb + ABb + Ab + AaBb + Aab + AaB + AaBb + aB + aBb + AaBb + Aab + aBb + ab = AB + Ab + aB + ab + 2ABb + 2aBb + 2AaB + 2Aab + 4AaBb.$[20]

In precisely similar fashion is the developmental series of hybrids exhibited when three kinds of differentiating characters are conjoined in them. The hybrids form eight various kinds of egg and pollen cells—ABC,

ABc, AbC, Abc, aBC, aBc, abC, abc—and each pollen form unites itself again on the average once with each form of egg cell.

The law of combination of different characters which governs the development of the hybrids finds therefore its foundation and explanation in the principle enunciated, that the hybrids produce egg cells and pollen cells which in equal numbers represent all constant forms which result from the combinations of the characters brought together in fertilisation.

Experiments with Hybrids of Other Species of Plants

It must be the object of further experiments to ascertain whether the law of development discovered for *Pisum* applies also to the hybrids of other plants. To this end several experiments were recently commenced. Two minor experiments with species of *Phaseolus* have been completed, and may be here mentioned.

An experiment with *Phaseolus vulgaris* and *Phaseolus nanus* gave results in perfect agreement. *Ph. nanus* had together with the dwarf axis, simply inflated, green pods. *Ph. vulgaris* had, on the other hand, an axis 10 feet to 12 feet high, and yellow-coloured pods, constricted when ripe. The ratios of the numbers in which the different forms appeared in the separate generations were the same as with *Pisum*. Also the development of the constant combinations resulted according to the law of simple combination of characters, exactly as in the case of *Pisum*. There were obtained

Constant combinations	Axis	Colour of the unripe pods	Form of the ripe pods
1	long	green	inflated
2	"	"	constricted
3	"	yellow	inflated
4	"	"	constricted
5	short	green	inflated
6	"	"	constricted
7	"	yellow	inflated
8	"	"	constricted

The green colour of the pod, the inflated forms, and the long axis were, as in *Pisum*, dominant characters.

Another experiment with two very different species of *Phaseolus* had only a partial result. *Phaseolus nanus*, L., served as seed parent, a perfectly constant species, with white flowers in short racemes and small white seeds in straight, inflated, smooth pods; as pollen parent was used *Ph. multiflorus*, W., with tall winding stem, purple-red flowers in very long racemes, rough, sickle-shaped crooked pods, and large seeds which bore black flecks and splashes on a peach-blood-red ground.

The hybrids had the greatest similarity to the pollen parent, but the flowers appeared less intensely coloured. Their fertility was very limited; from seventeen plants, which together developed many hundreds of flowers, only forty-nine seeds in all were obtained. These were of medium size, and were flecked and splashed similarly to those of *Ph. multiflorus*, while the ground colour was not materially different. The next year forty-four plants were raised from these seeds, of which only thirty-one reached the flowering stage. The characters of *Ph. nanus*, which had been altogether latent in the hybrids, reappeared in various combinations; their ratio, however, with relation to the dominant plants was necessarily very fluctuating owing to the small number of trial plants. With certain characters, as in those of the axis and the form of pod, it was, however, as in the case of *Pisum*, almost exactly 1 : 3.

Insignificant as the results of this experiment may be as regards the determination of the relative numbers in which the various forms appeared, it presents, on the other hand, the phenomenon of a remarkable change of colour in the flowers and seed of the hybrids. In *Pisum* it is known that the characters of the flower- and seed-colour present themselves unchanged in the first and further generations, and that the offspring of the hybrids display exclusively the one or the other of the characters of the original stocks. It is otherwise in the experiment we are considering. The white flowers and the seed-colour of *Ph. nanus* appeared, it is true, at once in the first generation [*from* the hybrids] in one fairly fertile example, but the remaining thirty plants developed flower-colours which were of various grades of purple-red to pale violet. The colouring of the seed-coat was no less varied than that of the flowers. No plant could rank as fully fertile; many produced no fruit at all; others only yielded fruits from the flowers last produced, which did not ripen. From fifteen plants only were well-developed seeds obtained. The greatest disposition to infertility was seen in the forms with preponderantly red flowers, since out of sixteen of these only four yielded ripe seed. Three of these had a similar seed pattern to *Ph. multiflorus*, but with a more or less pale ground colour; the fourth plant yielded only one seed of plain brown tint. The forms with preponderantly violet-coloured flowers had dark brown, black-brown, and quite black seeds.

The experiment was continued through two more generations under similar unfavourable circumstances, since even among the offspring of fairly fertile plants there came again some which were less fertile or even quite sterile. Other flower- and seed-colours than those cited did not subsequently present themselves. The forms which in the first generation [bred from the hybrids] contained one or more of the recessive characters remained, as regards these, constant without exception. Also of those plants which possessed violet flowers and brown or black seed, some did not vary again in these respects in the next generation; the majority, however, yielded, together with offspring exactly like themselves, some which displayed white flowers and white seed-coats. The red flowering plants remained so slightly fertile, that nothing can be said with certainty as regards their further development.

Despite the many disturbing factors with which the observations had to contend, it is nevertheless seen by this experiment that the development of the hybrids, with regard to those characters which concern the form of the plants, follows the same laws as in *Pisum*. With regard to the colour characters, it certainly appears difficult to perceive a substantial agreement. Apart from the fact that from the union of a white and a purple-red colouring a whole series of colours results [in F_2], from purple to pale violet and white, the circumstance is a striking one that among thirty-one flowering plants only one received the recessive character of the white colour, while in *Pisum* this occurs on the average in every fourth plant.

Even these enigmatical results, however, might probably be explained by the law governing *Pisum* if we might assume that the colour of the flowers and seeds of *Ph. multiflorus* is a combination of two, or more entirely independent colours, which individually act like any other constant character in the plant. If the flower-colour A were a combination of the individual characters $A_1 + A_2 + \ldots$ which produce the total impression of a purple coloration, then by fertilisation with the differentiating character, white colour, a, there would be produced the hybrid unions $A_1a + A_2a + \ldots$ and so would it be with the corresponding colouring of the seed-coats.[21] According to the above assumption, each of these hybrid colour unions would be independent, and would consequently develop quite independently from the others. It is then easily seen that from the combination of the separate developmental series a complete colour-series must result. If, for instance, $A = A_1 + A_2$, then the hybrids A_1a and A_2a form the developmental series—

$A_1 + 2A_1a + a$
$A_2 + 2A_2a + a.$

The members of this series can enter into nine different combinations, and each of these denotes another colour—

1 A_1A_2	2 A_1aA_2	1 A_2a
2 A_1A_2a	4 A_1aA_2a	2 A_2aa
1 A_1a	2 A_1aa	1 aa.

The figures prescribed for the separate combinations also indicate how many plants with the corresponding colouring belong to the series. Since the total is sixteen, the whole of the colours are on the average distributed over each sixteen plants, but, as the series itself indicates, in unequal proportions.

Should the colour development really happen in this way, we could offer an explanation of the case above described, viz. that the white flowers and seed-coat colour only appeared once among thirty-one plants of the first generation. This colouring appears only once in the series, and could therefore also only be developed once in the average in each sixteen, and with three colour characters only once even in sixty-four plants.

It must, nevertheless, not be forgotten that the explanation here attempted is based on a mere hypothesis, only supported by the very imperfect result of the experiment just described. It would, however, be well worth while to follow up the development of colour in hybrids by similar experiments, since it is probable that in this way we might learn the significance of the extraordinary variety in the colouring of our ornamental flowers.

So far, little at present is known with certainty beyond the fact that the colour of the flowers in most ornamental plants is an extremely variable character. The opinion has often been expressed that the stability of the species is greatly disturbed or entirely upset by cultivation, and consequently there is an inclination to regard the development of cultivated forms as a matter of chance devoid of rules; the colouring of ornamental plants is indeed usually cited as an example of great instability. It is, however, not clear why the simple transference into garden soil should result in such a thorough and persistent revolution in the plant organism. No one will seriously maintain that in the open country the development of plants is ruled by other laws than in the garden bed. Here, as there, changes of type must take place if the conditions of life be altered, and the species possesses the capacity of fitting itself to its new environment. It is willingly granted that by cultivation the origination of new varieties is favoured, and that by man's labour many varieties are acquired which, under natural conditions, would be lost; but nothing justifies the assumption that the tendency to the formation of varieties is so extraordinarily increased that the species speedily lose all stability, and their offspring diverge into an endless series of extremely variable forms. Were the change in the conditions the sole cause of variability we might expect that those cultivated plants which are grown for centuries under almost identical conditions would again attain constancy. That, as is well known, is not the case, since it is precisely under such circumstances that not only the most varied but also the most variable forms are found. It is only the *Leguminosae*, like *Pisum, Phaseolus*[22], *Lens,* whose organs of fertilisation are protected by the keel, which constitute a noteworthy exception. Even here there have arisen numerous varieties during a cultural period of more than 1000 years under most various conditions; these maintain, however, under unchanging environments a stability as great as that of species growing wild.

It is more than probable that as regards the variability of cultivated plants there exists a factor which so far has received little attention. Various experiments force us to the conclusion that our cultivated plants, with few exceptions, are *members of various hybrid series,* whose further development in conformity with law is varied and interrupted by frequent crossings *inter se.* The circumstance must not be overlooked that cultivated plants are mostly grown in great numbers and close together, affording the most favourable conditions for reciprocal fertilisation between the varieties present and the species itself. The probability of this is supported by the fact that among the great array of variable forms solitary examples are always found, which in one character or another remain constant, if only foreign influence be carefully excluded. These forms behave precisely as do those which are known to be members of the compound hybrid series. Also with the most susceptible of all characters, that of colour, it cannot escape the careful observer that in the separate forms the inclination to vary is displayed in very different degrees. Among plants which arise from *one* spontaneous fertilisation there are often some, whose offspring vary widely in the constitution and arrangement of the colours, while that of others shows little deviation, and among a greater number solitary examples occur which transmit the colour of the flowers unchanged to their offspring. The cultivated species of *Dianthus* afford an instructive example of this. A white-flowered example of *Dianthus caryophyllus,* which itself was derived from a white-flowered variety, was shut up during its blooming period in a greenhouse; the numerous seeds obtained therefrom yielded plants entirely white-flowered like itself. A similar result was obtained from a sub-species, with red flowers somewhat flushed with violet, and one with flowers white, striped with red. Many others, on the other hand, which were similarly protected, yielded progeny which were more or less variously coloured and marked.

Whoever studies the coloration which results in ornamental plants from similar fertilisation can hardly escape the conviction that here also the development follows a definite law which possibly finds its expression *in the combination of several independent colour characters.*

Concluding Remarks

It can hardly fail to be of interest to compare the observations made regarding *Pisum* with the results arrived at by the two authorities in this branch of knowledge, Kölreuter and Gärtner, in their investigations. According to the opinion of both, the hybrids in outward appearance present either a form intermediate between the original species, or they closely resemble either the one or the other type, and sometimes can hardly be discriminated from it. From their seeds usually arise, if the fertilisation was effected by their own pollen, various forms which differ from the normal type. As a rule, the majority of individuals obtained by one fertilisation maintain the hybrid form, while some few others come more like the seed parent, and one or other individual approaches the pollen parent. This, however, is not the case with all hybrids without exception. Sometimes the offspring have more nearly approached, some the one and some the other of the two original stocks, or they all incline more to one or the other side; while in other cases *they remain perfectly like the hybrid* and continue constant in their offspring. The hybrids of varieties behave like hybrids of species, but they possess greater variability of form and a more pronounced tendency to revert to the original types.

With regard to the form of the hybrids and their development, as a rule an agreement with the observations made in *Pisum* is unmistakable. It is otherwise with the exceptional cases cited. Gärtner confesses even that the exact determination whether a form bears a greater resemblance to one or to the other of the two original species often involved great difficulty, so much depending upon the subjective point of view of the observer. Another circumstance could, however, contribute to render the results fluctuating and uncertain, despite the most careful observation and differentiation. For the experiments plants were mostly used which rank as good species and are differentiated by a large number of characters. In addition to the sharply defined characters, where it is a question of greater or less similarity, those characters must also be taken into account which are often difficult to define in words, but yet suffice, as every plant specialist knows, to give the forms a peculiar appearance. If it be accepted that the development of hybrids follows the law which is valid for *Pisum,* the series in each separate experiment must contain very many forms, since the number of the terms, as is known, increases with the number of the differentiating characters as the powers of three. With a relatively small number of experimental plants the result therefore could only be approximately right, and in single cases might fluctuate considerably. If, for instance, the two original stocks differ in seven characters, and 100 and 200 plants were raised from the seeds of their hybrids to determine the grade of relationship of the offspring, we can easily see how uncertain the decision must become, since for seven differentiating characters the combination series contains 16,384 individuals under 2187 various forms; now one and then another relationship could assert its predominance, just according as chance presented this or that form to the observer in a majority of cases.

If, furthermore, there appear among the differentiating characters at the same time *dominant* characters, which are transmitted entire or nearly unchanged to the hybrids, then in the terms of the developmental series that one of the two original parents which possesses the majority of dominant characters must always be predominant. In the experiment described relative to *Pisum,* in which three kinds of differentiating characters were concerned, all the dominant characters belonged to the seed parent. Although the terms of the series in their internal composition approach both original parents equally, yet in this experiment the type of the seed parent obtained so great a preponderance that out of each sixty-four plants of the first generation fifty-four exactly resembled it, or only differed in one character. It is seen how rash it must be under such circumstances to draw from the external resemblances of hybrids conclusions as to their internal nature.

Gärtner mentions that in those cases where the development was regular, among the offspring of the hybrids the two original species were not reproduced, but only a few individuals which approached them. With very extended developmental series it could not in fact be otherwise. For seven differentiating characters, for instance, among more than 16,000 individuals—offspring of the hybrids—each of the two original species would occur only once. It is therefore hardly possible that these should appear at all among a small number of experimental plants; with some probability, however, we might reckon upon the appearance in the series of a few forms which approach them.

We meet with an *essential difference* in those hybrids which remain constant in their progeny and propagate themselves as truly as the pure species. According to Gärtner, to this class belong the *remarkably fertile hybrids Aquilegia atropurpurea canadensis, Lavatera pseudolbia thuringiaca, Geum urbano-rivale,* and some *Dianthus* hybrids; and, according to Wichura, the hybrids of the Willow family. For the history of the evolution of plants this circumstance is of special importance, since constant hybrids acquire the status of new species. The correctness of the facts is guaranteed by eminent observers, and cannot be doubted. Gärtner had an opportunity of following up *Dianthus Armeria deltoides* to the tenth generation, since it regularly propagated itself in the garden.

With *Pisum* it was shown by experiment that the hybrids form egg and pollen cells of *different* kinds, and that herein lies the reason of the variability of their offspring. In other hybrids, likewise, whose offspring behave similarly we may assume a like cause; for those, on the other hand, which remain constant the assumption appears justifiable that their reproductive cells are all alike and agree with the foundation-cell [fertilised ovum] of the hybrid. In the opinion of renowned physiologists, for the purpose of propagation one pollen cell and one egg cell unite in Phanerogams[23] into a single cell, which is capable by assimilation and formation of new cells to become an independent organism. This development follows a constant law, which is founded on the material composition and arrangement of the elements which meet in the cell in a vivifying union. If the reproductive cells be of the same kind and agree with the foundation cell [fertilised ovum] of the mother plant, then the development of the new individual will follow the same law which rules the mother plant. If it chance that an egg cell unites with a *dissimilar* pollen cell, we must then assume that between those elements of both cells, which determine opposite characters, some sort of compromise is effected. The resulting compound cell becomes the foundation of the hybrid organism, the development of which necessarily follows a different scheme from that obtaining in each of the two original species. If the compromise be taken to be a complete one, in the sense, namely, that the hybrid embryo is formed from two similar cells, in which the differences are *entirely and permanently accommodated* together, the further result follows that the hybrids, like any other stable plant species, reproduce themselves truly in their offspring. The reproductive cells which are formed in their seed vessels and anthers are of one kind, and agree with the fundamental compound cell [fertilised ovum].

With regard to those hybrids whose progeny is *variable* we may perhaps assume that between the differentiating elements of the egg and pollen cells there also occurs a compromise, in so far that the formation of a cell as foundation of the hybrid becomes possible; but, nevertheless, the arrangement between the conflicting elements is only temporary and does not endure throughout the life of the hybrid plant. Since in the habit of the plant no changes are perceptible during the whole period of vegetation, we must further assume that it is only possible for the differentiating elements to liberate themselves from the enforced union when the fertilising cells are developed. In the formation of these cells all existing elements participate in an entirely free and equal arrangement, by which it is only the differentiating ones which mutually separate themselves. In this way the production would be rendered possible of as many sorts of egg and pollen cells as there are combinations possible of the formative elements.

The attribution attempted here of the essential difference in the development of hybrids to *a permanent or temporary union* of the differing cell elements can, of course, only claim the value of an hypothesis for which the lack of definite data offers a wide scope. Some justification of the opinion expressed lies in the evidence afforded by *Pisum* that the behaviour of each pair of differentiating characters in hybrid union is independent of the other differences between the two original plants, and, further, that the hybrid produces just so many kinds of egg and pollen cells as there are possible constant combination forms. The differentiating characters of two plants can finally, however, only depend upon differences in the composition and grouping of the elements which exist in the foundation-cells [fertilised ova] of the same in vital interaction.[24]

Even the validity of the law formulated for *Pisum* requires still to be confirmed, and a repetition of the more important experiments is consequently much to be desired, that, for instance, relating to the composition of the hybrid fertilising cells. A differential [element] may easily escape the single observer[25], which although at the outset may appear to be unimportant, may yet accumulate to such an extent that it must not be ignored in the total result. Whether the variable hybrids of other plant species observe an entire agreement must also be first decided experimentally. In the meantime we may assume that in material points an essential difference can scarcely occur, since the unity in the developmental plan of organic life is beyond question.

In conclusion, the experiments carried out by Kölreuter, Gärtner, and others with respect to *the transformation of one species into another by artificial fertilisation* merit special mention. Particular importance has been attached to these experiments, and Gärtner reckons them among "the most difficult of all in hybridisation."

If a species *A* is to be transformed into a species *B*, both must be united by fertilisation and the resulting hybrids then be fertilised with the pollen of *B*; then, out of the various offspring resulting, that form would be selected which stood in nearest relation to *B* and once more be fertilised with *B* pollen, and so continuously until finally a form is arrived at which is like *B* and constant in its progeny. By this process the species *A* would change into the species *B*. Gärtner alone has effected thirty such experiments with plants of genera *Aquilegia, Dianthus, Geum, Lavatera, Lychnis, Malva, Nicotiana,* and *Oenothera*. The period of transformation was not alike for all species. While with some a triple fertilisation sufficed, with others this had to be repeated five or six times, and even in the same species fluctuations were observed in various experiments. Gärtner ascribes this difference to the circumstance that "the specific [*typische*] power by which a species, during reproduction, effects the change and transformation of the maternal type varies considerably in different plants, and that, consequently, the periods within which the one species is changed into the other must also vary, as also the number of generations, so that the transformation in some species is perfected in more, and in others in fewer generations." Further, the same observer remarks "that in these transformation experiments a good deal depends upon which type and which individual be chosen for further transformation."

If it may be assumed that in these experiments the constitution of the forms resulted in a similar way to that of *Pisum,* the entire process of transformation would find a fairly simple explanation. The hybrid forms as many kinds of egg cells as there are constant combinations possible of the characters conjoined therein, and one of these is always of the same kind as that of the fertilising pollen cells. Consequently there always exists the possibility with all such experiments that even from the second fertilisation there may result a constant form identical with that of the pollen parent. Whether this really be obtained depends, in each separate case upon the number of the experimental plants, as well as upon the number of differentiating characters which are united by the fertilisation. Let us, for instance, assume that the plants selected for experiment differed in three characters, and the species *ABC* is to be transformed into the other species *abc* by repeated fertilisation with the pollen of the latter; the hybrids resulting from the first cross form eight different kinds of egg cells, viz.:

ABC, ABc, AbC, aBC, Abc, aBc, abC, abc.

These in the second year of experiment are united again with the pollen cells *abc*, and we obtain the series

AaBbCc + AaBbc + AabCc + aBbCc + Aabc + aBbc + abCc + abc.

Since the form *abc* occurs once in the series of eight terms, it is consequently little likely that it would be missing among the experimental plants, even were these raised in a smaller number, and the transformation would be perfected already by a second fertilisation. If by chance it did not appear, then the fertilisation must be repeated with one of those forms nearest akin, Aabc, aBbc, abCc. It is perceived that such an experiment must extend the farther *the smaller the number of experimental plants and the larger the number of differentiating characters* in the two original species; and that, furthermore, in the same species there can easily occur a delay of one or even of two generations such as Gärtner observed. The transformation of widely divergent species could generally only be completed in five or six years of experiment, since the number of different egg cells which are formed in the hybrid increases as the powers of two with the number of differentiating characters.

Gärtner found by repeated experiments that the respective period of transformation varies in many species, so that frequently a species A can be transformed into a species B a generation sooner than can species B into species A. He deduces therefrom that Kölreuter's opinion can hardly be maintained that "the two natures in hybrids are perfectly in equilibrium." It appears, however, that Kölreuter does not merit this criticism, but that Gärtner rather has overlooked a material point, to which he himself elsewhere draws attention, viz. that "it depends which individual is chosen for further transformation." Experiments which in this connection were carried out with two species of *Pisum* demonstrated that as regards the choice of the fittest individuals for the purpose of further fertilisation it may make a great difference which of two species is transformed into the other. The two experimental plants differed in five characters; while at the same time those of species A were all dominant and those of species B all recessive. For mutual transformation A was fertilised with pollen of B, and B with pollen of A, and this was repeated with both hybrids the following year. With the first experiment $\frac{B}{A}$ there were eighty-seven plants available in the third year of experiment for selection of the individuals for further crossing, and these were of the possible thirty-two forms; with the second experiment $\frac{A}{B}$ seventy-three plants resulted, which *agreed throughout perfectly in habit with the pollen parent*; in their internal composition, however, they must have been just as varied as the forms in the other experiment. A definite selection was consequently only possible with the first experiment; with the second the selection had to be made at random, merely. Of the latter only a portion of the flowers were crossed with the A pollen, the others were left to fertilise themselves. Among each five plants which were selected in both experiments for fertilisation there agreed, as the following year's culture showed, with the pollen parent:

1st Experiment	2nd Experiment	
2 plants	—	in all characters
3 "	—	" 4 "
—	2 plants	" 3 "
—	2 "	" 2 "
—	1 plant	" 1 character

In the first experiment, therefore, the transformation was completed; in the second, which was not continued further, two more fertilisations would probably have been required.

Although the case may not frequently occur in which the dominant characters belong exclusively to one or the other of the original parent plants, it will always make a difference which of the two possesses the majority of dominants. If the pollen parent has the majority, then the selection of forms for further crossing will afford a less degree of certainty than in the reverse case, which must imply a delay in the period of transformation, provided that the experiment is only considered as completed when a form is arrived at which not only exactly resembles the pollen plant in form, but also remains as constant in its progeny.

Gärtner, by the results of these transformation experiments, was led to oppose the opinion of those naturalists who dispute the stability of plant species and believe in a continuous evolution of vegetation. He perceives[26] in the complete transformation of one species into another an indubitable proof that species are fixed within limits beyond which they cannot change. Although this opinion cannot be unconditionally accepted we find on the other hand in Gärtner's experiments a noteworthy confirmation of that supposition regarding variability of cultivated plants which has already been expressed.

Among the experimental species there were cultivated plants, such as *Aquilegia atropurpurea* and *canadensis, Dianthus caryophyllus, chinensis,* and *japonicus, Nicotiana rustica* and *paniculata,* and hybrids between these species lost none of their stability after four or five generations.

NOTES

This is the version of Mendel's paper that appeared in W. Bateson. 1909. *Mendel's Principles of Heredity*. Cambridge: Cambridge University Press. This is the same version that Fisher used. To that end, all notes, including the editorial notes in brackets written by Bateson, have been retained.

1. [This translation was made by the Royal Horticultural Society, and is reprinted with modifications and corrections, by permission. The original paper was published in the *Verh. naturf. Ver. in Brünn, Abhandlungen,* iv. 1865, which appeared in 1866.]

2. [It is to the clear conception of these three primary necessities that the whole success of Mendel's work is due. So far as I know this conception was absolutely new in his day.]

3. [Mendel uses the terms "albumen" and "endosperm" somewhat loosely to denote the cotyledons, containing food-material, within the seed.]

4. One species possesses a beautifully brownish-red coloured pod, which when ripening turns to violet and blue. Trials with this character were only begun last year. [Of these further experiments it seems no account was published. Correns has since worked with such a variety.]

5. [This is often called the Mummy Pea. It shows slight fasciation. The form I know has white standard and salmon-red wings.]

6. [In my account of these experiments (*R.H.S. Journal*, vol. xxv. p. 54). I misunderstood this paragraph and took "axis" to mean the *floral* axis, instead of the main axis of the plant. The unit of measurement, being indicated in the original by a dash ('), I carelessly took to have been an *inch,* but the translation here given is evidently correct.]

7. [It is somewhat surprising that no mention is made of Thrips, which swarm in Pea flowers. I had come to the conclusion that this is a real source of error and I see Laxton held the same opinion.]

8. [This also happens in Sweet Peas.]

9. [Mendel throughout speaks of his cross-bred Peas as "hybrids," a term which many restrict to the offspring of two distinct *species*. He, as he explains, held this to be only a question of degree.]

10. [Note that Mendel, with true penetration, avoids speaking of the hybrid-character as "transmitted" by either parent, thus escaping the error pervading the older views of heredity.]

11. [Gärtner, p.223.]

12. [This refers to the coats of the seeds borne by F_1 plants.]

13. [This is true also of the dwarf or "Cupid" Sweet Peas.]

14. [This paragraph presents the view of the hybrid-character as something incidental to the hybrid, and not "transmitted" to it—a true and fundamental conception here expressed probably for the first time.]

15. [This statement of Mendel's in the light of present knowledge is open to some misconception. Though his work makes it evident that such varieties may exist, it is very unlikely that Mendel could have had seven pairs of varieties such that the members of each pair differed from each other in *only* one considerable character *(wesentliches Merkmal)*. The point is probably of little theoretical or practical consequence, but a rather heavy stress is thrown on *"wesentlich."*]

16. [Note that Mendel does not state the cotyledon-colour of the first crosses in this case; for as the coats were thick, it could not have been seen without opening or peeling the seeds.]

17. [This and the preceding paragraph contain the essence of the Mendelian principles of heredity.]

18. [To prove, namely, that both were similarly differentiated, and not one or other only.]

19. [Whether segregation by such units is more than purely fortuitous may perhaps be determined by seriation.]

20. [In the original the sign of equality (=) is here represented by +, evidently a misprint.]

21. [As it fails to take account of factors introduced by the albino this representation is imperfect. It is however interesting to know that Mendel realized the fact of the existence of compound characters, and that the rarity of the white recessives was a consequence of this resolution.]

22. [*Phaseolus* nevertheless is insect fertilised.]

23. In *Pisum* it is placed beyond doubt that for the formation of the new embryo a perfect union of the elements of both reproductive cells must take place. How could we otherwise explain that among the offspring of the hybrids both original types reappear in equal numbers and with all their peculiarities? If the influence of the egg cell upon the pollen cell were only external, if it fulfilled the *rôle* of a nurse only, then the result of each artificial fertilisation could be no other than that the developed hybrid should exactly resemble the pollen parent, or at any rate do so very closely. This the experiments so far have in no wise confirmed. An evident proof of the complete union of the contents of both cells is afforded by the experience gained on all sides that it is immaterial, as regards the form of the hybrid, which of the original species is the seed parent or which the pollen parent.

24. "*Welche in den Grundzellen derselben in lebendiger Wechselwirkung stehen.*"

25. "*Dem einzelnen Beobachter kann leicht ein Differenziale entgehen.*"

26. ["Es sieht" in the original is clearly a misprint for "Er sieht."]

CHAPTER 3

■ Has Mendel's Work Been Rediscovered?[1]

BY R. A. FISHER, M.A., SC.D., F.R.S.

Galton Professor of Eugenics, University College, London

1. The Polemic Use of the Rediscovery

The tale of Mendel's discovery of the laws of inheritance, and of the sensational rediscovery of his work thirty-four years after its publication and sixteen after Mendel's death, has become traditional in the teaching of biology. A careful scrutiny can but strengthen the truth in such a tradition, and may serve to free it from such accretions as prejudice or hasty judgment may have woven into the story. Few statements are so free from these errors as that which I quote from H. F. Roberts' valuable book *Plant Hybridisation before Mendel* (p. 286):

The year 1900 marks the beginning of the modern period in the study of heredity. Despite the fact that there had been some development of the idea that a living organism is an aggregation of characters in the form of units of some description, there had been no attempts to ascertain by experiment, how such supposed units might behave in the offspring of a cross. In the year above mentioned the papers of Gregor Mendel came to light, being quoted almost simultaneously in the scientific contributions of three European botanists, De Vries in Holland, Correns in Germany, and Von Tschermak in Austria. Of Mendel's two papers, the important one in this connection, entitled "Experiments in Plant Hybridization", was read at the meetings of the Natural History Society of Brünn in Bohemia (Czecho-Slovakia) at the sessions of February 8 and March 8, 1865. This paper had passed entirely unnoticed by the scientific circles of Europe, although it appeared in 1866 in the Transactions of the Society. From its publication until 1900, Mendel's paper appears to have been completely overlooked, except for the citations in Focke's 'Pflanzenmischlinge', and the single citation of Hoffmann, elsewhere referred to.

When the History of Science is taken seriously the number of enquiries which such a story suggests is somewhat formidable. We want to know first: What did Mendel discover? How did he discover it? And what did he think he had discovered? Next, what was the relevance of his discoveries to the science of his time, and what was its reaction to them? In the case of Mendel these last questions must be duplicated, for we are concerned not only with the period following the reading of his principal paper in 1865, but with that following the widespread publicity it received in 1900. This will be considered first.

Seeing how often it is taken for granted that all clouds were cleared away at the rediscovery in 1900, it is singularly difficult to ascertain exactly how Mendel's experiments were conducted and, indeed, what experiments he carried out. We have, of course, his paper, principally devoted to garden peas, entitled "Versuche über Pflanzenhybriden", printed in the transactions of the Natural History Society of Brünn, in Bohemia, in 1866, and reprinted in 1910. In 1901 it was also twice reprinted, in *Flora*, and in Ostwald's *Kiassiker der exakten Wissenscizaften* (No. 121). A valuable English translation, prepared for the Royal Horticultural Society, was published in 1901, and reprinted with modifications by Bateson on several occasions. I shall refer to its appearance in Bateson's book *Mendel's Principles of Heredity* (Cambridge, 1909).

It cannot be denied that Bateson's interest in the rediscovery was that of a zealous partisan. We must ascribe to him two elements in the legend which seem to have no other foundation: (1) The belief that Darwin's influence was responsible for the neglect of Mendel's work, and of all experimentation with similar aims; and (2) the belief that Mendel was hostile to Darwin's theories, and fancied that his work controverted them. On the first point we may note a paragraph from Bateson's preface (p. 2):

While the experimental study of the species problem was in full activity the Darwinian writings appeared. Evolution, from being an unsupported hypothesis, was at length shown to be so plainly deducible from ordinary experience that the reality of the process was no longer doubtful. With the triumph of the evolutionary idea curiosity as to the significance of specific differences was satisfied. The *Origin* was published in 1859. During the following decade, while the new views were on trial, the experimental breeders continued their work, but before 1870 the field was practically abandoned.

It should be noted that Bateson here identifies experimental breeding with the hybridization of species. He ignores the fact that Mendel's advance over his predecessors was due to crossing closely allied varieties, not different species, which, as Mendel actually recognized, would differ in a large number of different factors. It is a consequence of Darwin's doctrine that the nature of the hereditary differences between species can be elucidated by studying heredity in crosses within species. So far were the new evolutionary ideas from discouraging experimental breeding that Darwin, himself, apart from other work, devoted eleven years prior to 1876 to the great series of experiments of which his book on *The Effects of Cross- and Self-fertilisation in the Vegetable Kingdom* is a report. Had his example been followed there would have been no such lull as succeeded his death. Like Mendel's experiments a few years earlier they seemed to lead to nothing more at the time. To-day, in the light of genetic analysis, we can go further towards appreciating their significance.

Bateson's eagerness to exploit Mendel's discovery in his feud with the theory of Natural Selection shows itself again in his misrepresentation of Mendel's own views. Although he was in fact not among those responsible for the rediscovery, his advocacy created so strong an impression that he is still sometimes so described. In the biographical notice which Bateson prefixes to his reprint of Mendel's papers he writes (p. 311):

> With the views of Darwin which were at that time coming into prominence Mendel did not find himself in full agreement, and he embarked on his experiments with peas, which as we know he continued for eight years.

The suggestion that Mendel was prompted by disagreement with Darwin's views to undertake his experiments is easily disproved. Mendel's experiments cannot have commenced later than 1857. Darwin's views on evolution were known only to a few friends prior to the papers which he communicated, jointly with Wallace, to the Linnean Society in 1858. That Mendel had heard of Darwin, as a geologist or an explorer, at the time his experiments with peas were commenced is, indeed, possible. More probably he knew nothing of Darwin's existence, and certainly nothing of the theory of Natural Selection, at this date. When, in 1865, Mendel reported his experiments, the situation had doubtless changed. Mendel now recognizes that the study of inheritance has a special importance in relation to evolutionary theory. He alludes to the subject, in his introductory remarks, in words which suggest not doubts, but rather a simple acceptance of the theory of evolution (79):

> It requires indeed some courage to undertake a labour of such far-reaching extent; this appears, however, to be the only right way by which we can finally reach the solution of a question the importance of which cannot be overestimated in connection with the history of the evolution of organic forms.

In this paper the only other mention of evolution occurs in the concluding remarks, in which the results and opinions of Gartner are discussed. It will be seen that Mendel expressly dissociates himself from Gartner's opposition to evolution, pointing out on the other hand that Gartner's own results are easily explained by the Mendelian theory of factors (114):

> Gärtner, by the results of these transformation experiments, was led to oppose the opinion of those naturalists who dispute the stability of plant species and believe in a continuous evolution of vegetation. He perceives in the complete transformation of one species into another an indubitable proof that species are fixed within limits which they cannot change. Although this opinion cannot be unconditionally accepted we find on the other hand in Gärtner's experiments a noteworthy confirmation of that supposition regarding variability of cultivated plants which has already been expressed.

It is seen from these, the only two allusions to evolution in Mendel's paper, that he did not regard his work as a direct contribution to that subject. What he does claim for the laws of inheritance he established is that they make sense of many of the results of the hybridists, and that they form a necessary basis for the understanding of the evolutionary process. On this point he shows himself fully aware of the importance of what he had done. Had he considered that his results were in any degree antagonistic to the theory of selection it would have been easy for him to say this also.

2. Should Mendel Be Taken Literally?

Bateson raised a point of great interest as to the conduct of Mendel's experiments in a footnote to a passage in the translation he used. After describing his first seven experiments Mendel opens his eighth (unnumbered) section with the words (90):

> In the experiments described above plants were used which differed only in one essential character *(wesentliches Merkmal)*.

Bateson notes:

> This statement of Mendel's in the light of present knowledge is open to some misconception. Though his work makes it evident that such varieties may exist, it is very unlikely that Mendel could have had seven pairs of varieties such that the members of each pair differed from each other in *only* one considerable character. The point is probably one of little theoretical or practical consequence, but a rather heavy stress is laid on the word *wesentlich*.

Most practical experimenters will feel the weight of this difficulty. Unless Mendel had known in advance of the separate inheritance of the characters he was studying he could scarcely have used seven such pairs of varieties. More probably, perhaps, he would have used fewer varieties, say four or five, and crossed these in all, six or ten, possible ways. In any case, we should expect that some or all of the crosses would have involved more than one contrasted pair of characters. Each progeny would then have segregated in more than one factor, and the question arises as to what Mendel did with these additional data. Two courses seem possible:

(i) He might, for each cross, have chosen arbitrarily one factor, for which that particular cross was regarded as an experiment, and ignored segregation in other factors.

(ii) He might have scored each progeny in all the factors segregating, assembled the data for each factor from the different crosses in which it was evolved, and reported the results for each factor as a single experiment.

The first course seems incredibly wasteful of data. This objection is not so strong as it might seem, since it can be shown that Mendel left uncounted, or at least unpublished, far more material than appears in his paper. He evidently felt no anxiety lest his counts should be regarded as insufficient to prove his theory. But, apart from being wasteful, to have adopted this course would seem to imply as much foreknowledge of the outcome as if he had deliberately chosen unifactorial crosses. It would seem in any case an extremely arbitrary course to take. The second course is in effect what most modern geneticists would do, unless they were discussing either the linkage or the interaction of more than one factor. Mendel nowhere gives summaries of the aggregate frequencies from different experiments, and this would be intelligible if the "experiments" reported in the paper were fictitious, being in reality themselves such summaries. Mendel's paper is, as has been frequently noted, a model in respect of the order and lucidity with which the successive relevant facts are presented, and such orderly presentation would be much facilitated had the author felt himself at liberty to ignore the particular crosses and years to which the plants contributing to any special result might belong. Mendel was an experienced and successful teacher, and might well have adopted a style of presentation suitable for the lecture-room without feeling under any obligation to complicate his story by unessential details. The style of didactic presentation, with its conventional simplifications, represents, as is well known, a tradition far more ancient among scientific writers than the more literal narratives in which experiments are now habitually presented. Models of the former would certainly be more readily accessible to Mendel than of the latter.

The great objection to the view suggested by Bateson's hint, that Mendel's "experiments" are fictitious, and that his paper is a didactic exposition embodying his accumulated data, lies in the words which Mendel himself used in introducing the successive steps of his account, *e.g.*, at the beginning of the eighth section (90) "The next task consisted in ascertaining....", and the opening sentence of the ninth section (95) "The results of the experiments previously described led to further experiments". It is true that the different experiments described are not numbered in a single series; those described in any one section are numbered afresh 1, 2, 3, ..., so that these numbers were certainly assigned when the account was written; also we are never told in what year different plants were grown; yet, if Mendel is not to be taken literally, when he implies that one set of data was available when the next experiment was planned, he is taking, as *redacteur,* excessive and unnecessary liberties with the facts. Moreover, the style throughout suggests that he expects to be taken entirely literally; if his facts have suffered much manipulation the style of his report must be judged disingenuous. Consequently, unless real contradictions are encountered in reconstructing his experiments from his paper, regarded as a literal account, this view must be preferred to all alternatives, even though it implies that Mendel had a good understanding of the factorial system, and the frequency ratios which constitute his laws of inheritance, before he carried out the experiments reported in his first and chief paper. Such a reconstruction is attempted in the next section.

3. An Attempted Reconstruction

A framework for dating the experiments is afforded by the statement (79):

This experiment was practically confined to a small plant group, and is now, after eight years, concluded in all essentials.

Mendel's paper was presented on the 8th of February, 1865; if he first grew his experimental peas in 1857 he could then be reporting on eight seasons' work.[2] His monastery had sent him for two years to the University at Vienna, where he had studied mathematics, physics, and biology. He returned and took up teaching duties in the Technical High School in 1853; he may then have undertaken work in the monastery garden for three years before starting his investigation of peas.

On this basis parts of the experiment can be definitely dated (80):

In all, thirty-four more or less distinct varieties of Peas were obtained from several seedsmen and subjected to a two-years' trial. . . . For fertilisation twenty-two of these were selected and cultivated during the whole period of the experiments.

It was evidently in the second trial year (1858) that the first cross-pollinations were made, namely, crosses for the two seed characters *wrinkled* and *green*, and the two plant characters *white flowers* and *dwarf*.[3] Of these the two first are said (*89*) to have shown segregation for six years, which must be 1859–64, the two named plant characters for five (1860–64), while the three other plant characters used by Mendel, *constricted pods, yellow pods*, and *terminal flowers*, for which only four segregating generations are mentioned, may have been first crossed a year later (1859).

In 1858 the recessiveness of the two seed characters must have appeared in the ripe seeds from the flowers cross-pollinated, for these would be round (or yellow) irrespective of the shape (or colour) of the self-fertilized seeds borne by the same plants. From the cross round by wrinkled sufficient seed was sown to raise 253 plants in 1859, while from the cross yellow by green 258 plants were raised. It is not improbable that about 250 plants heterozygous for each of the other two factors were also grown in 1859, but we are only told the numbers of plants raised from their seed in 1860, and these do not exceed what could have been bred from forty plants of each kind. In any case, ground for some 600 to 1000 crossbred plants must have been needed in 1859, and it may be noted that in this year the number of self-fertilized lines was reduced from 38 to 22, releasing probably the ground occupied by sixteen rows.[4] The area of the experiments may well have been the same in the three years 1857, 1858, and 1859.

The heterozygous plants grown in 1859 from white-flowered parents, and those from dwarf parents, must have established the recessiveness of these characters, and so confirmed the fact of dominance in reciprocal crosses observed with the seed characters in the previous year. In 1859 too, when the pods were ripe, seeds on plants heterozygous for *wrinkled* and *green* showed segregation in 3:1 ratios. For wrinkled seeds 253 plants gave 7324 seeds, an average of 29 to a plant. 5474 were round and 1850 wrinkled. The deviation from the expected 3:1 is less than its standard error of random sampling. For green seeds 258 plants gave 8023 seeds, an average of 31 to a plant. 6022 were yellow and 2001 green. The agreement with expectation is here even closer. Mendel does not test the significance of the deviation, but states the ratios as 2.96:1 and 3.01:1, without giving any probable error. The yield per plant seems low. Possibly only four or five pods on each plant were left to ripen, the remainder being consumed green; it is possible again that little room was allowed for each plant.

The discovery, or demonstration, whichever it may have been, of the 3:1 ratio was evidently the critical point in Mendel's researches. The importance of the work was demonstrated, if not to Mendel himself, at least to his associates, and, in the following years, the area of the experimental site must have been greatly enlarged. Perhaps for the same reason, in this year also three new crosses were initiated, using the factors for *constricted pods, yellow pods,* and *terminal flowers.*

That Mendel was satisfied with the two approximate ratios so far obtained would be intelligible if, either previously or immediately upon reviewing the 1859 results, he had convinced himself as to their explanation, and framed the entire Mendelian theory of genetic factors and gametic segregation. His confidence and lack of scepticism shows itself in three distinct ways.

(a) He has numerous opportunities in subsequent years of testing on a large scale whether or not the ratios really remained constant from year to year. If he made any such verification he does not record the data.

(b) The test of significance of deviations from expectation in a binomial series had been familiar to mathematicians at least since the middle of the eighteenth century. Mendel's mathematical studies in Vienna may have given little attention to the theory of probability; but we know that he was engaged in other researches of a statistical character, in meteorology, and in connection with sun-spots, so that it is scarcely conceivable, had the matter caused him any anxiety, that he knew of no book or friend that would enable him to examine objectively whether or not the observed deviations from expectation conformed with the laws of chance. He goes so far as to give "by way of illustration" the classification of the seeds from "the first ten individuals" of each of these two series (86). In both cases the variations are no larger than the deviations to be expected, but Mendel does not say so. The average numbers of seeds from these two samples are above those for the whole series, being 44 against 29 in the first case and 48 against 31 in the second. Indeed, only three of the twenty plants give less than the average number for its experiment. Possibly some poor-yielding plants were rejected when the list was made up, in which case Mendel's statement, though it may be entirely honest, cannot be entirely literal. Possibly, again, the first ten plants had happened in each case to have been grown in more favourable conditions than the majority of the rest.[5]

Mendel also gives examples of extreme deviations in both directions from each series. These extreme cases, again, cannot be judged more extreme than would be expected among samples of about 250 plants, but Mendel gives no grounds for this opinion, and, indeed, does not express it.

(c) The third point on which Mendel seems more incurious than we could imagine him being, were he not already satisfied, is in not comparing the outcome of reciprocal crosses. He alludes to the point at issue in a footnote to his concluding remarks (*116n23*):

> In *Pisum* it is placed beyond doubt that for the formation of the new embryo a perfect union of the elements of both reproductive cells must take place. How could we otherwise explain that among the offspring of the hybrids both original types reappear in equal numbers and with all their peculiarities? If the influence of the egg-cell upon the pollen-cell were only external, if it fulfilled the role of a nurse only, then the result of each artificial fertilization could be no other than that the developed hybrid should exactly resemble the pollen parent, or at any rate do so very closely. This the experiments have in nowise confirmed. An evident proof of the complete union of the contents of the two gametes is afforded by the experience gained on all sides that it is immaterial, as regards the form of the hybrid, which of the original species is the seed parent and which the pollen parent.

If, in 1859, any doubt as to the equivalence of the contributions of the two parents had entered Mendel's mind, he would surely have made a separate enumeration of the seeds borne by the two types of heterozygous plants derived from reciprocal pollinations. Their equivalence as regards dominance had been indicated in the previous year. Their equivalence in genic content Mendel seems early to have felt very sure of.

In 1930, as a result of a study of the development of Darwin's ideas, I pointed out that the modern genetical system, apart from such special features as dominance and linkage, could have been inferred by any abstract thinker in the middle of the nineteenth century if he were led to postulate that inheritance was particulate, that the germinal material was structural, and that the contributions of the two parents were equivalent. I had at that time no suspicion that Mendel had arrived at his discovery in this way. From an examination of Mendel's work it now appears not improbable that he did so and that his ready assumption of the equivalence of the gametes was a potent factor in leading him to his theory. In this way his experimental programme becomes intelligible as a carefully planned demonstration of his conclusions.

In 1860 the obstacles to the extension of his experimental programme had been overcome. In this year the two experiments with seed characters were completed by demonstrating that the 3:1 ratios observed in the previous year were genetically 1:2:1 ratios. In addition to an unknown number of wrinkled seeds, which came true for this character, 565 plants were raised from round seeds, of which 193 yielded round seeds only, while 372 behaved like their parents. Although at least a couple of pods from each of these 372 plants must have been allowed to ripen, the seed numbers are not reported and, perhaps, were not counted. In the second experiment some green seeds were sown, which duly gave green seeds only, while of 519 plants raised from yellow seeds 166 yielded yellow only and 353 were heterozygous. Again, no seed counts are reported from the 353 heterozygous plants. The ratios in both cases show deviations from the expected 2:1 ratio of less than their standard errors. This pair of experiments occupied the space of something more than 1084 plants. They were continued with smaller numbers for the next four years, but no further counts are given.

For the two plant characters *white flowers* and *dwarf* which in this year (1860) first showed segregation, provision was made on a larger scale. Of 929 plants 224 bore white flowers, while of 1064 plants 277 were dwarfed. In both cases the deviation is less than the standard error of random sampling. In addition to making provision for over 3000 plants from the crosses made in 1858 Mendel must in this year have raised perhaps 250 heterozygous plants from each of the three crosses started in 1859. His cultures were therefore probably increased this year by about 3000 plants.

In 1861 provision was made for 1000 plants each for completing the experiments with the first two plant characters, these being families of 10 plants each from a hundred of the 1860 crop, chosen as showing the dominant characters, coloured flowers, and tall stems respectively. The families from 36 plants had only coloured flowers, while those from 64 contained one or more white-flowered plants. The proportionate numbers among the 640 plants of these families was apparently not counted. Again, the families from 28 plants were exclusively tall, while 72 showed segregation of dwarfs. We are not told what was the frequency of dwarfs among these 720 plants. In neither case does the ratio depart significantly from the 2:1 ratio expected, although in the second case the deviation does exceed the standard deviation of random sampling.

In this year also the three crosses of plant characters started in 1859 required provision for nearly 1000 plants each. Of 1181 plants counted 299 had constricted pods, of 580 plants 152 had yellow pods, and of 858 plants 207 had terminal inflorescences. The deviation is below the standard in every case. Apart from progenies grown from recessive plants, these experiments account in all for 4619 plants. The total was thus probably greater than in the previous year, but the increase was not great.

So far as this, the first series of experiments, is concerned, there only remained in 1862 to provide for 3000 plants to establish the 2:1 ratios among the progenies of plants segregating for constricted pods, yellow pods, and terminal flowers. Out of a hundred parents tested there were respectively 29, 40, and 33 homozygous. Of these the first and third conform well with expectation. In the second case the observed frequencies, 40 homozygous to 60 heterozygous, shows a relatively large, but not a significant, deviation. It is remarkable as the only case in the record in which Mendel was moved to verify a ratio by repeating the trial. A second series of a hundred progenies, presumably grown in 1863, gave 65:35, as near to expectation as could be desired. Although in 1861 only 580 plants had been available to display the 3:1 ratio for yellow pods, and in these two trials respectively 600 and 650 more must have appeared, they do not seem to have been counted, and are not reported in the paper.

In connection with these tests of homozygosity by examining ten offspring formed by self-fertilization, it is disconcerting to find that the proportion of plants misclassified by this test is not inappreciable. If each offspring has an independent probability, .75, of displaying the dominant character, the probability that all ten will do so is $(.75)^{10}$ or .0563. Consequently, between 5 and 6 per cent of the heterozygous parents will be classified as homozygotes, and the expected ratio of segregating to non-segregating families is not 2:1 but 1.8874:1.1126 or approximately 377.5:222.5 out of 600. Now among the 600 plants tested by Mendel 201 were classified as homozygous and 399 as heterozygous. Although these numbers agree extremely closely with his expectation of 200:400, yet, when allowance is made for the limited size of the test progenies, the deviation is one to be taken seriously. It seems extremely improbable that Mendel made any such allowance, or that the numbers he records as segregating are "corrected" values, rounded off to the nearest integer, obtained by dividing the numbers observed to segregate by .9437. We might suppose that sampling errors in this case caused a deviation in the right direction, and of almost exactly the right magnitude, to compensate for the error in theory. A deviation as fortunate as Mendel's is to be expected once in twenty-nine trials. Unfortunately the same thing occurs again with the trifactorial data.

These seven experiments of the first series require, as we have seen, a total of four or five thousand plants in the years 1860 and 1861. Apart from the continuation of heterozygous series they account for only 3000 in 1862 and for 1000 in 1863. The pollinations for his second series of experiments were, therefore, probably carried out in 1861. The large trifactorial experiment could not indeed have been finished had it started later, and, as the factor for white flowers first showed segregation in 1860, it is difficult to place it earlier. The bifactorial experiment took a year less, and might have been started in 1860, since the ripened seeds of 1859 had established the 3:1 ratios of the two factors. I shall suppose that both were initiated in 1861, and that the same is true of the important but smaller experiments devoted to determining the gametic ratios.

To 1862, then, are ascribed the fifteen doubly heterozygous plants of the bifactorial experiment, of which the 556 seeds displayed the first 9:3:3:1 ratio reported. All these were sown in 1863, even the thirty-two wrinkled-green seeds, which suggests that in this year space was abundant. (It was, indeed, in this same year that we have supposed Mendel to depart from his usual practice, and repeat the determination of a frequency ratio, at the expense of growing 1000 additional plants. Even with these additions the summary [Table VI] shows 1863 as less crowded than most of the other years.) The plants from these seeds, classified by the seeds they bore, exhibited independent segregation of the two factors. Mendel's classification of the 529 plants which came to maturity is shown in Table I.

TABLE I — Classification of plants grown in the bifactorial experiment

	AA	Aa	aa	Total
BB	38	60	28	126
Bb	65	138	68	271
bb	35	67	30	132
Total	138	265	126	529

The numbers are close to expectation at all points, but they are not very large. In relation to possible linkage, for example, they may be regarded as excluding, at the 5 per cent level of significance, recombination fractions less than 44.9 per cent, which is not very strong negative evidence; yet on this point also Mendel evidently felt that further data would be superfluous, for he certainly could have obtained many more for the counting. The 138 plants, for example, recorded in the table above as being doubly heterozygous, doubtless bore over 4000 seeds segregating in the 9:3:3:1 ratio, and, even if the bulk of the crop were needed when green, at least ten seeds from each plant must have been allowed to ripen in order to classify the plant on which they grew.

The trifactorial experiment required 24 hybrid plants grown in 1862, which gave 639 offspring in 1863. In order to distinguish heterozygotes from homozygotes among the plants with coloured flowers progenies from at least 473 of these must have been grown. If, as in other cases, Mendel used a progeny of ten plants for such discrimination the experiment must have needed 4730 plants in 1864. Of this experiment Mendel says (93):

> Among all the experiments this demanded the most time and trouble,

and the extent of the third filial generation explains this remark. It was evidently on the completion of this extensive work that Mendel felt that his researches were ripe for publication. It may have constituted the whole of his experimental work with peas in the last year before his paper was read. Even so, probably this year saw more experimental plants than were grown in any previous year. Since the factor for coloured flowers used in this experiment obscures the cotyledon-colour of unopened seeds, not all of the vast number of seeds borne by these three generations was easily available to supplement the bifactorial and trifactorial data reported, yet even what was easily available must have been much more extensive than any data which Mendel published. Mendel's trifactorial classification of the 639 plants of the second generation is shown in Table II, which follows Mendel's notation, in which *a* stands for *wrinkled* seeds, *b* for *green* seeds, and *c* for *white flowers*.

In order to discriminate *CC* from *Cc* plants progenies from these, which are seen to number 463 together, must have been grown on in 1864. In addition to abundant new unifactorial data the additional bifactorial data supplied by the experiments is seen to be large. 175 of the plants were heterozygous for both of the two seed characters, and, if 30 seeds from each had been classified, these would have given 5250 seeds, nearly ten times as many as the 556 reported from the bifactorial experiment. The classification of these plants as double heterozygotes must indeed have required that about half this number of seeds from each plant were exam-

TABLE II—Classification of plants grown in the trifactorial experiment

	CC				Cc				cc				Total			
	AA	*Aa*	*aa*	Total	*AA*	*Aa*	*aa*	Total	*AA*	*Aa*	*aa*	Total	*AA*	*Aa*	*aa*	Total
BB	8	14	8	30	22	38	25	85	14	18	10	42	44	70	43	157
Bb	15	49	19	83	45	78	36	159	18	48	24	90	78	175	79	332
bb	9	20	10	39	17	40	20	77	11	16	7	34	37	76	37	150
Total	32	83	37	152	84	156	81	321	43	82	41	166	159	321	159	639

ined. In the following year also nine-sixteenths of the progeny of 127 F_2 plants, or about 815 F_3 plants, must have borne seeds segregating in the 9:3:3:1 ratio, so that a further 24,000 seeds could have been so classified in 1864. Evidently, however, Mendel felt that the complete classification of 529 plants in the bifactorial experiment was sufficient; he does not even add, for the simultaneous segregation of *Aa* and *Bb*, the 639 plants completely classified in the trifactorial experiment, which suffice to raise the recombination fraction significantly higher than 46.56 per cent (from 44.9 per cent).

TABLE III—Comparison of numbers reported with uncorrected and corrected expectations

	Number of plants tested	Number of non-segregating progenies observed	Number expected		Deviation	
			Without correction	Corrected	Without correction	Corrected
1st group of experiments	600	201	200.0	222.5	+1.0	−21.5
Trifactorial experiment	473	152	157.7	175.4	−5.7	−23.4
Total	1073	353	357.7	397.9	−4.7	−44.9

In the case of the 600 plants tested for homozygosity in the first group of experiments Mendel states his practice to have been to sow ten seeds from each self-fertilized plant. In the case of the 473 plants with coloured flowers from the trifactorial cross he does not restate his procedure. It was presumably the same as before. As before, however, it leads to the difficulty that between 5 and 6 per cent of heterozygous plants so tested would give only coloured progeny, so that the expected ratio of those showing segregation to those not showing it is really lower than 2:1, while Mendel's reported observations agree with the uncorrected theory.

The comparisons are shown in Table III. A total deviation of the magnitude observed, and in the right direction, is only to be expected once in 444 trials; there is therefore here a serious discrepancy.

If we could believe that Mendel changed his previous practice, and in 1863 went to the great labour of back-crossing the 473 doubtful plants, the data could be explained, for in such progenies misclassification would be only about one-fiftieth part as frequent as in progenies by self-fertilization. Equally, if we could suppose that larger progenies, say fifteen plants, were grown on this occasion, the greater part of the discrepancy would be removed. However, even using families of 10 plants the number required is more than Mendel had assigned to any previous experiment, and there is no reason for thinking that he ever grew so many as 7000 experimental plants in one year, apart from his routine tests.[6] Such explanations, moreover, could not explain the discrepancy observed in the first group of experiments, in which the procedure is specified, without the occurrence of a coincidence of considerable improbability.

An explanation of a different type is that the selection of plants for testing favoured the heterozygotes. In the first series of experiments the selection might have been made in the garden, or, if the whole crop was harvested, on the dry plants. In either case the larger plants might have been unconsciously preferred. It is also not impossible that, in some crosses at least, the heterozygotes may have been on the average larger than sister homozygotes. The difficulties to accepting such an explanation as complete are three. (i) In the tri-factorial experiment there was no selection, for all plants grown must have been tested. The results here do not, however, differ in the postulated direction from those of the first series. On the contrary, they show an even larger discrepancy. (ii) It is improbable that the supposed compensating selection of heterozygotes should have been equally effective in the case of five different factors. (iii) The total compensation for all five factors (21.5 plants) must be supposed to be greater than would be needed (16.8 plants) if families of 11 had been grown, and less than would be needed (30.0) if 9 only had been grown, though nearly exactly right for the actual number 10 of F_3 plants in each progeny (22.5).

The possibility that the data for the trifactorial experiment do not represent objective counts, but are the product of some process of sophistication, is not incapable of being tested. Fictitious data can seldom survive a careful scrutiny, and, since most men underestimate the frequency of large deviations arising by chance, such data may be expected generally to agree more closely with expectation than genuine data would. The twenty-seven classes in the trifactorial experiment supply twenty-six degrees of freedom for the calculation of χ^2. The value obtained is 15.3224, decidedly less than its average value for genuine data, 26, though this value by itself might occur once in twenty genuine trials.

This total may be subdivided in various ways; one relevant subdivision is to separate the nine degrees of freedom created by the discrimination of homozygous and heterozygous plants with coloured flowers from the remaining seventeen degrees of freedom based on discriminations made presumably in the previous year. To the total the 9 supply 6.3850, leaving only 8.9374 for the remaining 17. If anything, therefore, the subnormality in the deviations from expectation is more pronounced among the seventeen degrees of freedom than among the nine. If there has been sophistication there is no reason to think that it was confined to the final classification made in 1864.

To 1863 belong probably the bifactorial experiment and the five comparisons, each of four equal expected frequencies, supplied by the experiments on gametic ratios. The bifactorial experiment, having nine classes, supplies eight degrees of freedom for comparison, and gives a χ^2 of only 2.811—almost as low as the 95 per cent point. The fifteen degrees of freedom of gametic ratios supply only 3.6730, which is beyond the 99 per cent point. In the same year also should be included the verified 2:1 ratio for yellow pods, giving 0.125 for one degree of freedom

Putting together the comparisons available for 1863 we have:—

TABLE IV—Measure of deviation expected and observed in 1863

	Expectation	χ^2 observed
Trifactorial experiment	17	8.9374
Bifactorial experiment	8	2.8110
Gametic ratios	15	3.6730
Repeated 2:1 test	1	0.1250
Total	41	15.5464

The discrepancy is strongly significant, and so low a value could scarcely occur by chance once in 2000 trials. There can be no doubt that the data from the later years of the experiment have been biased strongly in the direction of agreement with expectation.

One natural cause of bias of this kind is the tendency to give the theory the benefit of doubt when objects such as seeds, which may be deformed or discoloured by a variety of causes, are being classified. Such an explanation, however, gives no assistance in the case of the tests of gametic ratios and of other tests based on the classification of whole plants. For completeness it may be as well to give in a single table the χ^2 values for all the experiments recorded.

TABLE V—Deviations expected and observed in all experiments

		Expectation	χ^2	Probability of exceeding deviations observed
3 : 1 ratios	Seed characters	2	0.2779	
	Plant characters	5	1.8610	
		— 7	— 2.1389	.95
2 : 1 ratios	Seed characters	2	0.5983	
	Plant characters	6	4.5750	
		— 8	— 5.1733	.74
Bifactorial experiment		8	2.8110	.94
Gametic ratios		15	3.6730	.9987
Trifactorial experiment		26	15.3224	.95
Total		64	29.1186	.99987
Illustrations of plant variation		20	12.4870	.90
Total		84	41.6056	.99993

The bias seems to pervade the whole of the data, apart, possibly, from the illustrations of plant variation. Even the 14 degrees of freedom available before 1863 give only 7.1872, a value which would be exceeded about 12 times in 13 trials.

TABLE VI—Approximate numbers of plants grown in different years[7]

	1857	1858	1859	1860	1861	1862	1863	1864
Stock lines	2280	2280	1320	1320	1320	1320	1320	1320
1st group	—	—	1011	3927	4719	3200	1350	350
2nd group	—	—	—	—	—	65	1719	4730
Total	2280	2280	2331	5247	6039	4585	1389	6400

What I have inferred respecting the extent of Mendel's cultures is summarized by years in Table VI. I have arbitrarily allowed sixty plants for each of the stock lines and fifty for each segregating line which was continued with smaller numbers after the completion of the main experiments. I have included also in 1862 and 1863 the two small experiments devoted to the demonstration of gametic ratios. Some of the totals for years may be correct to the nearest hundred, but I do not expect all to be so. I feel justified in concluding only that the experiment was greatly enlarged after the first three years and that, with only ten plants to a family, the year 1864 was probably the fullest of all.

4. The Nature of Mendel's Discovery

The reconstruction has been undertaken in order to test the plausibility of the view that Mendel's statements as to the course and procedure of his experimentation are to be taken as an entirely literal account, or whether, on the other hand, there is evidence that data have been assembled from various sources, or the same data rediscussed from different standpoints in different sections of his account. There can, I believe, now be no doubt whatever that his report is to be taken entirely literally, and that his experiments were carried out in just the way and much in the order that they are recounted. The detailed reconstruction of his programme on this assumption leads to no discrepancy whatever. A serious and almost inexplicable discrepancy has, however, appeared, in that in one series of results the numbers observed agree excellently with the two to one ratio, which Mendel himself expected, but differ significantly from what should have been expected had his theory been corrected to allow for the small size of his test progenies. To suppose that Mendel recognized this theoretical complication, and adjusted the frequencies supposedly observed to allow for it, would be to contravene the weight of the evidence supplied in detail by his paper as a whole. Although no explanation can be expected to be satisfactory, it remains a possibility among others that Mendel was deceived by some assistant who knew too well what was expected. This possibility is supported by independent evidence that the data of most, if not all, of the experiments have been falsified so as to agree closely with Mendel's expectations.

The importance of the conclusion, if it is well established, that Mendel's statements are to be taken literally, lies in the inferences which flow from this view. *First,* that prior to the reported experiments Mendel was sufficiently aware of the independent inheritance of seven factors in peas to have chosen seven pairs of varieties, each pair differing only in a single factor. If it be thought that out of thirty-eight varieties he could not by deliberate choice have found the material for seven such crosses, it should be remembered also that at this stage he was choosing not only the varieties but, perhaps, also the factors to use in his experiment, and that he may have known of other factors in peas in addition to those with which his experiments are concerned, which, however, could not have been introduced without bringing in an undesirable complication.[8] *Next,* it appears that Mendel regarded the numerical frequency ratios, in which the laws of inheritance expressed themselves, simply as a ready method of demonstrating the truth of his factorial system, and that he was never much concerned to demonstrate either their exactitude or their consistency. It may be that the seed counts of 1859 were a revelation to him of the precision with which his system worked, and could be demonstrated; they may also possibly have given him an exaggerated impression of the precision with which the theoretical ratios should be verified, but from that moment it is clear, from the form his experiments took, that he knew very surely what to expect, and designed them as a demonstration for others rather than for his own enlightenment. That the hereditary contribution of the two parents might be unequal he did not seriously consider, although his first experiments provided splendid evidence on this important question, which it does not occur to him to present. It seems also not to have occurred to him that the inheritance of different factors might not be wholly independent. He asserts independence for all his factors, but gives evidence for only three of them, and for these much less than he might have given. A feature such as linkage would have been a complication extraneous to his theory, as he conceived it, which he would only have taken seriously had the observations forced it under his notice.

The theoretical consequences of his system he had thought out thoroughly, and in this respect his thought is considerably in advance of that of the first generation of geneticists which followed his rediscovery. He pointed out that n factors would give rise to 3^n different genotypes, of which 2^n would be capable of breeding true. He realized that even in intra-specific crosses n would be sufficiently great for these to be very large numbers, and that even more factors must be involved when crosses are made between different species, when minor in addition to major differences are considered. This understanding of the consequences of the factorial system contrasts sharply with many of the speculations of the earlier geneticists, such as that new species might be formed by the mutation of a single factor, or that the mimetic groups, found among butterflies and other insects, might be explained by the paucity of the genetic factors controlling the pattern and coloration of the wings. In these respects it has taken nearly a generation to rediscover Mendel's point of view.

Mendel seems also to have realized that the factorial system resolved one of the chief difficulties felt and discussed by Darwin, namely that, if the wide variation observable in cultivated plants were caused by the changed conditions and increased nourishment experienced on being brought into cultivation, then this cause of variation must continue to act, as Darwin had written, "for an improbably long time", since anciently cultivated species are not less but rather more variable than others. With segregating, heritable factors, on the other hand, the variability is easily explained by the preservation in culture of variants which, apart from man, would have been eliminated by natural selection. This, indeed, seems to have been Mendel's view (106–7):

> It will be willingly granted that by cultivation the origination of new varieties is favoured, and that by men's labour many varieties are acquired which, under natural conditions, would be lost; but nothing justifies the assumption that the tendency to the formation of varieties is so extraordinarily increased that the species speedily lose all stability, and their offspring diverge into an endless series of extremely variable forms. Were the change in the conditions the sole cause of variability you might expect that those cultivated plants which are grown for centuries under almost identical conditions would again attain constancy. That, as is well known, is not the case, . . .

The reflection of Darwin's thought is unmistakable, and Mendel's comment is extremely pertinent, though it seems to have been overlooked. He may at this time have read the *Origin*, but the point under discussion may equally have reached his notice at second hand.

5. The Contemporary Reaction to Mendel's Work

The peculiarities of Mendel's work, to which attention has been called in the previous sections, seem to contribute nothing towards explaining why his paper was so generally overlooked. The journal in which it was published was not a very obscure one, and seems to have been widely distributed. In London, according to Bateson, it was received by the Royal Society and by the Linnean Society. The paper itself is not obscure or difficult to understand; on the contrary, the new ideas are explained most simply, and amply illustrated by the experimental results. In view of the parallel failure of the biological world to appreciate and follow up Darwin's experiments, it is difficult to suppose that, had Mendel's paper been more widely read, there would have been many mentally prepared to appreciate its significance. Some there certainly were; and, had the new facts and methods come to the knowledge of Francis Galton, the experimental analysis of heredity might well have been established twenty-five years earlier than it was in fact; but minds equally receptive were certainly rare.

Among German biologists the one with whom Mendel is known to have corresponded is von Nägeli. From his writings it is apparent either that Mendel's researches made no impression on his mind or that he was anxious to warn students against paying attention to them. In a paper published December 15, 1865, only ten months after the delivery of Mendel's paper on peas, and before its appearance in print, he seems to reprove observers who venture to think for themselves and to plan their own experiments instead of using the results of Gartner and Kölreuter (p. 190):

> The knowledge of hybridization would in recent times have made more progress, if many observers, instead of beginning anew, had made use of the results of the two first-named German investigators, who applied the labour of their lives to the solution of this problem.

In the beginning of his paper Mendel had, with modest confidence, contrasted his method of procedure with that of these two distinguished predecessors. In his final discussion, also, he reinterprets the results of Gartner in terms of the factorial system, showing that Gartner's observations agreed with Mendel's theory, while dissenting from Gartner's opinion that they were opposed to the theory of evolution.

In spite of his correspondence von Nägeli does not refer to Mendel's recent paper, and the following passage seems designed positively to ignore it (p. 231):

> Variability of the hybrids, that is to say, the diversity of forms which belong to the same generation, and their behaviour on propagation once or many times by self fertilization, form two points in the study of hybridization which are still least ascertained, and which appear to be the least subject to strict rules.

Mendel had claimed to have established precisely such strict rules. Another passage in the same paper seems designed directly to contradict Mendel's claims as to the dominance and independence of genetic factors (p. 222):

> The characters of the parental forms are, as a rule, so transmitted that, in each individual hybrid both influences make themselves felt. It is not that one character is transmitted, as it were, unchanged from the one parent, a second unchanged from the other but there occurs an interpenetration of the paternal and the maternal character, and a union between their characters.

It is difficult to suppose that these remarks were not intended to discourage Mendel personally, without drawing attention to his researches.

No such dishonourable intention can be ascribed to W. O. Focke, who, in his *Pflanzenmischlinge*, makes no less than fifteen references to Mendel. As in the case of other voluminous compilers, most of these references, though doubtless relevant to the different topics Focke had in mind, ignore the point of Mendel's work. The nearest Focke comes to giving any idea of what Mendel had done is found in the following sentence. This may stand as a good example of the limitations of even the best intentioned compilers of comprehensive treatises (p. 110):

> Mendel's numerous crossings gave results which were quite similar to those of Knight, but Mendel believed that he found constant numerical relationships between the types of the crosses.

The fatigued tone of the opening remark would scarcely arouse the curiosity of any reader, and in all he has to say Focke's vagueness and caution have eliminated every point of scientific interest. Could any reader guess that the "constant numerical relationships" were the universal and concrete ratios of 1:1 and 3:1, or even that Focke was speaking of the frequency ratios of a limited number of recognizable genotypes?

It is not an accident that Focke was vague. In this case, as perhaps in others, he had not troubled to understand the work he was summarizing. Mendel's discovery of dominance and the great use he had made of seed characters had escaped him altogether. His comment continues:

> In general, the seeds produced through a hybrid pollination preserve also, with peas, exactly the colour which belongs to the mother plant, even when from these seeds themselves plants proceed, which entirely resemble the father plant, and which then also bring forth the seeds of the latter.

H. F. Roberts makes an instructive comment on Focke's book:

> A careful study of Focke's report brings into interesting relief the reason for his having failed to appraise the Mendel paper at its present value. In the first place, Focke was especially interested in the works of those who produced more extended contributions. The work of Kölreuter, Gartner, Wichura and Wiegmann, whose works were much more voluminous in the field which they occupied, receive appropriate consideration, as do also Naudin's and Godron's prize contributions; but Mendel's paper evidently appeared to Focke simply in the guise of one of the numerous, apparently similar, contributions to the knowledge of the results of crossing within some single group.... It was supposedly not at all conceivable that the laws of hybrid breeding could be compassed within a series of experiments upon a single plant.

Roberts ends his comment on a note of appreciation:

> The details of his (Focke's) data are laborious, exact, well classified and scientifically arranged, comprising 79 families of dicotyledons, 13 families of monocotyledons, 2 families of gymnosperms, 2 of pteridophytes, one of the musci and one of the algae.

It is very well to be reminded that the high qualities catalogued in the sentence last quoted are yet compatible with the learned author having overlooked, in his chosen field, experimental researches conclusive in their results, faultlessly lucid in presentation, and vital to the understanding not of one problem of current interest, but of many.

The peculiar incident in the history of biological thought, which it has been the purpose of this study to elucidate, is not without at least one moral—namely, that there is no substitute for a careful, or even meticulous, examination of all original papers purporting to establish new facts. Mendel's contemporaries may be blamed for failing to recognize his discovery, perhaps through resting too great a confidence on comprehensive compilations. It is equally clear, however, that since 1900, in spite of the immense publicity it has received, his work has not often been examined with sufficient care to prevent its many extraordinary features being overlooked, and the opinions of its author being misrepresented. Each generation, perhaps, found in Mendel's paper only what it expected to find; in the first period a repetition of the hybridization results commonly reported, in the second a discovery in inheritance supposedly difficult to reconcile with continuous evolution. Each generation, therefore, ignored what did not confirm its own expectations. Only a succession of publications, the progressive building up of a *corpus* of scientific work, and the continuous iteration of all new opinions seem sufficient to bring a new discovery into general recognition.

NOTES

This paper was originally published in *Annals of Science* 1 (1936): 115–37. This version of Fisher's paper is from the R. A. Fisher Digital Archive (http://digital.library.adelaide.edu.au/dspace/bitstream/2440/15123/1/144.pdf). Only footnotes 5, 6, and 8 are from Fisher's original paper. The other footnotes are editorial comments. Fisher's paper has been reproduced without any alterations; any instances of variations in Mendel's quotations are Fisher's own.

1. For further commentary on Mendel's work written by Fisher in 1955, see *Experiments in Plant Hybridisation: Gregor Mendel* (Ed. J. H. Bennett). Edinburgh: Oliver & Boyd, 1965. As indicated there, all of the years given in Fisher's (1936) reconstruction of the timing of Mendel's experimental programme must be reduced by one.

2. See note 1.

3. This refers to the second year of the experiment and not of the trial mentioned in the previous paragraph. The two-years' trial was completed in 1855 and the eight years' experiment began in 1856.

4. The number of self-fertilised lines was reduced from 34 to 22 (i.e., by 12 rather than 16) after the second year of the trials.

5. I am obliged to Dr. J. Rasmusson, who has extensive experience of genetical work with *Pisum,* for the following explanation of Mendel's probable method of selection:

"It is my impression that the classification was made throughout on dry plants in Winter. That is to say, that Mendel harvested his plants in Autumn, probably tied them up plot by plot, and for scoring loosened up the bunch of plants and picked out from it one plant after another. This is the method which first presents itself in work of this kind it is also the method I am accustomed to use. The fact is that, working in this way, one will unconsciously choose the best plant first. This happens to me, whether I do the work myself or have other people picking out the plants from the bunch."

In respect to the average yield Dr. Rasmusson also says: "About 30 good seeds per plant is, under Mendel's conditions (dry climate, early ripening, and attacks of *Bruchus pisi*) by no means a low number. It seems to me, indeed, rather a good one, and I feel convinced that Mendel classified all the seeds from these plants."

6. The area available is given by Iltis as only 7 m. by 35 m. Dr. Rasmusson estimates that he might have grown 4000–5000 plants in this area.

7. The entries for 1857 and 1858 should be 1320 (see note 4).

8. It is particularly gratifying that this conclusion is supported by Dr. Rasmusson, basing his opinion upon existing types of garden peas, and on the development of those types since Mendel's time. He writes:

"From the most probable assortment of varieties available to Mendel there would be no difficulty whatever in making unifactorial crosses in all characters. Indeed, the assortment at hand seems to have been much better fitted for such crosses than for other combinations."

BIBLIOGRAPHY

W. Bateson. *Mendel's Principles of Heredity,* Cambridge University Press, 1909.
C. Darwin. *The Origin of Species,* London, John Murray, 1859.
C. Darwin. *The Effects of Cross- and Self-Fertilisation in the Vegetable Kingdom,* London, John Murray, 1876.
R. A. Fisher. *The Genetical Theory of Natural Selection,* Oxford, Clarendon Press, 1930.
W. O. Focke. *Die Pflanzenmschlinge,* Berlin, 1881.
H. Iltis. *Gregor Johann Mendel Leben, Werk und Wirkung,* Berlin, Julius Springer, 1924.
G. Mendel. *Versuche uber Pflanzenhybriden (Verhandlungen Naturforschender Vereines in Brünn.* 1866, 10, 1).
C. Von Nägeli. *Die Bastardbildung im Pflanzenreiche BotanischeMittheilungen,* 1865, 2, 187–235.
H. F. Roberts. *Plant Hybridization before Mendel,* Princeton, University Press, 1929.

CHAPTER 4

■ Are Mendel's Results Really Too Close?

A. W. F. EDWARDS

I. Introduction

Gregor Mendel read his famous paper 'Experiments in plant hybridization' to the Brünn Natural History Society in 1865 (Mendel, 1866). It attracted wide attention only from 1900 onwards, and in 1911 R. A. Fisher, then starting his third year as an undergraduate in Gonville and Caius College, Cambridge, noted in a talk to the Cambridge University Eugenics Society:

It is interesting that Mendel's original results all fall within the limits of probable error; if his experiments were repeated the odds against getting such good results is about 16 to one. It may have been just luck; or it may be that the worthy German abbot, in his ignorance of probable error, unconsciously placed doubtful plants on the side which favoured his hypothesis. (Norton & Pearson, 1976; Bennett, 1983)

Twenty-five years later Fisher published his analysis of Mendel's data (Fisher, 1936), admitting, in a covering letter to the editor of the journal, 'I had not expected to find the strong evidence which has appeared that the data had been cooked' (Bennett, 1965). Nearly 20 years later still he prepared an introduction and commentary to Mendel's paper which was published posthumously in 1965 (Bennett, 1965), and his final verdict was, 'The data have evidently been sophisticated systematically, and after examining various possibilities, I have no doubt that Mendel was deceived by a gardening assistant, who knew too well what his principal expected from each trial made.' A letter from Fisher to E. B. Ford dated 2 January 1936 (Bennett, 1983) amplifies what Fisher called his 'abominable discovery': 'I cannot conceive that Mendel himself had any hand in it'. From his comments in 1936 it seems that Fisher had by then forgotten what he had written in 1911.

Fisher's remark about the odds of 16 to 1 was probably taken directly from Weldon (1902), but that is not to suggest that Weldon himself harboured any suspicions about Mendel's results. He subjected them to a statistical analysis using probable errors but concluded only that Mendel's statements were 'admirably in accord with his experiment'. Fisher had encountered Mendel's paper itself in the English translation in Bateson's newly published *Mendel's Principles of Heredity* (Bateson, 1909) which he had bought as a freshman (Fisher, 1952). In 1936 he referred to a footnote of Bateson's as raising 'a point of great interest' because it hinted that Mendel's paper should not be taken as a literal account of his experiments, and this may be said to have been the first such suggestion.

Fisher's 1936 conclusion slowly became the received wisdom, but his painstaking analysis and his defence of Mendel's integrity have sometimes been incorrectly reported as having exposed a scientific fraud of major proportions, and the name of Mendel is in danger of acquiring the connotations of Piltdown or Burt. Recently there has been a spate of papers seeking to defend Mendel against the imaginary charges. Thus Pilgrim (1984) writes 'The purpose of this [i.e. his own] paper is to demonstrate that Fisher's reasoning was faulty and to clear the name of an honest man', and 'There is no evidence that Mendel did anything but report his data with impeccable fidelity. It is to the discredit of science that it did not recognize him during his lifetime. It is a disgrace to slander him now'. An answer to Pilgrim is Edwards (1986). A similar example of missionary zeal is afforded by Monaghan & Corcos (1985), who write that 'suggestions that Mendel's data were biased in support of an hypothesis about inheritance in peas are not supported'. Their paper is, however, completely untrustworthy in its statistical analysis (see Section III below). A recent biography of Mendel (Orel, 1984) concludes, on the basis of Weiling's analysis to which I refer below, 'This would seem to have removed the last shadow of doubt as to the credibility of Mendel's experiments and his published results'.

As I will show, such views are not supported by the evidence. In the present paper, after reviewing the many 'explanations' of Mendel's data, I shall let the facts speak for themselves as plainly as possible: there is no point in addressing the issue with statistical techniques of such complexity that they are more difficult to interpret than the original data. I shall assume that readers are familiar with Mendel's paper and that they have an English translation to hand.

II. A Survey of Explanations

The critical re-evaluation of experiments and observations is an important part of scientific endeavour. ('There is no substitute for a careful, or even meticulous, examination of all original papers purporting to establish new facts', Fisher, 1936.) Though in certain notorious cases this has led to the exposure of fraudulent work, it generally results in an increased appreciation of the original material or even a fresh discovery from it. Occasionally some puzzling aspects of the data surface on a close examination, and may even lead to the suggestion that the experiments did not take place exactly as the author reported them. Thus Galileo's unrepeatable results in his experiments on acceleration down an inclined plane mentioned in the *Dialogo* led Mersenne (1637) to doubt whether he had actually performed them. In that example it was the incomprehensibly bad data which raised doubts, but long before the era of modern statistics it was understood that surprisingly good data could equally well create justifiable suspicion. Babbage (1830), writing before the time of Mendel, classified some of the methods used to produce data too close to expectation:

Trimming consists in clipping off little bits here and there from those observations which differ most in excess from the mean, and in sticking them on to those which are too small.

Cooking . . . One of its numerous processes is to make multitudes of observations, and out of these to select those only which agree, or very nearly agree.

Many commentators on Mendel's results share Fisher's belief that some kind of 'cooking' has taken place (for example: de Beer, 1964; Sturtevant, 1965; Crew, 1966), but most have sought alternative explanations for the surprisingly good fit of the data to the hypothesis.

A popular one is that the data are the result of an imperfectly understood sampling process. An early example of such a misunderstanding occurs in the work of Francis Galton, who, by the standards of his time, had a remarkably clear head for statistical enquiry. In *Hereditary Genius* (Galton, 1892), however, he concluded that since judges were all men and came from families of average size five, each must have (on average) 2 ½ sisters and 1½ brothers. Galton is here wrongly assuming binomial sampling, with probability one half, *including* the judge; might not we be assuming the wrong sampling process for Mendel's data? (Galton, 1904, later realized his error.)

The natural suggestion is that Mendel stopped counting the peas in each experiment when he was satisfied with the result, and this has been put forward numerous times (Dunn, 1965a, b; Olby, 1966; Beadle, 1967; van der Waerden, 1968; C. D. Darlington, reported to me by J. H. Edwards; J. Kříženecký, reported by Orel, 1968). The difficulty is that there is no evidence whatsoever that Mendel did this, and in many of the experiments his description clearly excludes the possibility, so that to accept it as a means of exonerating Mendel from having reported data which had been adjusted in some way is to saddle him with the charge of not having reported his experimental method accurately. 'The style throughout suggests that he expects to be taken entirely literally; if his facts have suffered much manipulation the style of his report must be judged disingenuous' (Fisher, 1936). Van der Waerden (1968) is the only one of the authors quoted above who pursued the suggestion in detail, and although he found it possible to account for some of the earlier instances of surprisingly good fits in the experiments if he assumed sequential sampling, he concluded 'All in all my impression is that some data from later years were probably biased.'

Another favourite suggestion has been that the data exhibit non-binomial variability not because of the sampling process, but because of the reproductive physiology of the pea (Sturtevant, 1965; Weiling, 1966; Thoday; 1966; Beadle, 1967). Sturtevant and Beadle 'explored this possibility to see if it was sufficient to account for the apparent bias. It works in the right direction but is not sufficient' (Beadle, 1967). Weiling (1966, 1971, 1985) made a detailed comparison of Mendel's data with data from other authors and concluded that the results were all consistent with the hypothesis that the variance of the segregation was smaller than the binomial variance by a factor of between 0.6 and 1.0. In the next section I give my reasons for rejecting this suggestion: the evidence is that the pea is in fact an excellent randomizer!

To help explain the infra-binomial variance in the experiments in which Mendel progeny-tested F_2 plants which displayed one of the dominant plant-characteristics to see if they were homozygous or heterozygous (*88, 89*), Weiling (1966) suggested that on average only 8 of the 10 seeds sown to test each parent for heterozygosity will have produced a scorable plant, thus making the probability $(3/4)^8$ that no recessive was raised, rather than the $(3/4)^{10}$ assumed by Fisher (1936). Taking into account the resulting misclassification of some heterozygotes as homozygotes, the expectations are then even further from Mendel's data, and the fit consequently *worsened*. By contrast, Sturtevant (1965) thought that 'if the experiment included at least 10 seeds but often more than 10, then the correction to the 2:1 expectation will be less, and Fisher's most telling point will be weakened', and Wright (1966; also reported by Dunn, 1965b) advanced the same idea: 'if the average was 12, the probability of misclassifying [the heterozygote] falls from .056 [$(3/4)^{10}$] to .031 [$(3/4)^{12}$]'. This would *improve* the fit of Mendel's results. As a matter of fact this suggestion had already been made by Fisher (1936) in connection with the 473 plants with coloured flowers from the trifactorial experiment which needed progeny-testing to determine whether they were homozygous or heterozygous, for which Mendel did not restate his procedure. 'It was presumably the same as before', wrote Fisher, adding:

if we could suppose that larger progenies, say fifteen plants, were grown on this occasion, the greater part of the discrepancy would be removed. However, even using families of 10 plants the number required is more than Mendel had assigned to any previous experiment.... Such explanations, moreover, could not explain the discrepancy observed in the first group of experiments, in which the procedure is specified. (*130–31*)

Nothing could illustrate better than these two opposing theories the ingenuity that has been expended on accounting for Mendel's surprisingly good data. Here are two explanations, one postulating that *fewer* than ten plants were scored, and the other that *more* than ten were scored, both with a view to accounting for results judged too close to a 2:1 ratio. In the next section I shall comment further on Fisher's handling of this question, and show that Weiling's suggestion is of little importance statistically, but once again we should note that there is no evidence that Mendel did either of the things suggested. He stated explicitly (*88*) that 'ten seeds of each were cultivated', implying that there were exactly ten which grew. In Sherwood's translation (Stern & Sherwood, 1966) '*angebaut*'—'cultivated'—is rendered as 'sown', but I am advised (K. Martin, personal communication) that this is an unwarranted change.

One puzzling aspect of the problem of Mendel having apparently overlooked the fact that progeny-testing with only ten plants will have led to some heterozygotes being misclassified as homozygotes is that towards the end of his paper (*113*) he was completely clear that a form with low probability (one-eighth in the case he was treating) might by chance not appear in a particular experiment.

Wright (1966) also suggested that Mendel might often have been able to distinguish a 'segregating group from a nonsegregating one even in the absence of any recessives.' In other words, for some characters Mendel might have been able to spot the heterozygote, and Wright mentioned seed-coat colour especially. This was indeed often true for seed-coat colour (*106*), though Mendel never stated he used the fact. Indeed, he explicitly employed progeny-testing (*88*) or 'further investigation' (*93*), which all commentators have assumed to mean progeny-testing, in the case of seed-coat colour as with the other plant characters. Mendel (*85*) also mentioned that F_1 hybrids from short × tall parents were usually taller than the tall parent, but he thought this possibly due to the 'greater luxuriance which appears in all parts of plants when stems of very different length are crossed' and did not mention its occurrence in the F_2. It is one thing to notice that tall offspring are generally taller than their tall parent, but another thing altogether to notice that a tall population of plants actually consists of two sub-populations. Even if there were some truth in the suggestion that in respect of seed-coat colour Mendel could distinguish between a mixed group of ten plants and a homozygous group, the explanation would only affect one character and the statistical consequences would be small. But the most telling objection to this hypothesis is, again, the fact that Mendel explicitly described a quite different procedure (*88*).

In addition to the papers cited above, there have been many relevant reviews, though mostly of a somewhat inconclusive nature, and all lacking any extensive statistical analysis of the scope of Fisher (1936) or even Weiling (1966). Mention may, however, be made of the wide-ranging commentaries by Piegorsch (1983), who studied Weiling's analysis with care, and Root-Bernstein (1983), who concluded his review of earlier work by saying 'The issue remains unresolved,' and then, accepting the existence of bias, suggested that some plants were difficult to classify and on that account were tallied last in a direction to fit the ratios (cf. Fisher, 1911, quoted above in Section I). In respect of the others, I believe that I have read all the published material on this question, and that no useful purpose would be served by referencing papers which, though they may have fulfilled a valuable educational role at the time, are of no lasting value. Similarly, I make no reference to the many excellent commentaries on Mendel's paper that do not touch on the goodness-of-fit issue.

III. Previous Statistical Analyses

Fisher (1936) gave the results of the most extensive statistical analysis of the data so far. On recomputing his χ^2 values (1936, table V) which he calculated to four places of decimals, I find them all correct, though Piegorsch (1983) noted that the famous 'P-value' for $\chi^2 = 41.6056$ on 84 d.f. should be .99997 and not .99993!

There are, however, two minor criticisms of Fisher's analysis. First, the 20 d.f. *Illustrations of plant variation* are not independent of the 2 d.f. 3:1 *ratios – seed characters* since the data for the former are included in the latter. Subtracting them out increases χ^2_2 from 0.2779 to 0.6232. Secondly, in those experiments in which, as Fisher was the first to point out, 2:1 ratios are not to be expected (see Section II above), he did in fact use them in his calculations of χ^2, presumably because they were what he thought Mendel expected. Root-Bernstein (1983) regarded this as a 'ploy' of Fisher's (a choice of word more revealing of Root-Bernstein's thought than Fisher's), but there are arguments both ways.

Fisher was testing the divergence between Mendel's counts and their expectations, and the problem is *which* expectations, Mendel's or nature's? The difficulty is that any goodness-of-fit test confounds two sources of departure from expectation, the systematic (due to using the incorrect expectation) and the random (due to binomial variability). Usually one is assuming binomial variability and testing an expectation, in which case the null hypothesis under test clearly involves that expectation, but if one is assuming an expectation and testing the variability against the binomial model, as when the implied alternative is that the data have been adjusted towards the expectation, it is arguable that one should then assume *that* expectation. However, we need not explore this statistical byway further (for which see Edwards, 1972, chapter 9) since even assuming the alternative expectations on Mendel's data, Fisher's overall conclusion is barely affected: Weiling (1966) computed that the χ^2_{84} is increased from 41.6056 to 48.910 and remarked correctly, 'This value is also highly significant.' Finally, on Fisher's statistical analysis, we should note that van der Waerden's (1968) suggestion of a numerical error seems to be a mistake.

After Fisher, Weiling (1966) has given the next most extensive statistical treatment. He squarely addressed the question of whether the variance might not be binomial by postulating that it is infra-binomial by a factor c, for the reasons given in Section II above, and estimating c. Exactly the same thought occurred to me at a later date (Edwards, 1972, chapter 9) in a general discussion of χ^2. My k^2 is his c, my figure 25 highly reminiscent of Weiling's figures, and my final comment indicative of an intention to do what, unbeknown to me, Weiling had already done:

> Since the alternative hypothesis [in considering the goodness-of-fit of Mendel's data] is that the variance of the results is less than expected, rather than that the means are in doubt, the use of χ^2 seems entirely appropriate. It would be interesting to rework Fisher's analysis using the justification of χ^2 offered by the Method of Support.

Of course, once one has estimated c (for which Weiling found the broad limits 0.6–1.0) there is nothing left to test, and Weiling persuaded himself that Mendel's results were compatible with those of other workers who repeated his experiments around the turn of the century. However, the facts indicate otherwise. The overwhelming majority of the post-Mendelian data on segregation in the pea comes from two sources, Bateson & Kilby (1905) and Darbishire (1908, 1909), for which Weiling calculated $\chi^2_{408} = 411.101$ and $\chi^2_{654} = 597.689$ respectively, or $\chi^2_{1062} = 1008.79$ in aggregate. The estimate of c is therefore $1008.79 / 1062 = 0.9500$. In other words these massive data exhibit a standard deviation of about 97.5% of that expected on a binomial model, and far from this lending support to the biological hypothesis that the pea does not segregate randomly, it fills one with admiration for the perfection of the randomizing mechanism! The compatibility which impressed Weiling is but a reflection of the wide interval estimate he obtained for c on Mendel's data.

Moreover, the conclusion that the pea is a good randomiser is actually supported by an alternative analysis Weiling gave which he misinterpreted. If the data really do depart from the binomial model in the way he suggested, then not only should the observed χ^2 values be less than their expectations (= their degrees of freedom), but more than half of the component χ^2_1's should be less than their theoretical median value of 0.4549. He found (his table 4) that for Bateson & Kilby's data 116 were less and 120 more, and for Darbishire's 347 were less and 307 more (I assume that only 236 d.f. in Bateson & Kilby's data refer to single-factor segregations). These figures, 463 vs. 427, lend no real support to the infra-binomial-variance hypothesis, but Weiling did not appreciate this because he wrote:

> In spite of the consequent results of our investigations into all the analyzed pea crosses (yielding an appreciable deviation of the value c from 1), the χ^2-test may, without hesitation, at least be applied to the evaluation of the individual segregations. Then the test of frequency distribution of all other inquiries into the underlying individual probabilities shows no noticeable deviation in the direction of P = 1.0, as [his] Table 4 illustrates.

Weiling here conceded 'no noticeable deviation' but thought this consistent with an appreciable overall departure of c from 1. He thus assumed that a collective of individual χ^2_1 can exhibit no systematic departure from their theoretical distribution even when their sum does, but this is a misunderstanding of the nature of χ^2. I consider the logic of χ^2 further in the next section.

Weiling also analysed the much less extensive data of Correns and Tschermak, and the fact that I have concentrated my remarks on the data of Bateson & Kilby and Darbishire is explained by the circumstance that I had already chosen these data-sets for a complete re-analysis because of their extent and their ready availability to me. When I started the re-analysis I was not aware of the existence of the other smaller data-sets in the German literature, and since I do not read German I must leave their reconsideration to others. I only labour this point in order to deflect any suggestion that I have deliberately ignored data which supports this or that hypothesis. I hope to publish a full analysis of the very large English data sets in due course.

Weiling is to be congratulated for his determined attempt to rescue the Mendelian experiments from the Fisherian conclusion. He left no stone unturned in his search for explanations, increasing the value of χ^2 by using the true instead of the 2:1 ratio and by postulating that not all the 10 seeds sown in some of the experiments yielded mature plants, and then accounting for the too-small χ^2 still remaining by the hypothesis of a diminished variance. But in the end the attempt fails: there is too much to explain. The diminished-variance hypothesis is not supported by other more extensive data, and, as I show in my own analysis below, simply reducing the variance is in any case not enough to explain the peculiarities of Mendel's data.

No such congratulations can be offered to Monaghan & Corcos (1985). All their four tables contain substantial errors. At the trivial level, many of the χ^2 values they give are simply wrong to the accuracy claimed. Even the simple addition of these values is wrong in table I, and this departure alone from Fisher's value should have alerted the authors to the fact that either they or Fisher were in error. Table III contains the astonishing assertion that $0.95 < P < 0.99$ for $\chi^2_{67} = 32.57$, on which obvious error the authors hang their conclusions. Table IV *Results of other workers compared to those of Mendel* leads to the observation 'that the results of Bateson for cotyledon shape *[sic]* and Tschermak for cotyledon color deviate very widely from the other workers', which is not surprising given that Monaghan & Corcos have copied them both down wrongly from the literature. The list of references is a bibliographic disaster area. I spare the reader further criticism.

My overall impression from reading all the commentaries since Fisher (1936) is that a good deal of special pleading, not to mention downright advocacy, has failed to make any substantial impact on Fisher's conclusion. Moreover, as Fisher was the first to point out, much of the special pleading requires that Mendel's account itself would have to be judged 'disingenuous'. Certainly, there is every reason for each student of Mendel to keep an open mind and to make his own analysis and form his own conclusion. I give my own attempt in Section V, preceded in Section IV by a discussion of the precise nature of goodness-of-fit χ^2, the statistical procedure on which the whole issue hinges and which has often been misunderstood.

IV. Goodness-of-Fit χ^2

Karl Pearson published his famous χ^2 goodness-of-fit test in the year that Mendel's paper was rediscovered (Pearson, 1900), and it is of interest that he warned against its use with large numbers of degrees of freedom:

Thus, if we take a very great number of groups our test becomes illusory. We must confine our attention in calculating P to a finite [*sic*] number of groups, and this is undoubtedly what happens in actual statistics. *n* will rarely exceed 30, often not be greater than 12.

The basis of Pearson's concern seems to have been that in a large number of dimensions practically the whole of the probability in a multivariate normal distribution is to be found in the 'tails', or regions remote from the central region of high probability *density*. He seems to have sensed the dilemma facing anyone who ponders the question of what a 'typical' sample-point might be from a standardized multivariate normal distribution in, say, 84 dimensions (the number of degrees of freedom in Mendel's data). If, for example, the choice falls somewhere in the region of high probability density within one unit (= standard deviation in one dimension) of the centre, or mean, then the corresponding χ^2_{84} is less than one and, by this criterion, the point is surprisingly close to the centre. Yet if the choice falls about nine units from the centre, to make χ^2_{84} about 81 and thus near its expectation, it is in a region where the probability density has fallen to about 2.6×10^{-18} of its maximum value.

It is interesting to find this implicit concern for the logic of 'tail-area' significance tests so early in their history. More recent explicit criticisms are well-known, often in the writings of Bayesians (for example, Jeffreys,

1939) but not always (for example, Edwards, 1972). What is the 'rejection region' to be? Taking the 5% significance level as an example, is it to be the region delimited by the line (or surface, etc.) of constant probability density which divides the sample space into probabilities of 0.95 and 0.05 such that the probability density everywhere in the latter part is less than the probability density everywhere in the former part? Or is it to be the line which delimits that part of the sample space beyond a certain distance from the mean, the distance chosen so that just 5% of the probability is in it? In either case, what transformations of the sample space are allowed (since they manifestly affect both criteria)?

The χ^2 distribution is a transformation of the multivariate normal sample space, many dimensions being mapped into one by combining all points of equal probability density. On either of the above criteria the rejection region for the standardized multivariate normal will be the outer skirts delimited by a hypersphere, but in the case of χ^2, which is not a symmetrical univariate distribution, it will be the two tails, consisting either of equal probability content 0.025 or cut off at equal probability *densities* and totalling 0.05 in probability. The corresponding region in the multivariate normal is not that described above, but rather two disjoint regions delimited by hyperspheres, one surrounding the centre (the image of the left-hand tail of χ^2) and one far removed from it (the image of the right-hand tail). A χ^2 value much below expectation (such as 41.6056 on 84 d.f.) is represented in the multivariate normal space by a point within the hyperspherical region surrounding the centre, the region of highest probability density, yet in the χ^2 space it is in a region of low probability density. The criticism of Mendel's results is that they are too close to expectation in the normal space as judged by being too far from expectation in the χ^2 space. *Quis custodiet ipsos custodes?*

We may emphasize this *Paradox of the Probable* in another way. Suppose the χ^2 test had antedated Mendel, and that in his paper he had reported a value of 84.0000 on 84 d.f. The reaction of a latter-day Fisher might well have been to conclude that Mendel's assistant had known that what Mendel really needed for his paper was not good Mendelian ratios but a good value of χ^2.

I have discussed these questions at length in my book *Likelihood*: '*The quantification of surprise in terms of probability is likely to tell only half the story*' (Edwards, 1972). The other half depends on the fact that the more improbable a result on a particular hypothesis as judged on *some* criterion, the greater the possibility of an alternative hypothesis leading to the result with a substantially higher probability. In other words there is a greater possibility of a hypothesis of higher likelihood, and hence one which, other things being equal, is more satisfactory. In particular, any *pattern* which we recognize in some data and which is unexplained on the current hypothesis is a signal that we should seek an alternative hypothesis, because an alternative which accounts for it is almost bound to have a higher likelihood. This may be the case even where the actual data are more probable than any other that might have occurred given the hypothesis, for we can then often more readily think of an alternative explanation than would be the case for some more arbitrary outcome. Thus numerical equality of the sexes at a children's party (see *Likelihood,* p. 190) is readily explained although, being the most probable outcome of a random selection of children, it might be thought to be the least in need of explanation. It is surely for this reason that attempts to base a theory of statistical estimation on maximising probabilities (Barndorff-Nielsen, 1976) have failed.

To sum up, we must not allow our judgment to be dominated by tests of significance and other calculations of probability which are at best pointers for further thought and at worst misleading. This is especially important when, as in the case of Mendel's experiments, the suggestion is that the data are 'too good' in the sense of being *too probable,* too close to the target. If it were just a question of having hit the bull's eye with a single shot we might conclude, as some commentators have (e.g., Pilgrim, 1984), that Mendel was simply lucky, but when a whole succession of shots comes close to the bull's eye we are entitled to invoke skill or some other factor. A χ^2 of 41.6056 on 84 d.f. may look like a single lucky shot, but in reality it is a succession of 84 shots which must be manifesting some kind of pattern calling out for investigation. This will be the basis of my own analysis of Mendel's data in the next section.

Finally, there is one technical point about χ^2 which should be noted. It has been implied (Weiling, 1966) that one of the reasons for doubting the χ^2 analysis is that it is based on the assumption of normally distributed data, whereas genetic segregations are (at least on the simplest hypothesis) binomial. But it is a fact that the *expectation* of χ^2 on a binomial model is *exactly* equal to its degrees of freedom (Pearson, 1900). This is true *by definition,* for Pearson obtained his goodness-of-fit criterion by assuming that the normal distribution in question had a mean and variance given by the binomial distribution. In modern notation:

Let x be normally distributed with mean μ and variance σ^2. Then by definition $\chi^2_1 = (x - \mu)^2 / \sigma^2$. Now put $\mu = np$ and $\sigma^2 = npq$, the mean and variance of a binomial distribution, and let the observed number of successes x be a and of failures be $b = n - a$. Then, substituting for μ and σ^2,

$$\chi^2_1 = (a-np)^2 / npq,$$
$$= (a-np)^2 / np + (a-np)^2 / nq,$$
$$= (a-np)^2 / np + (b-np)^2 / nq,$$
$$= \Sigma \;[\;(\text{obs}-\text{exp})^2 / \text{exp}\;]\text{ exactly.}$$

It is obvious that the expectation is exactly 1 because $E(a-np)^2 = npq$ by the definition of variance, and if all the χ^2_1's are independent then the expectation of their sum will exactly equal the degrees of freedom.

It follows that although the fine detail of the χ^2 distribution may be wrong its mean and variance are right, and there can be no doubt over its applicability save when sample sizes are very small, much smaller than Mendel's, and even then the expectation is still correct.

V. A Fresh Analysis

The 84 d.f. in Mendel's data may be individually identified in a number of different ways depending on the order in which the simultaneous segregations are considered. I shall arbitrarily order the loci according to Mendel's own classification. Thus in analysing the results of selfing *AaBb* where *A, B* is the order of the loci in Mendel's list, I shall take the segregation at the *A* locus first and then the two segregations at the *B* locus for each of the two *A*-locus phenotypes, making three independent comparisons in all. With this arrangement we will then have 84 independent variates to analyse.

For our purposes it will be better to work with χ rather than χ^2, partly because it preserves the information about the direction in which the segregation departs from expectation, but also because χ is expected to have a normal distribution to an excellent approximation, and methods for analysing a sequence of normal variables are commonplace. By the argument of the previous section the expectation of each χ on the binomial hypothesis is exactly 0 and its variance exactly 1, so that its distribution is $N(0, 1)$ for all practical purposes. It may also be noted that since in the binomial case χ is given exactly by dividing the observed deviation from the hypothetical binomial mean by the hypothetical standard deviation, whether we derive it as the square-root of Pearson's χ^2 criterion or directly as a standardized deviation is immaterial.

There is, of course, also an arbitrariness as to which departures from expectation we regard as positive and which as negative, but we will consistently regard a deviation in favour of the first of the two segregating classes as positive, and the first class will always be taken to be the one with the larger expectation. In the case of the experiments on gametic ra-

TABLE 1 — Notation for Mendel's seven characters

Seed characters	
A, a	Form of seed
B, b	Colour of albumen
Plant characters	
C, c	Colour of the seed-coats
D, d	Form of pods
E, e	Colour of the unripe pods
F, f	Position of flowers
G, g	Length of stem

tios, where 1:1 is expected, the first class will always be the one with the larger number of dominant genes, so the order will be either $AA:Aa$ or $Aa:aa$ as the case may be.

The notation for the characters is given in Table 1, and agrees with Mendel's for the major part of his paper. Table 2 gives the 84 values of χ and their origins. The page references are to Mendel's paper in Bennett (1965), and 'A' stands for $AA + Aa$ and 'a' for aa in accordance with the usual convention. As mentioned at the beginning of our Section III, the data for the illustration of plant variation (84) should be subtracted out from the data in the first two experiments (84–85), and this has been done. Similarly, modified expectations based on the ratio 0.629124:0.370876 have been used instead of 2:1 where appropriate. All calculations have been rounded to four significant figures after completion.

It will be seen that of the fifteen χ values for the segregation ratio 0.6291:0.3709 (rows 30–35 and 61–69), 13 are positive and only 2 negative, suggesting something of a bias towards the larger class. The total segregation is 720:353, or 0.6710:0.3290, with an associated χ value of +2.8408, indicating a very poor fit indeed. Mendel (88–89) explicitly accepted this type of experiment as establishing the 2:1 ratio, and as an aid to judging the extent to which the data conform to this rather than the true ratio we can compute the likelihood ratio for the 2:1 hypothesis *vs.* the 0.6291:0.3709 hypothesis on the assumption of binomial variation (for likelihood techniques, see Edwards, 1972). The result is 57.90, or a support value (taking logs) of just over 4, which is large enough to substantiate what Sturtevant (1965) called 'Fisher's most telling point' in his argument that the data have been biased in the direction of agreement with what Mendel expected.

Having concluded from an examination of the *means* for these fifteen segregations that possibly all is not well, we now turn to an examination of the *variances* of the other 69 segregations. For we know from the analyses considered in Section III that it is the infra-binomial variance that raises doubts about the data. As far as the means are concerned, of their 69 χ values, 38 are positive, 30 negative, and 1 zero. Their sum is 4.0880 which, dividing by $\sqrt{69}$ to make it a standard normal deviate, is equal to 0.4291, as unexceptionable a value as one could wish for. By contrast, the variance of the χ values is certainly exceptional. χ^2 is now the appropriate statistic, and the sum of the 69 χ^2 values from Table 2 is 30.8138, highly remarkable on any interpretation of tests of significance.

TABLE 2 — Mendel's segregations and their χ values

		Segregation			
	Character	Expected	Observed		Total	χ
		F_2, Seed characters			
1	A	3 : 1	5138	1749	6887	−0.7583
2	A	3 : 1	5667	1878	7545	+0.2193
		F_2, Illustrations of plant variability			
3	A	3 : 1	45	12	57	+0.6882
4	A	3 : 1	27	8	35	+0.2928
5	A	3 : 1	24	7	31	+0.3111
6	A	3 : 1	19	10	29	−1.1793
7	A	3 : 1	32	11	43	−0.0880
8	A	3 : 1	26	6	32	+0.8165
9	A	3 : 1	88	24	112	+0.8729
10	A	3 : 1	22	10	32	−0.8165
11	A	3 : 1	28	6	34	+0.9901
12	A	3 : 1	25	7	32	+0.4082
13	B	3 : 1	25	11	36	−0.7698
14	B	3 : 1	32	7	39	+1.0170
15	B	3 : 1	14	5	19	−0.1325

(table continues)

(continued)

	Character	Expected	Observed		Total	χ
			F_2, Illustrations of plant variability			
16	B	3 : 1	70	27	97	−0.6448
17	B	3 : 1	24	13	37	−1.4237
18	B	3 : 1	20	6	26	+0.2265
19	B	3 : 1	32	13	45	−0.6025
20	B	3 : 1	44	9	53	+1.3482
21	B	3 : 1	50	14	64	+0.5774
22	B	3 : 1	44	18	62	−0.7332
			F_2, Plant characters			
23	C	3 : 1	705	224	929	+0.6251
24	D	3 : 1	882	299	1181	−0.2520
25	E	3 : 1	428	152	580	−0.6712
26	F	3 : 1	651	207	858	+0.5913
27	G	3 : 1	787	277	1064	−0.7788
			F_3, Seed characters			
28	A	2 : 1	372	193	565	−0.4165
29	B	2 : 1	353	166	519	+0.6518
			F_3, Plant characters			
30	C	0.63 : 0.37	64	36	100	+0.2252
31	D	0.63 : 0.37	71	29	100	+1.6743
32	E	0.63 : 0.37	60	40	100	−0.6029
33	F	0.63 : 0.37	67	33	100	+0.8462
34	G	0.63 : 0.37	72	28	100	+1.8813
35	E	0.63 : 0.37	65	35	100	+0.4322
			F_3, Bifactorial experiment			
36	A	3 : 1	423	133	556	+0.5876
37	B, among 'A'	3 : 1	315	108	423	−0.2526
38	B, among 'a'	3 : 1	101	32	133	+0.2503

(table continues)

(continued)

	Character	Expected	Observed		Total	χ
			F_3, Bifactorial experiment			
39	A, among 'AB'	2 : 1	198	103	301	−0.3261
40	A, among 'Ab'	2 : 1	67	35	102	−0.2100
41	B, among 'aB'	2 : 1	68	28	96	+0.8660
42	B, among Aa 'B'	2 : 1	138	60	198	+0.9045
43	B, among AA 'B'	2 : 1	65	38	103	−0.7664
			F_2, Trifactorial experiment, seed characters			
44	A	3 : 1	480	159	639	+0.0685
45	B, among 'A'	3 : 1	367	113	480	+0.7379
46	B, among 'a'	3 : 1	122	37	159	+0.5037
			F_3, Trifactorial experiment, seed characters			
47	A, among 'AB'	2 : 1	245	122	367	+0.0369
48	A, among 'Ab'	2 : 1	76	37	113	+0.1330
49	B, among 'aB'	2 : 1	79	43	122	−0.4481
50	B, among Aa 'B'	2 : 1	175	70	245	+1.5811
51	B, among AA 'B'	2 : 1	78	44	122	−0.6402
			F_2, Trifactorial experiment, plant character			
52	C, among AaBb	3 : 1	127	48	175	−0.7419
53	C, among AaBB	3 : 1	52	18	70	−0.1380
54	C, among AABb	3 : 1	60	18	78	+0.3922
55	C, among AABB	3 : 1	30	14	44	−1.0445
56	C, among Aabb	3 : 1	60	16	76	+0.7947
57	C, among AAbb	3 : 1	26	11	37	−0.6644
58	C, among aaBb	3 : 1	55	24	79	−1.1043
59	C, among aaBB	3 : 1	33	10	43	+0.2641
60	C, among aabb	3 : 1	30	7	37	+0.8542
			F_3, Trifactorial experiment, plant character			
61	C, among AaBb	0.63 : 0.37	78	49	127	−0.3488
62	C, among AaBB	0.63 : 0.37	38	14	52	+1.5174

(table continues)

(continued)

	Character	Expected	Observed		Total	χ
		F_3, Trifactorial experiment, plant character				
63	C, among AABb	0.63 : 0.37	45	15	60	+1.9384
64	C, among AABB	0.63 : 0.37	22	8	30	+1.1816
65	C, among Aabb	0.63 : 0.37	40	20	60	+0.6020
66	C, among AAbb	0.63 : 0.37	17	9	26	+0.2610
67	C, among aaBb	0.63 : 0.37	36	19	55	+0.3903
68	C, among aaBB	0.63 : 0.37	25	8	33	+1.5276
69	C, among aabb	0.63 : 0.37	20	10	30	+0.4257
		Gametic ratios, seed characters, first experiment				
70	A	1 : 1	43	47	90	−0.4216
71	B, among AA	1 : 1	20	23	43	−0.4575
72	B, among Aa	1 : 1	25	22	47	+0.4376
		Gametic ratios, seed characters, second experiment				
73	A	1 : 1	57	53	110	+0.3814
74	B, among Aa	1 : 1	31	26	57	+0.6623
75	B, among aa	1 : 1	27	26	53	+0.1374
		Gametic ratios, seed characters, third experiment				
76	A	1 : 1	44	43	87	+0.1072
77	B, among AA	1 : 1	25	19	44	+0.9045
78	B, among Aa	1 : 1	22	21	43	+0.1525
		Gametic ratios, seed characters, fourth experiment				
79	A	1 : 1	49	49	98	0.0000
80	B, among Aa	1 : 1	24	25	49	−0.1429
81	B, among aa	1 : 1	22	27	49	−0.7143
		Gametic ratios, plant characters				
82	G	1 : 1	87	79	166	+0.6209
83	C, among Gg	1 : 1	47	40	87	+0.7505
84	C, among gg	1 : 1	38	41	79	−0.3375

We now go further than any previous analysis of Mendel's data by looking at the distribution of the 69 χ values themselves to see whether the unexpectedly low variance is a reflection of any particular anomaly. Fig. 1 displays the values on normal probability paper. It is immediately obvious that the reduced variance is not characteristic of the whole data, as Weiling's theory would require, but is confined to the tails of the distribution, where the extreme variates are not extreme enough to conform to expectation. Between the upper and lower quartiles (the 25-percentiles) of the expected distribution the slope of the plot is acceptable, but outside these points it is not. It must be remembered that a normal probability plot is not a linear regression, so that intuitive judgments of the departure from expectation may be misleading; for comparison, a simulated plot from a standard normal distribution is given in Fig. 2, using the first 69 random normal deviates from Lindley & Scott (1984), reading columnwise.

Another way to judge the apparent lack of extreme deviates is by examining the range, which is from −1.4237 (row 17) to +1.5811 (row 50), a total of 3.0048, whereas the expected range for a sample of 69 random normal deviates is 4.7440. Indeed, the expected absolute values of the first three most extreme deviates out of 69 are 2.6216, 2.2755, and 2.0828, and nothing on Mendel's data approaches these. The inescapable conclusion is that some segregations beyond the outer 5-percentiles (approximately) have been systematically biased towards their expectations so as to fall between the 5-percentiles and the 25-percentiles. Further analysis shows that the effect is not confined to particular sample sizes or segregation ratios, but is quite general.

It is relevant to recall here that Mendel thought it worthwhile to repeat an experiment in which a segregation of 60:40 had appeared where he had expected 2:1 (*88–89*) a segregation with a corresponding χ-value of only −1.4142. As Fisher (1936) observed, this is not a significant value; yet only twice out of 69 times does Mendel record more extreme segregations (those quoted above). We should not forget, however, that when Mendel actually singled out the most extreme segregations in the first two experiments for comment (*86*) they were no different from what one might expect (Fisher, 1936).

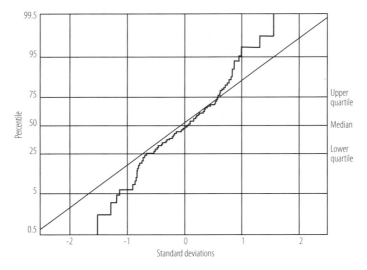

FIG. 1 The values of χ for the 69 segregations with undisputed expectations plotted on normal probability paper

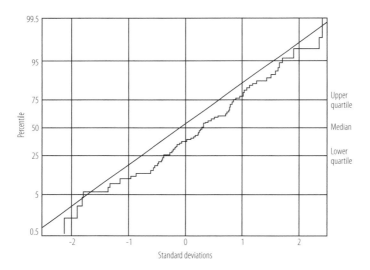

FIG. 2 The values of 69 simulated standard normal deviates plotted for comparison with fig. 1

VI. Conclusions

We thus see that Mendel's data exhibit two independent peculiarities both of which point to the same conclusion, namely, that in general the segregations agree more closely with what Mendel expected than chance would dictate. Where Mendel expected a 2:1 ratio even though the true expectation must have been different it was the former ratio which was achieved, and throughout the rest of his results there is a persistent lack of extreme segregations.

In Sections II and III above I reviewed earlier analyses and explanations and concluded that only Fisher (1936) could withstand detailed criticism. It may be helpful if I admit at this point that for many years I supposed that Fisher's analysis was going to be able to be faulted because of its total reliance on the 'repeated sampling' logic of the χ^2 goodness-of-fit test which I had come to mistrust, but a complete review of the whole problem has now persuaded me that his 'abominable discovery' must stand. I agree with Sturtevant (1965): 'In summary, then, Fisher's analysis of Mendel's data must stand essentially as he stated it.' As to the precise method by which the data came to be adjusted I would rather not speculate, though it does seem to me that any criticism of Mendel himself is quite unwarranted. Even if he were personally responsible for the biased counts his actions should not be judged by today's standards of data recording. In 1967 Dobzhansky reviewed six separate volumes celebrating the centenary of Mendel's paper and had this to say:

I believe that a far simpler explanation [than Fisher's suggestion of an over-zealous assistant] is at least as plausible. Few experimenters are lucky enough to have no mistakes or accidents happen in any of their experiments, and it is only common sense to have such failures discarded. The evident danger is ascribing to mistakes and expunging from the record perfectly authentic experimental results which do not fit one's expectations. Not having been familiar with chi-squares and other statistical tests, Mendel may have, in perfect conscience, thrown out some crosses which he suspected to involve contamination with foreign pollen or other accident.

With the reservation that in order to explain some of the results misclassification rather than simple throwing away of plants will have to be invoked, I agree with Dobzhansky. Perhaps after all the undergraduate Fisher came close to the correct solution 75 years ago: 'It may be that the worthy German abbot, in his ignorance of probable error, unconsciously placed doubtful plants on the side which favoured his hypothesis.'

VII. Summary

1. All the analyses and discussions of Mendel's data in his 1866 paper 'Experiments in plant hybridization' are reviewed, with special reference to the suggestion by Fisher (1936) that the segregations are in general closer to Mendel's expectations than chance would dictate. It is concluded that in spite of many attempts to find an explanation, Fisher's suggestion

that the data have been subjected to some kind of adjustment must stand.

2. A fresh analysis based on the individual consideration of the 84 independent segregations reported by Mendel confirms this conclusion in two separate ways. In the words of my title, Mendel's results really are too close.

NOTE

This paper was originally published as Edwards, A. W. F. 1986. *Biological Review* 61:295–312. Reprinted with permission of Cambridge University Press.

REFERENCES

Babbage, C. (1830). *Reflections on the Decline of Science in England and on Some of Its Causes*. Fellowes, London.
Bateson, W. (1909). *Mendel's Principles of Heredity*. Cambridge University Press.
Bateson, W. & Kilby, H. (1905). Experimental studies in the physiology of heredity: Peas (*Pisum sativum*). In *Royal Society Reports to the Evolution Committee* 2, 55–80.
Barndorff-Nielsen, O. (1976). Plausibility inference. *Journal of the Royal Statistical Society* B 38, 103–131.
Beadle, G. W. (1967). 'Mendelism, 1965'. In *Heritage from Mendel* (ed. R. A. Brink), pp. 335–350. University of Wisconsin Press, Madison.
Bennett, J. H. (ed.). (1965). *Experiments in Plant Hybridisation,* by Gregor Mendel, with a Commentary and Assessment by R. A. Fisher. Oliver & Boyd, Edinburgh.
Bennett, J. H. (ed.). (1983). *Natural Selection, Heredity, and Eugenics*. Clarendon Press, Oxford.
Crew, F. A. E. (1966). *The Foundations of Genetics*. Pergamon, Oxford.
Darbishire, A. D. (1908). On the result of crossing round with wrinkled peas, with especial reference to their starch-grains. *Proceedings of the Royal Society* B80, 122–135.
Darbishire, A. D. (1909). An experimental estimation of the theory of ancestral contributions in heredity. *Proceedings of the Royal Society* B81, 61–79.
De Beer, G. (1964). Mendel, Darwin, and Fisher (1865–1965). *Note and Records of the Royal Society* 19, 192–226.
Dobzhansky, T. (1967). Looking back at Mendel's discovery. *Science* 156, 1588–1589.
Dunn, L. C. (1965a). *A Short History of Genetics*. McGraw-Hill, New York.
Dunn, L. C. (1965b). Mendel, his work and place in history. *Proceedings of the American Philosophical Society* 109, 189–198.
Edwards, A. W. F. (1972). *Likelihood*. Cambridge University Press. (Reprinted 1984.)
Edwards, A. W. F. (1986). More on the too-good-to-be-true paradox and Gregor Mendel. *Journal of Heredity* 77, 138.
Fisher, R. A. (1936). Has Mendel's work been rediscovered? *Annals of Science* I, 115–137. (Reprinted in Bennett, 1965, pp. 59–87, and Stern & Sherwood, 1966, pp. 139–172.)
Fisher, R. A. (1952). Statistical methods in genetics. *Heredity* 6, 1–12.
Galton, F. (1892). *Hereditary Genius,* 2nd ed., Macmillan, London.
Galton, F. (1904). Average number of kinsfolk in each degree. *Nature* 70, 529 and 626.
Jeffreys, H. (1939). *Theory of Probability*. Clarendon Press, Oxford. (Reprinted 1983.)
Lindley, D. V. & Scott, W. F. (1984). *New Cambridge Elementary Statistical Tables*. Cambridge University Press.
Mendel, G. (1866). Versuche über Pflanzen-Hybriden. *Verhandlungen des naturforschen-*

den *Vereines in Brünn* 4, 3–47. (English translations in Bateson, 1909 [reprinted in Bennett, 1965, pp. 7–51] and Stern & Sherwood 1966, pp. 1–48.)

Mersenne, M. (1637). *L'Harmonie Universelle, Seconde Partie*. Ballard, Paris.

Monaghan, F. & Corcos, A. (1985). Chi-square and Mendel's experiments: Where's the bias? *Journal of Heredity* 76, 307–309.

Norton, B. & Pearson, E. S. (1976). A note on the background to, and refereeing of, R. A. Fisher's paper '[On] The correlation between relatives on the supposition of Mendelian inheritance'. *Notes and Records of the Royal Society* 31, 151–162.

Olby, R. C. (1966). *Origins of Mendelism*. Schocken, New York. (2nd ed. University of Chicago Press, 1985).

Orel, V. (1968). Will the story on 'too good' results of Mendel's data continue? *BioScience* 18, 776–778.

Orel, V. (1984). *Mendel*. Oxford University Press.

Pearson, K. (1900). On the criterion that a given system of deviations from the probable in the case of a correlated system of variables is such that it can be reasonably supposed to have arisen from random sampling. *London, Edinburgh, and Dublin Philosophical Magazine, Fifth series* 50, 157–175.

Piegorsch, W. W. (1983). The questions of fit in the Gregor Mendel controversy. *Communications in Statistics: Theory and Methods* 12, 2289–2304.

Pilgrim, I. (1984). The too-good-to-be-true paradox and Gregor Mendel. *Journal of Heredity* 75, 501–502.

Root-Berstein, R. S. (1983). Mendel and methodology. *History of Science* 21, 275–295.

Stern, C. & Sherwood, E. R. (1966). *The Origin of Genetics: A Mendel Source Book*. Freeman, San Francisco.

Sturtevant, A. H. (1965). *A History of Genetics*. Harper & Row, New York.

Thoday, J. M. (1966). Mendel's work as an introduction to genetics. *Advancement of Science* 23, 120–124.

Van der Waerden, B. L. (1968). Mendel's experiments. *Centaurus* 12, 275–288.

Weiling, F. (1966). Hat J. G. Mendel bei seinen Versuchen 'zu genau' gearbeitet?—Der χ^2-Test und seine Bedeutung für die Beurteilung genetischer Spaltungsverhältnisse *Der Züchter* 36, 359–365. (An English translation by Piegorsch is available as Paper BU-718-M of the Biometrics Unit, Cornell University, Ithaca, New York.)

Weiling, F. (1971). Mendel's 'too good' data in *Pisum* experiments. *Folia Mendeliana* 6, 75–77.

Weiling, F. (1985). What about R. A. Fisher's statement of the 'too good' data of J. G. Mendel's *Pisum*-paper? *45th Session of the International Statistical Institute*. Amsterdam.

Weldon, W. F. P. (1902). Mendel's laws of alternative inheritance in peas. *Biometrika* 1, 228–254.

Wright, S. (1966). Mendel's ratios. In Stern & Sherwood (1966), 173–175.

POSTSCRIPT TO CHAPTER 4

■ Alternative Hypotheses and Fisher's *The Design of Experiments*

A. W. F. EDWARDS

In my 1986 article on this subject, I did not pursue the link to Fisher's book *The Design of Experiments,* which was published in 1935, the year before his analysis of Mendel's data. This connection was, however, made in *Likelihood* (Edwards 1972), in section 9.3: Support Tests Based on the Normal and χ^2 Distributions. The long discussion of that section is heavily dependent on chapter 10 of Fisher's book, and particularly on his section 63, The χ^2 Test, which opens with the sentence, "What is meant by choosing a test of significance appropriate to a special purpose, may now be illustrated by considering what should be done if the experimenter were interested, not in whether the mean of the distribution could exceed a given value, or could lie in a given range, but in the value of the variance of the same distribution."

Fisher is throughout emphasizing that the choice of a test of significance will be influenced by the alternative hypotheses one has in mind, even if these are not explicitly formulated as in the Neyman-Pearson theory of hypothesis testing. In *Likelihood*, I argued that this is a natural result of working in terms of likelihoods, a method I called "the method of support," and I concluded with the comment already noted by Franklin: "It would be interesting to rework Fisher's analysis [of Mendel's data] using the justification of χ^2 offered by the Method of Support." This intention was realized in my 1986 article; here I wish to place further emphasis on the matter of the alternative hypothesis.

The point may be made most simply if we contemplate a test statistic that, on the null hypothesis, has a normal distribution. If a particular realization of the test leads to a value of the statistic far out into the tails of the distribution, we "reject" the null hypothesis essentially because we can easily see that an alternative null hypothesis—one predicting a different mean for the test statistic—would greatly increase the probability of the value actually achieved. In other, technical, language, the alternative hypothesis has a much higher *likelihood*. If, however, a particular realization leads to a value of the test statistic that is at or very near its expected value, we again easily see that an alternative hypothesis with a much smaller variance for the test statistic would have a much higher likelihood. In this case the null hypothesis is rejected because the fit is "too good." But if the test statistic falls somewhere unremarkable on the "shoulders" of the normal distribution, no alternative hypothesis suggests itself and the null hypothesis is accepted (always provisionally, of course). This way of looking at the problem stems from Fisher's 1935 discussion and is implicit in Weiling's 1966 analysis as well my 1986 work.

This approach is particularly pertinent when considering Mendel's data. It has often been remarked that since any extensive data-set will have an extremely small probability of occurrence whatever the null hypothesis, this cannot in itself be a reason for rejecting that hypothesis. Instead, the standard logic requires a test of significance to be based on the tail probability of the distribution of the test statistic and not just on the probability of the particular value realized by the data. But from a likelihood point of view the tail probabilities are irrelevant; what always matters is whether there is an alternative hypothesis that would confer substantially higher probability on what has actually been observed.

In the case of Mendel's data, this alternative hypothesis is, of course, that some adjustment of the data has happened. It is now important to recognize exactly what the two hypotheses are: either the assumption of the simple binomial and multinomial model (the null hypothesis of the goodness-of-fit test), or the assumption of model with a reduced variance. But the latter has been suggested *after examining the data* and on this account alone will "explain" it with an exaggerated probability. The two hypotheses are not being fairly compared.

To sum up, we are free to speculate about hypotheses that purport to explain departures from expectation, but we should not be too prescriptive. The statistical analysis may point to an alternative explanation, but in the end the judgment of its veracity is only in part a statistical one. As Fisher remarked in *The Design of Experiments,* "the 'one chance in a million' will undoubtedly occur, with no less and no more than its appropriate frequency, however surprised we may be that it should occur to *us.*"

REFERENCE

Fisher, R. A. 1935. *The Design of Experiments*. Edinburgh: Oliver and Boyd.

CHAPTER 5

■ Controversies in the Interpretation of Mendel's Discovery

VÍTĚZSLAV OREL AND DANIEL L. HARTL

Introduction

Gregor Mendel (1822–1884) is acknowledged as the founder of the science of heredity, given the name "genetics" by W. Bateson in 1905. The reason for Mendel's unique standing in the history of science is that his pioneering contribution was virtually ignored during his lifetime and, in fact, until the end of the century.[1] Not until 1900 did Hugo de Vries, Carl Correns, and Erich von Tschermak, working independently in different countries, publish results of experiments similar to Mendel's, draw attention to Mendel's paper, and attribute priority to him.[2]

Heredity as a scientific problem began to be investigated in European countries in the nineteenth century but, until the end of the century, only Mendel had elaborated an experimental approach for the study of the hereditary transmission of traits from parents to progeny that could explain the essence of the process. Soon after 1900, there appeared in the literature generalizations of Mendel's theory, encapsulated as the "laws of heredity," which were ascribed to him. Later, biologists and geneticists formulated different numbers of "Mendel's laws" and gave them quite different interpretations. Nevertheless, despite the late appreciation of Mendel and the "laws of heredity" fallaciously ascribed to him, the conventional wisdom, throughout most of the twentieth century, has been that Mendel understood the significance of his experiments and knew what he was talking about.

However, in recent years, some historians of science and a few geneticists have dismissed this interpretation as whiggish or orthodox. They have proposed that Mendel was so much in the tradition of plant hybridists that his overriding interest in the role of hybridization in the formation of new species blinded him to the significance of his results for the study of heredity. Among the most influential of the iconoclastic papers was that of Olby, who in 1979 argued that Mendel himself was no Mendelian because he purportedly failed to understand that alleles segregate in heterozygous genotypes as well as in homozygous genotypes.[3] The revised interpretation of Mendel is all the more remarkable in light of the conventional wisdom among geneticists, expressed by Sturtevant in 1965, that Mendel understood the significance of his work a good deal more deeply than any of the three men who, 34 years later, carried out similar experiments and rediscovered Mendel's paper.[4]

We have recently weighed in on the whiggish side of this argument.[5] The basis of our argument is the realization that, in oral presentation, the principal manner of emphasis is repetition. Since Mendel prepared his paper specifically for oral presentation, we have examined his paper for phrases repeated multiple times as clues to what Mendel himself might have regarded as the most significant points. One phrase is repeated no fewer than six times in almost identical words in his relatively short paper: "... *pea hybrids form germinal and pollen cells that in their composition correspond in equal numbers to all the constant forms resulting from the combination of traits united through fertilization*" (emphasis by Mendel).[6] This statement is not an experimental result but rather an inference that summarizes, in one clear and concise sentence, the implications of what are now called the "laws" of segregation and independent assortment. The statement conveys the essence of Mendelian heredity, and anyone who subscribes to it must be regarded as a Mendelian.

The purpose of the present paper is to review some of the historical background of the study of heredity in the late nineteenth century and to discuss various aspects in the interpretation of Mendel's work. We address several major points:

1. The emergence of the problem of heredity.
2. The methodology of the research.
3. The plausibility of the new theory.
4. The translation of innovative theory.
5. The allegation of data falsification.

A mild disclaimer: although we aspire to objectivity, we also realize that our final opinions must inevitably reflect, to an unknown degree, our own backgrounds. We both came to appreciate Mendel's paper from our teachers. For Orel, it was mainly Jaroslav Kříženecký (1896–1964) in Brno; and for Hartl, it was mainly Curt Stern (1902–1981) in Berkeley. Both men were on friendly terms and both were great admirers of Mendel. Their words convey their appreciation. They were unapologetic whigs. For example, Kříženecký remarked in 1964, "I like to return to Mendel's classic work, it produces repeatedly a new impression upon me and I find always new motivation in it for my own research."[7] In 1966, Stern wrote, "Mendel's paper is not solely a historical document. It remains alive as a supreme example of scientific experimentation and profound penetration of data."[8]

Emergence of the Problem of Heredity

In the second half of the nineteenth century biologists began to distinguish heredity as a biological process separate from the enigma of generation. Olby has examined the emergence of the concept of heredity in the context of nineteenth century medical science. He attributes to Francis Galton what geneticists conventionally attribute to Mendel: the first precise conception of the subject of heredity in terms of "the statistical relations between the distribution of characters in successive generations."[9] Concerns of medical science seem to have had little impact on Mendel's thinking, however; for him the relevant fields were agricultural science and the tradition of plant hybridization.

On the Continent, one of the pioneers in distinguishing heredity as a process was the celebrated cytologist August Weismann. As we will see, Weismann's perspectives changed through the years. In the context of Weismann's views, it is also instructive to examine those of his antecedent, R. Leuckart, as well as those of Leuckart's teacher, R. Wagner. The latter is directly relevant to Mendel because Wagner was on friendly terms with J. E. Purkyně, who repeatedly visited the Augustinian monastery in Brno. Later we will give evidence that the abbot of the monastery, F. C. Napp (1792–1867), had already made a distinction between heredity and generation in the first half of the century—though not, certainly, as sharp a distinction as Galton or Weismann.

In 1883, just prior to Mendel's death, Weismann rejected the then-prevailing view that experiments attempting to explain the traditional enigmas of heredity were premature and unlikely to be illuminating.[10] At that time, as pointed out by Churchill, many German cytologists already subscribed to an experimental approach for the investigation of heredity as distinct from the concept of generation *(Zeugung)*, meaning the origin and development of the new individual.[11] With his theory of germ plasm, Weismann replaced the prevailing concept that the hereditary material could be modified by extrinsic influences (a view denoted by Mayr[12] as "soft heredity") with a new concept of an unchanging hereditary material ("hard heredity") based on the transmission of physical entities from generation to generation. Stimulated by Darwin's theory of evolution, Weismann was emboldened to examine the problem of heredity in isolation:

Aber unter der oft gehörten Forderung, man solle so complicirte Erscheinungen, wie z. B. Vererbung, jetzt noch nich in Angriff nehmen, verbirgt sich noch eine andere Unklarheit, nämlich die, als sei eine Tatsache deshalb unsicherer, weil ihre Ursachen sehr verwickelte, für uns zunächst noch nicht übersehbare sind.[13] (In the often-heard claim that we are not yet ready to attack such complicated phenomena as, for instance, heredity, is hidden still another fallacy—namely, that facts are less certain because their causes are not yet understood.)

Considering the observation that "the participation of heredity from the side of the father and mother is completely or nearly equal," Weismann concluded that some active substance determining heredity must be in the germ cells and, therefore, he became convinced that the time was ripe for attacking the problem experimentally. In his view:

Die Wissenschaft von Lebendigen hat nicht zu warten, bis Physik und Chemie fertig sind und *die Erforschung der Vererbungsvorgänge* hat nicht zu warten, bis die Physiologie der Zelle fertig ist[14] (emphasis added). (Life science cannot wait until physics and chemistry are complete, and research on heredity cannot wait until cell physiology is complete.)

Weismann's recommendation was to look for experimental approaches to investigate heredity from different points of view. He himself, however, could not offer any new method. Later, in the introduction to his published papers, Weismann challenged his contemporaries to carry out research in heredity, admitting that his own papers were not popular treatises; indeed,

... sie wollen nicht bekannte Ansichten weiteren Kreisen zugänglich machen, sondern *neuen* Ansichten zur Geltung bringen. Es sind Untersuchungen, wenn auch nicht solche, welche durchaus auf Mikroskop und Lupe, ich meine auf neuen Beobachtungen beruhen, sondern solche, welche sich vielfach auf bereits bekannte Tatsachen stützen, diese aber in neuer Weise verbinden und beleuchten. (... they do not aim to make known views accessible to a wider audience but to make new views valid. They are investigations, not based on new observations obtained with the microscope or magnifying glass, but based on already known facts, which connect and illuminate them in a new way.[15])

In his biography of 1896, Weismann wrote that he had occupied himself with the idea of writing a theory of heredity during the previous eight years and had already written a manuscript of some hundred pages arising from the theory of epigenesis.[16] Ultimately, not being able to find any explanation of heredity, he gave up his plan. Thereafter, he paid renewed attention to evolution, resulting in the publication of his "Voträge über Deszendenztheorie." In the third edition in 1913, Weismann described Mendel's theory as one of "genius," to which attention has been drawn recently by Danailov.[17] Relative to Mendel's theory, Weismann wrote:

Wir haben also in den Vererbungserscheinungen, wie wir sie bei der Bastardisierung kennen gelernt haben, einen weiteren Beweis für die reale Existenz von Determinanten.[18] (Thus we have, in the phenomenon of heredity, which we learned from hybridization, further proof of the actual existence of determinants.)

In his research direction, Weismann was influenced by the theory of evolution put forth by Darwin in 1859. However, Weismann's preoccupation with heredity was stimulated directly by Carl Nägeli, who was a supporter of Darwin's concept of natural selection and wished to understand the role of hybridization in the origin of variation and new species. The idea of units of heredity was introduced by Spencer,[19] and Darwin's provisional hypothesis of pangenesis also postulated hypothetical hereditary units.[20] Nägeli, in 1884, drew attention to the hereditary substance, which he called idioplasm, present in all cells but transmitted by the germ cells. Nägeli also assumed the existence of a "law of heredity," analogous to the physical law of gravitation, and ultimately came to the more focused view of heredity as the transfer of traits from parents to offspring. From this point of view the question "what is inherited?" follows immediately.[21]

In a later period, influenced by the success of cytology, Hugo de Vries attributed sudden variation in plant traits to changes in units of heredity present in the structures of germ cells. As an enthusiastic proponent of Darwin's theory of evolution, de Vries soon came to a conclusion similar to Weismann's—that the continuous phenotypic variation caused by differences in environment during any individual's lifetime cannot be inherited. This reasoning led de Vries to his theory of intracellular pangenesis.[22] He rejected Darwin's hypothesis of pangenesis and replaced Darwin's term "gemmules" with the new term "pangenes." But Weismann's theory of the constancy of the germ plasm was also refuted by de Vries because de Vries was afraid that the segregation of determinants of heredity implicated in Weismann's theory would imply, in effect, the splitting of species. After publishing the theory of pangenesis, de Vries also investigated the occurrence of monstrosities, believing that hybrids would show higher frequencies of them. In crossing experiments with plants, he had noted the proportions of the parental traits occurring in the progeny of hybrids as early as 1893. However, according to Meijer, de Vries may not have appreciated the significance of regular numerical segregation ratios in the progeny of hybrids until after reading Mendel's paper.[23]

Although Weismann distinguished clearly between heredity and generation, this separation had antecedents. In his writings about heredity, Weismann took little cognizance of the work of R. Leuckart regarding parthenogenic fertilization in animals.[24] Weismann had come into contact with Leuckart during his stay in Giessen, to which attention has been drawn by Riesler.[25] Some years previous to Weismann's visit to Giessen, R. Wagner, Professor of Physiology at the University in Göttingen, had asked Leuckart, his former student, to elucidate the problem of generation and also that of heredity for publication in Wagner's dictionary of physiology.[26] At that time, Leuckart could find neither an explanation of the phenomena nor a method for experimental investigation. Wagner was dissatisfied and added two postscripts to Leuckart's treatment.

In his first postscript, Wagner examined the enigma of generation and heredity from a "historical and methodological point of view" and recommended crossing experiments with animals.[27] He defined six areas of the problem of heredity to be investigated and outlined a method for research in crossing animals with different traits that included the application of statistics. He believed that "a more exact ascertainment of the numerical data could furnish a reliable clue."

In his second postscript, Wagner mentioned the newly discovered penetration of rabbit sperm into the egg and concluded that the embryo came into being with the participation of both parents.[28] This idea Wagner attributed to J. E. Purkyně, with whom Wagner was on friendly terms. Purkyně (rendered Purkinje in German, French, and English publications) was Professor of Physiology at the University of Breslau (now Wroclaw) until 1850 and afterwards in Prague.[29] In this postscript Wagner also briefly wrote about his own crossing experiments with frogs and fish, concluding that such experiments are extensive and time-consuming undertakings. He recommended crossing experiments with sheep, horses, dogs, and even exotic animals at the London Zoo. One of Wagner's noteworthy conclusions concerned hybridization in plants: "teaching about plant crossing rests on a similar and more enlightened basis, and therefore research into generation and heredity might also be realized through the crossing of plants." In this context, Wagner also called attention to the then-current dispute among plant physiologists regarding Schleiden's preformationist view that the embryo was already contained inside the pollen tube. Wagner emphasized that Mohl and Hofmeister had shown the participation of both parents in the formation of the embryo and considered the epigenetic approach the starting point for further research into both generation and heredity.

Wagner was apparently unaware that the enigma of generation and heredity had been discussed about 40 years earlier in Mendel's homeland, Moravia, by the sheep breeders who had become organized in the Sheep Breeder's Association in Brno.[30] As early as 1819 an attempt at generalization of empirical knowledge regarding the transfer of traits from parents to offspring was published in Brno by E. Festetics under the heading "genetical laws," which dealt with the observed phenomena of the heredity of traits determining the quality of wool.[31] In 1827, J. K. Nestler, Professor of Natural History and of Agricultural Science at the University of Olomouc, included in his teaching a separate section on scientific breeding and heredity. Two years later, Nestler published his lecture under the title *On the Influence of Generation on the Characteristics of Progeny*.[32]

In his second postscript, Wagner did mentioned sheep. He may have been informed about sheep breeders' investigations of heredity by Purkyně, who had visited F. C. Napp, abbot of the Augustinian monastery in Brno, in 1835.[33] At that time, Napp took an active part in discussions with the sheep breeders and cooperated with Nestler. The published lectures of Nestler evoked new interest among practical breeders in theoretical explanations of the observed transmission of traits from parents to offspring. At their annual meeting in Brno, the breeders even began to discuss such problems as the origin of life by means of a "generation force" *(in der Zeugungskraft)*, also referred to as a "genetical force."[34] These discussions reached a climax in 1836–1837 when abbot Napp, summarizing the discussion, concluded that the problem was one of experimental physiology and that the key questions were these: What is inherited? How is it inherited?[35] Soon after, Nestler examined the parental contributions with reference to heredity and expressed his agreement with the viewpoint expressed by Napp in 1836 as follows: "I agree with Napp's further proposition, saying: heredity of characteristics from producers to offspring consists predominantly in the mutual elective kinship of paired animals."[36]

Just prior to Purkyně's 1835 visit to Brno, he had published a paper about generation and heredity, and so we assume that he and Napp spoke about the subject. By means of the generation substance, Purkyně tried to explain the reduction of traits in the parents in the germs *(Keime)* into the pure quality in the process he called *involution*.[37] According to Purkyně, the origin of the embryo results from the material dynamic interaction of the germs from both parents giving rise to a new individual in a process he called *evolution*. According to the manuscript of his lectures entitled *The Developmental History of Plant and Animal Organisms,* Purkyně postulated that the generation substance contained creative rudiments *(Gestaltungsanlagen),* which determined that the traits of both parents were present in the progeny.[38] Later, in his article in Czech on animal cells, Purkyně distinguished between an external cell substance by which each organic part persists physically and chemically and presents itself to the senses and an internal, germinal, life-forming generative part, nearly anticipating Weismann's influential concept of the constancy of the germ plasm.[39]

Purkyně visited the monastery in Brno for the second time in 1850 and undertook an excursion in the surroundings of Brno with Mendel's fellow friar, Matthew Klácel, and discussed problems of Naturphilosophy.[40] (At that time, Mendel was in Brno after his first unsuccessful examination at the university in Vienna.) In 1869, Klácel emigrated to the United States where he earned his living as a journalist, living outside the church and propagating his utopian notions about ideal society.[41] In 1882 he died in seclusion at Belle Plaine, Iowa. In his papers preserved in Prague, there is the manuscript of his lecture on Darwinism in which he wrote about the investigation of the laws of heredity going "down to the very seed of the animal or plant, which give rise to the offspring, even to those infinitesimal dimensions which no microscope has yet reached."[42] Mendel remarked in his autobiography about the importance of the advice of experienced men in his self-education in natural science, and it can be assumed that Klácel was among those who nurtured him.[43] From Purkyně, the prominent physiologist, through Klácel, Mendel had the opportunity to learn new views about generation as well as heredity. The year after Purkyně's second visit, abbot Napp sent Mendel to study natural sciences in Vienna and Mendel learned still more about these problems.

Heredity as a process was distinguished from the enigma of generation and treated separately at the annual meetings of the Sheep Breeders' Society in Brno in the years 1814–1840.[44] At that time, pure naturalists (with the exception of Nestler from the University of Olomouc) did not consider heredity as a separate scientific problem.[45] As Sandler and Sandler have emphasized, until the end of the nineteenth century, Mendel was virtually the only naturalist to carry out exact experiments into heredity as distinct from the enigma of generation.[46] Meanwhile, progress in cytology in the last quarter of the century was stimulating deeper interests of biologists in heredity, which culminated in 1900 in the rediscovery of Mendel's paper.

In 1900 heredity emerged as a separate scientific problem and soon became the starting point for a new science of heredity. The protagonists of the scientific discipline ascribed the basic laws of heredity to Mendel's work. Some of his admirers wished to explain in detail the new theory he had presented in the Pisum paper but, even by that time, biologists' views of hybridization, variation, and heredity were inevitably colored by Darwin's theory of evolution, in which an important challenge was to synthesize the understanding of heredity and evolution. The elucidation of Mendelian inheritance ultimately resulted in the evolutionary synthesis.[47] In this early period, there also appeared very different interpretations of the relation between Mendel's and Darwin's theories, to which attention has been drawn in several monographs on the history of genetics.[48] Sapp has written an insightful review of the manner in which each new generation of geneticists (and historians of genetics) has seen Mendel and the significance of his work primarily in the glow of its own reflected light.[49] Sapp concludes: "Since the emergence of genetics, Mendel has become a cultural resource to assert the truth about not only what it means to be a good scientist, a geneticist, but also what Mendelian genetics implies.... To understand geneticists' reconstructions of Mendel's intentions is to understand the divergent and sometimes conflicting definitions of what Mendelian genetics signifies or connotes."[50] We agree with this analysis and suggest further that it would be foolhardy to expect that the opinions about Mendel and Mendelian inheritance prevailing in the mid-1990s are either more objective or will be more permanent than those of previous generations.

Methodology of the Research

On the occasion of the Mendel Centennial, J. H. Bennett published a letter from R. A. Fisher to one of the editors of *Annals of Science* stressing that "the rediscoverers of 1900 were just as incapable as the non-rediscoverers of 1870 of assimilating from it [Mendel's paper] any ideas other than those which they were already prepared to accept."[51] After studying Mendel's paper in detail, and trying to reconstruct the experiments with Pisum, Fisher wrote in 1936 about "the tale of Mendel's discovery of the laws of inheritance, and of the sensational rediscovery of his work thirty-four years after its publication."[52] Assuming that Mendel had indeed intended to study heredity as a process, Fisher wished to explain the nature of Mendel's discovery and provoked geneticists and historians of science to study Mendel's paper anew to try to answer such questions as "What did Mendel discover?" and "How did he discover it?"

In his critique of Mendel's experiments in 1936, Fisher called attention to "the numerical frequency ratios, in which the laws of inheritance expressed themselves, as a ready method of demonstrating the truth of his factorial system." In discussing the experiments in 1955, Fisher made a similar point: "The simplicity of his plan, and the adequacy of the numbers of the first crosses reported, are indications that he knew in advance very much what he ought to expect."[53] Fisher is most often quoted for his criticism of the excessively good fit of the data in Mendel's paper. Perhaps more important is Fisher's assertion that Mendel carried out his experiments according to the theory developed *a priori*. Behold Gregor Mendel—some suggest that he failed to understand heredity; others that he understood it so well that his experiments were mere demonstrations?

In the 1930s, geneticists were preoccupied in fundamental research relating heredity to evolution and with the misuse of genetics by racial-supremacist ideologs, and Fisher's paper drawing attention to the methodological aspects of Mendel's research remained without an echo. In the 1950s, geneticists in the United States celebrated the fiftieth anniversary of the rediscovery of Mendel's paper as the Golden Jubilee of genetics. According to Julian Huxley, "In the fifty years since Mendel's laws were so dramatically rediscovered, genetics has been transformed from groping incertitude to a rigorous and many-sided discipline, the only branch of biology in which induction and deduction, theory and experiment, observation and comparison have come to interlock, in the sort of way they have for many years done in physics."[54] C. D. Darlington voiced the opinion that the original rediscoverers rediscovered only Mendelian segregation but that "it has taken 50 years to rediscover the determinants which he [Mendel] called elements and which we call genes."[55] *Primus inter parus* in the circle of whig geneticists, Darlington thusly credited Mendel with the concept of the gene, elevating it to the level of "rather the primary law of biology."[56] Only in the mid-1960s was Fisher's critical paper on Mendel brought into the limelight by Zirkle and de Beer.[57]

In 1966, Olby examined Mendel's research in the context of the tradition of plant hybridizers and showed the connection of Mendel's research program with early studies of plant hybridization.[58] In his classic paper in 1979, Olby criticized geneticists for the glorification of Mendel as a discoverer of the gene concept, concluding that Mendel did not clearly understand the paired nature of alleles in homozygous genotypes as evidenced by his use of a single-letter symbol—for example, *A*—for these genotypes instead of the use of a doubled letter—for example, *AA*. Use of the doubled letter is a practice introduced by early twentieth century geneticists.[59] The sociologist A. Brannigan, analyzing the social basis of scientific discovery in general, gave Mendel's research as an example of purely empirical research into plant hybridization at a time when hybridization was considered an experimental model of the process of organic evolution.[60] In this view, Mendel did not investigate heredity as a process in itself, nor did he intend to: His name came to be associated with the Mendelian "laws" of heredity only secondarily because his experiments were relevant; the more important reason was that the discovery was attributed to him by social consensus among the early Mendelians.

Coming to a similar conclusion about Mendel's lack of interest in heredity, Corcos and Monaghan also asserted that Mendel "was not a theorist, but an empirist."[61] In contradiction, in so far as studying inheritance is concerned, Falk and Sarkar expressed the opinion that "Mendel was studying and reporting on inheritance simply because hybridization is conceptually inseparable from inheritance."[62] Defending their own view, Corcos and Monaghan stated that they had scrutinized Mendel's paper carefully and analyzed it "sentence by sentence again and again for the last nine years." Acknowledging the historical context of intensive breeding activity in Moravia in the first half of the nineteenth century, they nevertheless concluded that the breeders in Moravia were not looking for the laws of heredity and "like most of the breeders of plants or animals before and after them, they were not concerned with the theoretical underpinnings of their practice."[63] The supposed lack of interest in genetical theory seems at odds with the publication of the previously mentioned "genetical laws" in 1819 and also with Napp's formulation of the fundamental research question of heredity as: What is inherited? How is it inherited?[64] Nevertheless, in the preface to their book *Gregor Mendel's Experiments on Plant Hybrids,* Corcos and Monaghan agree that "Mendel was a great scientist, a pioneer in integrating ideas across three scientific disciplines—botany, physics, and mathematics. Through this integration he created a mathematically precise quantitative theory of formation of hybrids and the development of their offspring through several generations."[65] It would appear, therefore, that the key issue pertaining to Mendel as the founder of the science of heredity is whether a "mathematically precise quantitative theory of formation of hybrids and the development of their offspring through several generations" is, in fact, a theory of heredity. Corcos and Monaghan also remark that "Mendel was a brilliant experimenter."[66] The integration of mathematics and experiments in Mendel's work was also stressed by I. Sandler and L. Sandler in their comment that a central element in his procedure "was the prior formulation of an hypothesis from which deductions were made and then tested experimentally.[67] From this conclusion, the Sandlers pose the question: "Was it this hypothetico-deductive method, pursued at a time when scientists stressed an empirical-inductive procedure, that was inimical to the reception of his work?"

A different approach to the analysis of scientific discovery was offered by M. Grmek, who set forth a definition of discovery as "a new statement, characterized by a novelty, where novelty is understood as what cannot be deduced but must be constructed within the sphere of logic, or observed in the course of experimental sciences, whether physical or biological."[68] From this viewpoint, Grmek rejected the idea that Mendel's discovery was an example of inductive empiricism because, in Mendel's work, "theorizing and experimenting went hand in hand: the essence of Mendel's discovery was acquired during the carrying out of the experiments." About two decades ago, Kříženecký also sought an explanation of the origin of Mendel's theory and denoted Mendel's main scientific inspiration as a *spiritual mutation*.[69] By this term Kříženecký meant a sudden moment of enlightenment in Mendel's reasoning. According to Grmek, such a moment was also described by C. Bernard as the culmination of a "subconscious maturation process marked by assimilation of new data" based on the observations of the researcher and leading through hypothesis to experiment.[70] Such a conceptual process can take place only within a suitable theoretical framework. It might therefore be profitable to try to define whatever theoretical framework guided Mendel's reasoning that resulted in the sequence of hypotheses and experiments in the Pisum research.

Returning from the University of Vienna in 1853, Mendel seems to have recognized a set of phenomena in plant hybridization and plant breeding for which no satisfactory explanation existed. With his background, gained from his studies of natural history and agriculture in Brno and of natural sciences in Vienna, Mendel was able to design systematic experiments to investigate these phenomena.[71] Indeed, Mendel's words and the flow of experiments in his paper, taken at face value, imply that he created, step by step, an experimental program that was new from the standpoint of methodology and unparalleled among his predecessors as well as contemporaries.[72] The model of discrete, contrasting trait-pairs was the initial theoretical framework that Mendel used for generating a series of specific hypotheses to be tested experimentally.

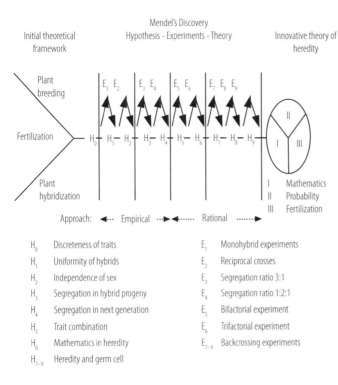

FIG. 1. Scheme of Mendel's experiment (E_i) and successive Hypotheses (H_i) combining the traditions of plant breeding and plant hybridization with mathematics and probability to account for the numerical proportions of the traits observed among offspring in successive generations.

Mendel's reasoning in creating hypotheses and testing their validity in specially conceived and separate experiments is indicated in Figure 1. Although there is some dispute as to whether Mendel actually carried out the experiments in as deliberate and systematic a manner as he described, there is nothing in the paper to suggest otherwise and, in any case, the presentation of the paper itself provides insight into Mendel's frame of mind when he wrote it.

In Figure 1, the idea of discreteness of traits is denoted H_0. Mendel may have arrived at this notion before initiating the experiments described in the paper. The hypotheses H_1 denotes the uniformity of hybrids, and the further hypotheses H_2 through H_9 denote equivalence of reciprocal crosses in producing hybrids, segregation of one pair of contrasting traits in hybrid progeny, segregation of these traits in next generation, segregation of two and three pairs of contrasting traits, and the transmission of trait determinants through the germ cells. Every hypothesis Mendel proved in separate experiments, denoted E_1–E_9 in Figure 1. Based on experiments E_3–E_6, Mendel explained the frequencies of segregating traits in the progeny of hybrids in symbolic terms as mathematical series; in this manner also, at the end of his first lecture presenting the paper (February 8, 1865), he demonstrated the use of mathematics in his research to explain the transmission of traits from generation to generation. In experiments E_7–E_9, Mendel demonstrated that the role of fertilization in heredity was the transmission of determinants of traits through the germ cells. In explaining his results, Mendel applied the theory of probability to the production of germ cells and the transmission of the determinants of heredity. Thus, Mendel's theoretical explanation was a synthesis stemming from three fields of knowledge:

- The application of mathematics to research in heredity.
- The elucidation of the role of fertilization in heredity.
- The application of probability theory to biological processes.

These components of explanation were brought together in imaginative conjectures proved by experiment. Putting together previously unrelated scientific and empirical ideas was the essence of Mendel's discovery. The explanation includes not only what, after 1900, were called the law of segregation and the law of independent assortment of traits, but also the demonstration that the traits of plants are inherited through physical determinants present in the germ cells. As noted earlier, Mendel summarized his discovery in words repeated almost verbatim six times in the text of his short paper: "... pea hybrids form germinal and pollen cells that in their composition correspond in equal numbers to all the constant forms resulting from the combination of traits united through fertilization." We have argued that Mendel himself regarded this statement as the essence of his discovery.[73]

Kalmus, in examining the origin of Mendel's concepts, also emphasized the association of multiple scientific ideas from different fields creating the basis of a new discovery or theory.[74] It may be objected that Mendel did not create his synthesis systematically but hit upon it by chance. Perhaps so, but Pasteur's famous slogan that "chance favors only those who are prepared" puts the situation in perspective. Mendel's preparation for his research while living and working in the creative cultural environment of the Augustinian monastery in Brno under the abbotship of Napp had unique opportunities for studies of culture and science: Th. Bratranek, as a "Naturphilosoph," occupied himself with, among other things, the study of esthetics in the plant kingdom; M. Klácel, also a Naturphilosoph, tried to explain social problems of society in terms of cosmological principles; P. Křížkovský, a composer, tried to express beauty and piety in musical compositions; and A. Keller, administrator of the monastery estates, had experience in agriculture and wished to create new plant varieties by applying artificial fertilization.

Insight into Mendel's creativity (and that of other scientists as well) can also be gleaned from some ideas expressed by the physicist and philosopher K. Popper. In his opening lecture on *Creative Critics in Science and Art*, on the occasion of the music festival in Salzburg in 1979, Popper explained that science and art are both motivated by the desire of human beings to perceive and understand the phenomena of nature and of themselves.[75] According to Popper, the mythos of the artist looks for beauty in art, rhythm in music, tension in drama, and the force of conviction. In the inner process of self-criticism, the mythos drives an artist to literature, music, painting, and sculpture as an expression of the art. The mythos of a scientist directs itself to a rational explanation of natural phenomena—for example, to explain the origin of species or the appearance of life on Earth. Scientific knowledge thus starts with naive ideas that are gradually developed into a scientific problem. The natural scientist strives to create hypotheses that can be tested by experiment. In this process, new theories are either refuted or further developed successively, forming the basis for progress in scientific understanding.

The creative process of Mendel is also illuminated by the much earlier ideas of the Moravian scholar, J. A. Comenius (1592–1670), who lived for some years in the vicinity of Mendel's birthplace. In his masterpiece, entitled *General Consultation on the Improvement of All Things Human*, which was written in Latin and published only in the Czech translation until 1992, Comenius attempted to describe the origin of scientific discovery as a component of his original pansophy. His was one of the first attempts at epistemology. Comenius explained the origin of discovery as follows: "The creator forms his new ideas about his task, first for himself, and for a certain period of time rejoices in it. When he wishes to show his work to another, he sketches it on paper, or he shapes it into other matter. Later he comes to the realization of his work in the final form, allowing all to see it in order to gain the real profit of it."[76] (This passage reminds one of Sapp's more blunt statement in recognition of that fact that most scientific papers are written neither exactly as the experiments were carried out nor with an accurate description of the ever-changing thought processes: "'The scientific paper' is not fraud; it is rhetoric. The structure of the narrative of the scientific paper plays an important persuasive role in science."[77])

Mendel was fully aware of the innovative character of his experimental plan and theoretical explanation. In the introduction of his paper, he states: "That no generally applicable law of the formation and development of hybrids has yet been successfully formulated can hardly astonish anyone who is acquainted with the extent of the task and who can appreciate the difficulties with which experiments of this kind have to contend."[78] In the last sentence of his introduction, Mendel continues: "Whether the plan by which the individual experiments were set up and carried out was adequate to the assigned task should be decided by a benevolent judgment." He welcomed repetition of his experiments by the audience who listened to his lectures as well as readers of his paper. Later, in his second letter to Nägeli on April 18, 1867, he wrote; "I knew that the results I obtained were not easily compatible with our contemporary scientific knowledge, and that under the circumstances publication of one such isolated experiment was doubly dangerous: dangerous for the experimenter and for the cause he represented."[79] Here we can see the self-consciousness of the scholar (as well as Mendel's characteristic modesty) in offering his theory for criticism by others. In his specific cultural environment, the mythos of Mendel motivated a desire to understand the essence of variation in traits of plants and to master the process of creating new varieties of cultivated plants. He looked for a scientific explanation of his results that could give a rational explanation of both the enigma of plant hybridization and the transmission of parental traits to progeny. The first observation—the discreteness of plant traits—became the starting point for the whole sequence of hypotheses and experiments in Figure 1, and the result was a new theory presented in new scientific language.

The proposed explanation of Mendel's reasoning during his experiments contrasts with an alternative view that he was a pure empiricist who used mathematics only to account formally for his results with no deeper theoretical insight or purpose.[80] Our view also contrasts with that of de Beer, according to which Mendel carried out the whole set of experiments guided by a preconceived theory elaborated earlier.[81] We are convinced that Mendel's ideas developed as he implied they did—in the course of the experiments. The Sandlers find this the key to Mendel's methodology: "central to Mendel's procedure was the prior formulation of an hypothesis from which deductions were made and then tested experimentally."[82] In other words, the process of Mendel's discovery arose from the formulation of an initial theoretical framework followed by a sequence of hypotheses that were tested in the experiments ultimately resulting in the construction of a new theory.

Plausibility of the New Theory

When Mendel began his experiments, contemporary naturalists were not much interested in research in heredity. The general view was that generation and heredity were inseparable aspects of the same problem.[83] Wagner, in recommending crossing experiments for elucidating generation and heredity, was one of the few who seemed to make a distinction between the processes.[84] Another exceptional view was expressed in 1864 by the botanist and plant breeder R. Geschwind, who recommended research in hybridization for creating new and more efficient varieties of forest trees; he noted that "the art of producing plant hybrids by means of forced mutual fertilization of plants was established in horticulture a hundred years ago. Hence, it is no wonder that men of science welcome the process of hybridization with open arms, because of the opportunity it provides of examining the mystery of its laws."[85] Mendel, who most probably did not know of Geschwind at the time of his experiments, nevertheless succeeded in finding a theory of plant hybridization that was at the same time a theory of heredity.

In the middle of the nineteenth century, some botanists were also interested in the problem of the transformation of species through hybridization, which inevitably connected plant breeding with heredity. In 1848, writing in the journal Flora, the acknowledged botanist Ch. F. Hornschuh (1793–1850) examined the differences between the seeds of vetch *(Vicia faba)*, pea *(Pisum sativum)*, and lentil *(Errus lens)*. At the time, it was claimed that many forms of vetch derived from forms of peas and lentils. According to Hornschuh, rigorous segregation of forms of flowers and plants was observed in crosses, without the appearance of any intermediate forms. In his view, these types of questions were only just starting to became the subject of research for a handful of scholars, while other scholars considered them almost ridiculous.[86] In the introduction of his lengthy serialized paper *Ueber Ausartung der Pflanzen (On the Degeneration of Plants)*, Hornschuh used 24 lines to quote interesting methodological comments by Ottweiler encouraging the readers to create new theories in botany for the explanation of the enigmas of species degeneration and species transformation under the influence of hybridization and environment. Botanists of the time accepted the constancy of species, but Hornschuh did not consider it to be proved and wrote that "Man würde daher Unrecht thun, Jemand ein Verbrechen daraus zu machen, wenn er den Autoritätsglauben zu erschüttern sucht." ("It would be wrong to accuse somebody of a crime for casting doubt on the authority of faith.")

While studying the monograph of Gärtner on plant hybridization,[87]

Mendel made a marginal note drawing attention to the citation of Hornschuh's paper. The journal *Flora* was delivered to the library of the Agricultural Society in Brno, and Mendel might have read it, in which case he would have found encouragement for his methodological approach to the research.

On the other hand, Mendel did not mention the term "heredity" either in the title of his paper or in the text. He stated his purpose in carrying out the experiments as to discover a "generally applicable law of the formation and development of hybrids." The audience attending Mendel's lectures received them warmly but could not understood the theoretical explanation. After 1867, Mendel's paper was quoted in the scientific literature, but no one interpreted the experiments as related to heredity, and Mendel's theoretical explanation remained incomprehensible. Not until after 1900 did geneticists begin to connect Mendel's name with the fundamental laws of heredity. Many years later, in 1936, Fisher tried to reconstruct the Pisum experiments and came to the conclusion that geneticists would have to study the paper anew in order to understand what Mendel really did discover.

Starting about 1950, geneticists began to write as if Mendel were the discoverer of the modern concept of the gene, which Olby rightly criticized as an unwarranted glorification of Mendel.[88] Historians of science soon began to disparage the interpretation of Mendel as the founder of genetics as the *traditional* or *orthodox* view. P. J. Bowler went so far as to write that "the traditional image of Mendel is a myth created by the early geneticists" who wished "to see Mendel as a pioneer student of their discipline."[89]

A fresh approach in the analysis of the Pisum experiments came from M. Campbell.[90] According to her, Mendel's theory was not accepted by naturalists until after 1900 because his explanation was deemed implausible. Although the experiments confirmed the predictions, Campbell considers confirmation and plausibility as two distinct criteria of acceptability and, of the two, plausibility as the more important. Hence, the formulation of Mendel's research problem should be examined from the standpoint of plausibility. Mendel presented the results of his experiments at meetings of the Natural Science Society in Brno and handed over the manuscript for publication without alteration. The published paper differs in both form and content from publications of his predecessors and contemporaries in plant hybridization. After the introduction and formulation of the problem, Mendel described the choice of experimental plant, the organization of the experimental crosses of plants differing in one trait or more than one trait, and the organization of the experiments on the reproductive cells of hybrids. In the concluding remarks, he offers his theoretical explanation. Although hybridization is explicitly in the title of Mendel's paper, it must be emphasized that, in 1865, the natural science view of hybridization had already changed from what it had been in 1854 when Mendel started his experiments with Pisum. Before the publication of Darwin's theory of evolution in 1859, hybridization was primarily a problem for botanists and plant breeders in the context of observing traits present in parental plants and offspring. Beginning in the 1860s, naturalists gave primary attention to the evolution of species, and the study of hybridization lost its former significance. Later, in 1877, Darwin himself remarked that the subject of hybridization was "one of the greatest obstacles to the general acceptance of the great principle of evolution."[91]

In the introduction to his lecture, Mendel first mentions the experiments of plant breeders who used artificial fertilization in order to obtain new color varieties, and then he calls attention to other hybridization experiments by botanists not related to the problem of heredity. Mendel set himself the task "to follow the development of hybrids in their progeny."[92] The key problem formulated was to discover a "generally applicable law of the formation and development of hybrids," which Mendel considered to be "the one correct way of finally reaching the solution to a question whose significance for the developmental history of organic forms must not be underestimated."[93]

The secretary of the Natural Science Society in Brno, G. Niessl, Professor of Geodesy, was also acknowledged as a botanist interested in plant hybrids. He also spoke with Mendel about his experiments. In Mendel's obituary in 1884, he drew attention to Mendel's "independent and special manner of reasoning."[94] Like everybody else, however, Niessl failed to understand Mendel's explanation of his experimental results. It may be argued that Niessl—together with biologists until the end of the century—regarded Mendel's explanation as implausible. The same may be said of biologists who appeared on the scene much later, as illustrated in the collected *Classic Papers in Genetics* edited by Peters, in which the two final sections of the Pisum paper were omitted from the reprint because, as explained in the introduction, "these paragraphs have little bearing on the principles that Mendel proposed" and because experience with students had indicated that "these pages serve primarily to confuse rather than to clarify."[95] It is in the omitted pages that Mendel explained his idea of the mechanism of transmission of traits in the process of fertilization.

In Brno, apparently only animal and plant breeders used the term "heredity." Mendel delivered his lecture to the Natural Science Society, the members of which were more interested in pure science than breeding practice. Mendel published his lecture six years after the appearance of Darwin's theory of evolution, according to which natural selection operates on heritable variation. Darwin had to admit that the laws governing inheritance were unknown. Later, in trying to explain heredity, Darwin introduced the provisional hypothesis of pangenesis, incorporating Lamarck's view that heredity is influenced by environment.[96] Even though his explanation was ultimately rejected, Darwin's emphasis on heredity stimulated biologists to undertake research in heredity as a new scientific problem distinct from generation.

An argument can be made that, in attempting to explain the origin and development of traits in hybrids, Mendel was also aware that he would be explaining the transmission of traits from parents to offspring in general. Such awareness is evident in the papers of Nägeli, who wrote that "the topic of the origin of hybrids is also very important from the reproductive point view," thus calling attention to the transmission of traits from parents to offspring.[97] Mendel's first letter to Nägeli stressed the importance of "new experiments in which the degree of kinship between the hybrid forms and their parental species are practically determined" (as opposed to estimated by subjective general impression).[98] By "degree of kinship between hybrid forms," Mendel seems to mean the similarity in their traits as determined by heredity. In his second letter to Nägeli, Mendel emphasized that segregated traits were unchanged, and there was "nothing that one of them has either inherited or taken over from the other."[99] This was the only occasion, as far as we know, in which Mendel explicitly used the term "inherited."

The originality of Mendel's thinking can be appreciated by comparison with some of the ideas of Nägeli. In explaining his theory of hybrid formation, Nägeli stated that the hybrid combinations, designated *AB* and *BA*, cannot be identical. In this case, the letters *A* and *B* represent the entirety of the parental plants, not merely the form of expression of some specified trait. In this paper, Nägeli also asserted that nobody had carried out self-fertilization of individual plants in successive generations.[100] In the progeny of hybrids, Nägeli had observed three types of forms, one corresponding to the hybrid plant and the other two similar to the parental forms. Nägeli added that the forms in hybrid progeny have low stability and readily change among themselves. Adhering to the concept of blending inheritance, Nägeli did not believe that the traits of parents might reappear unchanged in the offspring. He failed completely to understand Mendel's idea of unchanging hereditary determinants.

Until the end of the nineteenth century, only Mendel had analyzed quantitatively the appearance of traits in hybrid progeny and subsequent generations and demonstrated that the phenotypic segregation ratio 3:1 in the F_2 generation results from the genotypic ratio 2:1:1.[101] This demonstration was the definitive refutation of the concept of blending inheritance. Designating the traits with capital and small letters, and using combinatorial reasoning, Mendel explained the progeny of hybrids as series using a new kind of symbolic notation rediscovered in 1900 and still in use today. In using symbols in this manner, Mendel went beyond the realm of empiricism and provided a theoretical framework from which the relative proportions of traits observed among the offspring of crosses could be predicted.

According to a contemporary report, Mendel's second lecture was devoted to cell formation, fertilization, seed production in general, and seed production of hybrids. Apparently his explanation was not easy to follow at a time when the very process of fertilization in higher plants was still an enigma. Mendel brought to his research the latest findings of Pringsheim, who explained the fertilization process in algae and other lower plants and who ascribed determinants for the transmission of traits to the germ cells.[102] Mendel's reasoning is demonstrated in experiments with reciprocal backcrossing of peas differing in pod shape, pod color, and flower position, about which he asserted "the results obtained were in full agreement: all combinations possible through union of the different traits appeared when expected and in nearly equal numbers."[103] He also symbolized the behavior of germinal and pollen cells in the fertilization process in accordance with the law of probability, writing the designation for the types of germinal and pollen cells in the form of fractions with pollen cells above the line and germinal cells below. The average course of self-fertilization of hybrids when two differing traits are segregating was presented graphically as follows:

$$\frac{A}{A} + \frac{A}{a} + \frac{a}{A} + \frac{a}{a} = A + 2Aa + a$$

Mendel proved his premises about fertilization indirectly by examining the progeny of backcrosses, which yielded a phenotypic ratio of 1:1 in contrast to the ratio of 3:1 in the progeny of self-fertilized hybrids. In these experiments, Mendel also refuted Schleiden's idea that the pollen tube penetrating into the embryo sac gave rise to the embryo. In fact, Mendel's experiments proved that fertilization in higher plants results from the union of one pollen and one germinal cell.

In the concluding remarks of his paper, Mendel also compared his results with those of his botanical predecessors—Köreuter and Gärtner in particular. Gärtner's monograph described "simple hybrids," in the progeny of which the traits of the parental forms segregated, and "constant hybrids," which propagate themselves without change.[104] Mendel examined the simple hybrid forms known to him from breeding practice. Referring to the opinion of unnamed "famous physiologists," he explained the propagation of higher plants as "the union of one germinal and one pollen cell into one, which is able to develop into an independent organism through incorporation of matter and the formation of new cells." To these germinal and pollen cells Mendel ascribed the determinants of the transmission of traits and added: "this development proceeds in accord with a constant law based on the material composition and arrangement of the elements that attained a viable union in the cell."[105] Beyond that he could say nothing more about the character of these "elements." Many years later, Weismann understood these elements to be physical determinants that function in all hereditary transmission.[106]

In order to explain the difference between variable and constant hybrids, Mendel developed the idea of the "mediating cell," a sort of helper. Referring to constant hybrids, he says: "When a germinal cell is successfully combined with a dissimilar pollen cell, we have to assume that some compromise takes place between those elements of both cells that cause their differences. The resulting mediating cell becomes the basis of the hybrid organism whose development must necessarily proceed in accord with a law different from that for each of the two parental types. If the compromise be considered complete, in the sense that the hybrid embryo is made up of cells of like kind in which the differences are *entirely and permanently mediated,* then a further consequence would be that the hybrid would remain as constant in its progeny as any other stable plant variety. The reproductive cells formed in its ovary and anthers are the same and like the mediating cell from which they derive" (emphasis Mendel's).[107]

In the variable hybrids, Mendel assumed a compromise between the different elements of the germinal and pollen cell "great enough to permit the formation of a cell that became the basis for the hybrid, but this balance between the antagonistic elements is only temporary, and does not extend beyond the lifetime of the hybrid plant."[108] The conclusion was that the different elements of the hybrid form "succeed in escaping from the enforced association as late as at the stage at which the reproductive cells develop."[109]

According to Mendel, the "attempt to relate the important difference in the development of hybrids to *permanent or temporary association* of different cell elements can, of course, be of value only as a hypothesis which, for lack of well-substantiated data, still leaves some latitude" (emphasis Mendel's).[110] Mendel provided a justification for this hypothesis in his findings "that, in Pisum, the behavior of a pair of differing traits in hybrid union is independent of any other differences between the two parental plants and, furthermore, that the hybrid produces as many kinds of germinal and pollen cells as there are possible constant combination forms. The distinguishing traits of the two plants can, after all, be caused only by differences in the composition and grouping of elements existing in dynamic interaction in their primordial cells."[111] Mendel's musings about the nature of the hereditary determinants and whether the determinants are paired has aroused a series of learned discussions, which are briefly examined in the next section.

In applying the theory of probability and experimentally demonstrating his premises, Mendel came to the following conclusion: "Thus experimentation also justifies the assumption that pea hybrids form germinal and pollen cells that in their composition correspond in equal numbers to all the constant forms resulting from the combination of traits united through fertilization."[112] As noted, Mendel repeated this statement, or close variants of it, no fewer than six times in his paper;[113] we regard it as Mendel's best attempt to explain his theory in a single sentence, and we believe that Mendel himself regarded it as the essence of his discovery.[114] For biologists of the 19th century, Mendel's theory was evidently either incomprehensible or implausible.

Translation of Innovative Theory

Textbooks in biology or genetics often attribute to Mendel not only the basic laws of heredity but also the concept that genes come in pairs, one from each parent. In arguing that Mendel actually had no notion of the modern concept of pairs of factors that determine contrasting characters, Olby noted that Mendel's "overriding concern was with the role of hybrids in the genesis of new species."[115] Echoing Olby's view, Monaghan and Corcos assert that "Mendel's research was not concerned with heredity but with the behavior of hybrids" and that the research resulted only in empirical laws governing the formation and development of hybrids.[116] Later, Olby supported his interpretation with the statement that the goal of Mendel's paper "was surely to ascertain whether or not fertile hybrids can give rise to constant forms and thus to the multiplication of species."[117] This interpretation gives more emphasis to the evolutionary aspect of the problem than to heredity.

Bateson, in 1909, was the first to argue that the purpose of Mendel's work was to provide evidence against Darwin's theory of evolution.[118] Bateson's idea has recently been resurrected by Callender, who, in a new analysis of Mendel's theory, stressed that "Mendel was an opponent of the fundamental principle of evolution itself—that is to say, of descent with modification."[119] On the other hand, in the absence of any definitive statement by Mendel himself, who can say whether he was a supporter of Darwin's theory or opposed to it? Fisher may have had the deepest insight into Mendel's attitude toward the theory of evolution by noting that there are only "two allusions to evolution in Mendel's paper," indicating that Mendel did not regard his work as a direct contribution to evolution.[120] Perhaps Mendel was indifferent to the theory of evolution by means of natural selection. In any case, he certainly did not discuss how his theory provided a mechanism of heredity that would allow natural selection to work, in contrast to the concept of blending inheritance in vogue at the time.

Mendel brought to his research the ideas on developmental history (*Entwicklungsgeschichte*) that prevailed in the first half of the 19th century, and we agree with Fisher that, had Mendel "considered that his results were in any degree antagonistic to the theory of selection, it would have been easy for him to say this also." By analogy, we argue that, had Mendel considered that his research was primarily to examine the problem of the constancy of hybrids, it would have been easy for him to say so. In our opinion—although Mendel was constrained to use the terminology common among naturalists of the first half of the 19th century—his objective was to try to explain the transmission of traits as studied through hybridization, and he offered an innovative theory for explaining the essence of variable hybrids.

Another dispute exists as to whether Mendel developed the concept of the unit of heredity. Meijer[121] drew attention to a 1947 paper by Heimans,[122] who averred that the modern gene theory did not originate from Mendel's research but from ideas put forth by de Vries in his book *Intracellular Pangenesis*. Heimans later argued that Mendel concerned himself only with phenotypic differences—that is, with directly observable characters—for which Mendel almost invariably used the term "Merkmal."[123] Still later, Heimans argued that Mendel did not understand the physical nature of the elements determining the transmission of traits and did not conceive that these elements are paired.[124] Olby agreed that the presence of paired hereditary elements had escaped Mendel, but he also concluded that "Mendel was committed to a materialist and determinist explanatory framework, i.e., that the characters of living organisms are determined by material entities in the cell."[125]

In the context of paired hereditary elements, it is instructive to consider other evidence bearing on the level of Mendel's understanding of the nature of his "potentially formative elements" (*bildungsfähige Elemente*) that determine the transmission of traits. One issue is whether Mendel thought of them as physical particles rather than as fluids or emulsions. This issue has been raised by Meijer, among others.[126] In his paper, Mendel asserts only that the development of the fertilized egg "proceeds in accord with a constant law based on the material composition and arrangement of the elements that attained a viable union in the cell."[127] This phrase would seem to imply that Mendel envisaged particles or at least that he was thinking in terms of entities with a material reality. Even the weaker interpretation has been disputed by Kalmus, who dismisses Mendel's statement as "an afterthought" and argues instead that Mendel might have been thinking of elements in terms of scholastic metaphysics, perhaps unconsciously alluding to the Aristotelian concept of the potential.[128] However, Kalmus's opinion seems flatly contradicted by another of Mendel's statements, unselfconsciously materialistic, that "the distinguishing traits of two plants can, after all, be caused only by differences in the composition and grouping of the elements existing in dynamic interaction in their primordial cells."[129]

In a wider sense, we regard it as irrelevant whether Mendel was thinking in terms of particles or fluids. The key point is emphasized in his letter of April 18, 1867, to Nägeli, in which he explains the transmission and segregation of traits and says that the course of development "consists simply in this; that in each generation the two parental traits appear, separated and unchanged, and that there is nothing to indicate that one of them has either inherited or taken anything from the other."[130]

A major argument that Mendel failed to understand that the determinants of hereditary traits are paired is based on his use of the symbols *A* and *a*. Mendel routinely used either *A* or *a* to represent the constant forms whose genotypes would be written in modern symbolism as *AA* or *aa*. Mendel's use of symbols in this manner has been interpreted as meaning that he did not comprehend that the hereditary elements occur in pairs.[131] However, throughout most of his paper, Mendel used the symbols *A* and *a* in quite a different sense than used in modern genetics with reference to genes. In Mendel's usage, "*A*" simply refers to a plant that breeds true for the dominant trait, and "*a*" refers to the true-breeding recessive. Occasionally he uses the same symbols to refer to the hereditary determinants, as in a passage illustrating the transmission of traits in the process of fertilization; in the context of Mendel's discussion, it is quite clear that his expression

$$\frac{A}{A} + \frac{A}{a} + \frac{a}{A} + \frac{a}{a} = A + 2Aa + a$$

summarizes the expected genetic constitutions of the progeny on the left-hand side of the equal sign and gives their physical and breeding characteristics on the right-hand side. There are also other views of Mendel's use of symbols, including that of Kalmus. Drawing upon Jindra's analysis of the logical nature of Mendel's ideas,[132] Kalmus argues that, from the standpoint of the axiom of *idempotency* or *identity*, expressions like $A + A$ and $2Aa$ are superfluous, which Kalmus believes explains Mendel's writing $A + Aa + a$ rather than $AA + 2Aa + aa$.[133] A similar conclusion was reached by Meijer.[134] Before reading too much into Mendel's often inconsistent use of symbols, it is worthwhile bearing in mind that modern Drosophila geneticists routinely use unpaired gene symbols when referring to homozygous recessives—for example, in referring to Drosophila strains, the symbol *al* means the genotype *al/al*, and *cn bw* means the genotype *cn bw/cn bw*.

The issue whether Mendel considered segregation to take place in *AA* and *aa* genotypes, as well as in *Aa*, provides the principal argument that Olby puts forth for concluding that Mendel was no Mendelian.[135] Olby draws attention to Mendel's discussion of segregation in the formation of germ cells in hybrids: "In the formation of these cells all elements present participate in completely free and uniform fashion, and only those that differ separate from each other."[136] What meaning Mendel intended to convey by this passage should be interpreted in the context of his previous sentence. Mendel had noted that no change in the characteristics of variable hybrids can be noticed throughout the entire vegetative period, and he concluded that the "differing elements [in hybrids] succeed in escaping from the enforced association as late as at the stage at which the reproductive cells develop."

The concept of paired heredity elements Olby ascribes to Correns, who rephrased a key passage of the Pisum paper as follows: "In the hybrid, reproductive cells are produced in which the Anlagen for the individual parental characteristics are contained in all possible combinations, *but both Anlagen for the same pair of traits are never combined.*"[137] The emphasized text is not correctly translated into English in the book by Stern and Sherwood. Correns's original German text reads: "Der Bastard bildet Sexualkerne, die in allen möglichen Combinationen die Anlagen für die einzelne Merkmale der Eltern vereinigen, *nur die desselben Merkmalspaares nicht.*"[138] In English, the translation should be: "In the hybrids, reproductive cells are formed in which the Anlagen for the individual parental traits are contained in all possible combinations, but not those of the same trait pair." In this context "the same trait pair" might mean the combination *Aa*.

Strongly influenced by the critical views of Heimans and others, Monaghan and Corcos recently argued that Mendel did not even arrive at the notion of particulate hereditary determinants in spite of his statement that the germinal and pollen cells are endowed with Anlagen.[139] These authors go so far as to suggest a revision of the text of the Pisum paper. Mendel's text reads: "In our experience we find everywhere confirmed that constant progeny can be formed only when germinal cells and fertilizing pollen are alike, both endowed with the potential *(Anlage)* for creating identical individuals as in normal fertilization of pure strains. Therefore, we must consider it inevitable that in a hybrid plant also identical factors are acting together in the production of constant forms."[140]

Monaghan and Corcos recommend a revised text as follows: "In our experience we find everywhere confirmed that constant progeny can be formed only when germinal and pollen cells are alike, as in normal fertilization of pure strains. Therefore we must consider it inevitable that in a hybrid plant also identical germinal and pollen cells are acting in the production of constant forms."[141] Comparing with the original text, Mendel's phrase "both endowed with the potential *(Anlage)* for creating identical individuals" has been deleted, and Mendel's word "factors" has been replaced with "germinal and pollen cells." This is the first attempt we know of in which Mendel's paper has been rewritten to be more in line with the views of contemporary critics! Reasonable people may differ in their views about Mendel's motivation or understanding of his work, but most would agree that judgments must be based on the words in which Mendel chose to express himself.

From internal evidence in Mendel's paper and in his letters to Nägeli, we have argued against the view that Mendel was a purely empirical plant hybridist with little theoretical understanding of the implications of his discovery. There is also another extreme view that Mendel undertook his experiments after he had already elaborated a complete theory and intended only to demonstrate its validity. It cannot easily be argued that Mendel's theory would have been obvious to anyone who performed similar hybridization experiments and tabulated the results, because many plant hybridizers after Mendel did carry out similar experiments without hitting on his theoretical explanation. Serre,[142] emphasizing primarily Mendel's rejection of the concept of blending inheritance, also argued that some prior theoretical ideas are necessary to any scientific research and quoted the philosopher of science Canguilhelm: "'No practice can furnish a theory with data which are theoretically usable and valid, if the theory has not itself first been invented and the conditions of validity according to which the data will be received."[143] This statement is entirely in accord with our interpretation of the origin and essence of Mendel's theory as a sequence of hypotheses followed by experiments to validate them.

Allegation of Data Falsification

There is a widely held view that the excessive goodness of fit between Mendel's data and his theoretical expectations provides conclusive evidence that Mendel deliberately falsified or otherwise tampered with his data. This view has also been fostered by sensationalism in the controversy-driven popular press.

According to B. Norton and E. S. Pearson, the celebrated statistician, R. A. Fisher, had been fascinated with reconstructing Mendel's experiments and analyzing his data even as an undergraduate in 1911.[144] In his early analysis, Fisher concluded that the data fell within the limits of probable error and added "It may have been just luck; or it may have been that the worthy German abbot, in his ignorance of probable error, unconsciously placed doubtful plants on the side which favored his hypothesis." In his 1936 analysis, Fisher reached a different conclusion—that the data were systematically too much in agreement with expectation—but he suggested that perhaps an assistant had adjusted the numerical results to fit the expected segregation ratios.[145] In 1955, Fisher asserted that the data in the Pisum paper had been "systematically sophisticated."[146]

A number of other geneticists have also examined Mendel's data for their goodness of fit to the expectations. One of these is S. Wright, who provided an analysis for Stern to publish along with a reprint of Fisher's paper.[147] Stern writes: "Why Mendel's specific data are too good from a statistical point of view remain unknown, but comments which throw some light on this question have kindly been provided by Professor Wright." Wright's brief paper acknowledges some minor anomalies in the data but concludes: "Taking everything into account, I am confident, however, that there was no deliberate effort at falsificaton."[148]

Additional study of Mendel's data was carried out by the biometrician Weiling who, from 1965 onwards, published a series of papers vouching for the integrity of Mendel's results. Inspired by Weiling's analysis, Pilgrim came to the conclusion that "there is no evidence that Mendel did anything but report his data with impeccable fidelity."[149] Two years later, Pilgrim added that the chi-squared method is inadequate to answer the key question whether the data in Mendel's paper represent a random sample under the null hypothesis.[150] Weiling had also considered this issue and concluded that "Fisher's objection to the supposedly greater-than-chance fit between observed ratios and expected numbers is therefore based on the assumption that Mendel's data are ratios of random binomial samples while they represent in fact net sums of initially larger numbers of individual segregations of a hypergeometrical type in the fertilized seeds diminished by zygote, seedling, growth and/or other losses.... In other words, a too-good agreement between observation and expectation is seen only when the data of several single plant segregations or segregations of individual offspring (or their chi-squared values) are added."[151] Weiling's 1985 summary attributes the excessive goodness of fit to Fisher's assumption of a true binomial distribution of progeny types when, in Weiling's opinion, a hypergeometric distribution would be more appropriate.[152]

The most recent rigorous statistical analysis of Mendel's data is that of Edwards, a student of Fisher.[153] Edwards's analysis gives no support to Weiling's rejection of the assumption of a binomial distribution of progeny types: "the evidence is that the pea is in fact an excellent randomizer."[154] As for the excessive goodness of fit, "the segregations in [Mendel's paper] are in general closer to Mendel's expectation than chance would dictate."[155] However, it must be emphasized that the data from most of the individual experiments in Mendel's paper give no evidence of having been adjusted to fit expectation. The excessive goodness of fit results from the underrepresentation of extreme deviates, as would take place if the counting process had been terminated when the fits seemed satisfactory or if extreme deviates had been discarded as "bad data" (genteelly called "outliers"). With minor reservations, Edwards states his agreement with Dobzhansky's conclusion that "few experimenters are lucky enough to have no mistakes or accidents happen in any of their experiments, and it is only common sense to have such failures discarded. The evident danger is ascribing to mistakes and expunging from the record perfectly authentic experimental results which do not fit one's expectations. Not having been familiar with chi-squares and other statistical tests, Mendel may have, in perfect conscience, thrown out some crosses which he suspected to involve pollen contamination or other accident."[156]

Mendel's data also include what Fisher called his "abominable discovery" that some of the experiments yield an excellent fit to an incorrect expectation. Two series of experiments consisted of progeny tests in which plants with the dominant phenotype were self-fertilized and their progeny examined for segregation to ascertain whether each parent was heterozygous or homozygous. In the first series of experiments (tabulated as entries 30 through 35 in Edwards's Table 2), Mendel explicitly states that he cultivated 10 seeds from each plant.[157] Because $(3/4)^{10}$ of all such progenies from heterozygous parents will not exhibit segregation, the expected ratio of dominant homozygote to heterozygotes is $(2/3)[1 - (3/4)^{10}] : (1/3) + (2/3)(3/4)^{10}$, or 0.63 : 0.37. Among 600 plants tested, therefore, the true expected ratio is 377 : 223. However, Mendel reports 399 : 201, a ratio in much better agreement with 0.67 : 0.33 (that is, 2 : 1) than with 0.63 : 0.37. However, a χ^2 test of the reported ratio against the expected 377 : 223 yields $\chi^2 = 3.3$, for which P > 0.05. The observed result is, therefore, not significantly more deviant from the true expectations than could be expected by chance alone. In other words, this series of progeny tests yields no evidence that the data had been adjusted.

The second series of progeny tests (tabulated as entries 61 through 69 in Edwards's Table 2) is more problematical. In this case—assuming again 10 seeds from each plant—the $\chi^2 = 8.1$, for which P < 0.01. This is the discovery that Fisher regarded as his "abominable discovery"[158]: the reported data differ highly significantly (in statistical parlance) from the true expectation. How could this happen without deliberate fudging? In Fisher's analysis the key assumption is that, in these progeny tests, like those in the previous series, Mendel cultivated exactly 10 seeds from each plant. Mendel does not say this, however. He says: "This experiment was conducted in a manner quite similar to that used in the preceding one."[159] Mendel is usually very precise, and if he had done the experiment in exactly the same manner as before it would have been easy for him to say so. But he hedges by saying that the method was "quite similar." In this case the exact method makes all the difference: if Mendel had cultivated 12 seeds per plant rather than 10, then $\chi^2 = 3.0$, for which P > 0.05 and the insinuation of data tampering evaporates. In fairness to Fisher, he considered this very possibility: "If we could suppose that larger progenies, say fifteen plants, were grown on this occasion, the greater part of the discrepancy would be removed."[160] However, Fisher continues, "the number required is more than Mendel had assigned to any previous experiment...." and so he disregards this as a plausible explanation. But in dismissing this possibility, Fisher is being disingenuous. Mendel's first series of progeny tests examined 600 parents and therefore, at 10 seeds per parent, must have required the cultivation of 6,000 plants. It would therefore be quite plausible for Mendel to have examined 12 seeds per plant in the second series of progeny tests, since the total number of tested parents was 473; the second series would therefore have required the cultivation of 5,676 plants, which is fewer than in the first series. Perhaps what Mendel meant by saying that the method in the second series was "quite similar" to that in the first is that he allocated a certain plot of his garden for the purpose of progeny testing and cultivated as many plants per parent as he could to fill this space. If the space allocation was adequate for 6,000 plants then, in the second series of progeny tests, Mendel could have cultivated an average of 12.7 seeds per parent. He may well have cultivated more than 6,000 plants in the second series because he commented on the amount of work it required, saying that "of all experiments it required the most time and effort."[161] Hence, Fisher's dismissal of the explanation based on more than 10 progeny is too facile, especially in light of Mendel's vague specification of how similar the two experiments actually were and the plausible alternative interpretation of Mendel's text. Fisher's "abominable discovery" is therefore much less damaging than first appears. In short, although Mendel's expectations are certainly wrong, Fisher's expectations may be wrong as well. Thus, the uncertainties in the experiment and the ambiguities in the analysis discredit any inference of deliberate manipulation or falsification of data.

Mendel's data are some of the most extensive and complete "raw data" ever published in genetics. Additional examinations of the data will surely be carried out as new statistical approaches are developed. However, the principal point to be emphasized is that, up to the present time, no reputable statistician has proven, or even alleged, that Mendel knowingly and deliberately adjusted his data in favor of the theoretical expectation. Indeed, it is not clear even to statisticians what sort of statistical analysis is appropriate for a set of experiments as extensive as Mendel's. Edwards, in discussing this fact, emphasizes that "we must not allow our judgment to be dominated by tests of significance and other calculations of probability which are at best pointers for further thought and at worst misleading."[162] Although there have been allegations of falsification in Mendel's data, they have come from the popular press—for example, in Gardner's book, *Great Fakes of Science*, in which the statement is made: "Yes, even Brother Mendel lied."[163] This example also illustrates how easily great scientific achievement can be discredited by dilettantes.

A completely different issue bearing indirectly on Mendel's veracity has to do with his failure to find linkage and the coincidence between his studying seven pairs of traits and Pisum having seven pairs of chromosomes. This issue warrants only brief attention. For a long time after 1900, geneticists thought that the genes determining the traits Mendel investigated must have been distributed evenly, with one gene located in each of the chromosomes. Chance alone would render this coincidence quite unlikely, and so some textbook authors have implied that Mendel probably studied more than seven traits, some of them linked, but chose to discuss only those showing independent assortment. However, Blixt has shown that, among the seven loci that Mendel studied, none are on chromosomes 2 and 3 and that chromosomes 1 and 4 each carry more than one.[164] Soon after, Novitski and Blixt pointed out that the frequency of recombination between Mendel's linked loci would have made linkage difficult for Mendel to detect.[165] In any event, examining the various accusations against Mendel dealing with his experimental data, Piegorsch came to the conclusion that Mendel cannot be accused of data falsification especially with regard to the question of linkage.[166]

Another interpretation of Mendel's experimental approach, and especially Mendel's presentation of experimental data, has recently been revived by a philosopher, Di Trocchio.[167] Despite Mendel's seemingly careful description of his research methods and additional information provided in his letters to Nägeli, Di Trocchio has come to the conclusion that, after preliminary experiments, Mendel began his research program by crossing plants differing, not only in one trait pair as described by Mendel, but differing in at least three trait pairs. According to Di Trocchio, the monohybrid data with seven individual trait pairs reported by Mendel are fictitious in the sense that the numerical data were obtained by progressively disaggregating those from polyhybrid crosses. In other words, Di Trocchio alleges that Mendel reported marginal totals from polyhybrid crosses without saying so. While the allegation is not one of falsification, it is certainly one of misrepresentation inasmuch as Di Trocchio says that Mendel's monohybrid crosses were carried out only on paper. The possibility that Mendel reported marginal totals without saying so was first raised by Fisher in 1936 and dismissed on the grounds that "the style [of Mendel's paper] throughout suggests that he expects to be taken entirely literally; if his facts have suffered much manipulation the style of his report must be judged disingenuous. Consequently, unless real contradictions are encountered in reconstructing his experiments from his paper, regarded as a literal account, this view must be preferred to all alternatives...."[168] A literal account would also be consistent with Mendel's training in physics and with his complaint in his first letter to Nägeli "that this worthy man [Gärtner] did not publish a detailed description of his individual experiments, and that he did not diagnose his hybrid types sufficiently, especially those resulting from like fertilizations."[169] Because there are no serious contradictions in reconstructing the experiments from either the text of Mendel's paper or from his letters to Nägeli, Weiling[170] has concluded that the evaluation of Di Trocchio's interpretation must ultimately be based on judgments about the background of Mendel's work and on considerations of his surroundings and research objectives.

Conclusion

Attention has been drawn to contradictory interpretations of Mendel's theory by the majority of geneticists, on one side, and historians of science, on the other side. Students in either genetics or in the history of science might well be forgiven some confusion when coming across such opposite views in their textbooks.

Over the past few years we have examined several controversies dealing with Mendel and his work. These include the extent to which Mendel was investigating the problem of heredity as opposed to hybridization; to what extent his work should be considered empirical as opposed to theoretical; to what extent Mendel understood that the determinants of heredity are paired; the reasons why contemporary naturalists failed to comprehend or accept his explanation; and why some present-day historians of science and a few geneticists have rejected the conventional interpretation of Mendel's contribution.

We regard Mendel as a plant hybridist of the 19th century who explained his experimental results according to a theory based on the transfer of heredity determinants of traits through the germ cells in the process of fertilization. Even in Mendel's time, the problem of heredity was beginning to emerge as a problem separate from the enigma of generation, primarily among animal and plant breeders but later also among physiologists. We have stressed Mendel's Moravian connection because, in Brno and other parts of Moravia, motivated by practical considerations in breeding, the problem of heredity seems to have been developed more fully earlier than elsewhere. Surprisingly, the cultural context in which Mendel lived and worked has been largely ignored in modern reappraisals of Mendel's contribution. Mendel's research program was marked by innovative experimental design and deep theoretical insights in explaining the results. The emergence of a new scientific problem, the formulation of the problem in the language of naturalists of the first half of the nineteenth century, the innovative research methods of a physicist, and the presentation of the results in a new scientific language—all these conspired to render Mendel's explanation implausible to his contemporaries and resulted in the obscurity of Mendel's work until 1900 when the rediscoverers attributed priority to him.

Special attention is paid to Mendel's notation for traits and the determinants governing their transmission from parents to offspring. We conclude that Mendel is sometimes inconsistent in his use of symbols in describing his experiments and explaining his theory. An issue is whether, without the concept of paired hereditary determinants, Mendel could have discussed his theory and presented his results as he did. This issue seems to be what Pontecorvo had in mind when he noted that Mendel did not distinguish explicitly between factors and characters, and perhaps he did not even have this distinction clearly in his own mind, "yet we have no alternative but to read it in his paper, which would make no sense without it."[171] Our analysis, not only of Mendel's paper and his letters to Nägeli but also of new documents and information brought to light after 1965, supports the view that Mendel based his explanation on paired hereditary elements governing the transmission of traits from parents to offspring.

It must, however, be added that Mendel derived his theory from experiments with Pisum and assumed the validity of the theory only for variable hybrids. Later, Mendel wished to examine the validity of his theory in experiments with other plant species. Mendel's views in this regard have also become the subject of controversy, which will be treated separately.

Interpretations of Mendel's paper and his understanding of it are ever changing. Fisher pointed out that the evolution of opinion is partly a reflection of each generation's own biases and prejudices: "Each generation, perhaps, found in Mendel's paper only what it expected to find; in the first period a repetition of the hybridization results commonly reported, in the second a discovery in inheritance supposedly difficult to reconcile with continuous evolution. Each generation, therefore, ignored what did not confirm its own expectations."[172] To Fisher's two early periods may, perhaps, be added two later ones: the period of Mendel's glorification, which coincided with genetics becoming preeminent among the biological sciences; and the period of Mendel's diminution, reflecting the general iconoclasm and hero-bashing of more recent times.

NOTES

This paper was originally published as Orel, V. and D. L. Hartl. 1994. "Controversies in the Interpretation of Mendel's Discovery." *History and Philosophy of the Life Sciences* 16:423–64. Reprinted with permission of *History and Philosophy of Life Sciences*.

ACKNOWLEDGMENT. We thank Professor Ernst Mayr for his careful reading of the manuscript and his many helpful comments, Professor Gerhard Czihak for his information about the collection of Stern's papers and his kind offer of the book by Weismann from 1892, and Elisabeth Hauschteck-Jungen for help in translating certain key passages from the German. We are also grateful to Professor Federico Di Trocchio for his kind review and suggestions for improvement.

1. G. Mendel, "Experiments in Plant Hybridisation," 78.
2. H. de Vries, 'Das Spaltungsgesetz der Bastarde', *Berichte der Deutschen Botanischen Gesellschaft*, 18 (1900), 83–90; C. Correns, 'G. Mendels Regel über das Verhalten der Nachkommenschat der Bastarde', *Berichte der Deutschen Botanischen Gesellschaft*, 18 (1900), 158–168; E. v. Tschermak, 'Ueber künstliche Kreuzung bei *Pisum sativum*', *Berichte der Deutschen Botanischen Gesellschaft*, 18 (1900), 232–239. Note: English translation of the papers by de Vries and Correns are published in C. Stern and E. R. Sherwood (eds.), *The Origin of Genetics: A Mendel Source Book*, San Francisco: Freeman, 1966, 107–132.
3. R. Olby, 'Mendel No Mendelian?' *History of Science*, 17 (1979), 55–72.
4. A. H. Sturtevant, *A History of Genetics*, New York: Harper and Row, 1965.
5. D. L. Hartl and V. Orel, 'What Did Gregor Mendel Think He Discovered?' *Genetics*, 131 (1992), 245–253.
6. G. Mendel, *100*. Close variants of the statement occur elsewhere in the paper.
7. Both scientists were on friendly terms since the 1930s. Kříženecký's collection of reprints in the Mendelianum in Brno includes nearly all of Stern's papers. Conversely, Stern's collection of papers, preserved in the Institute of Evolutionary Biology and Genetics at the University of Salzburg, includes nearly all of Kříženecký's papers, preserved by G. Czihak, Professor of Genetics in the Institute. The quoted appreciations of Mendel's research were expressed around the time when geneticsts were preparing to commemorate the hundredth anniversary of the presentation of Mendel's paper. In 1962, Stern visited Kříženecký in Brno in an effort to gather information and documents relating to the Mendel Centennial. Kříženecký was a tragic victim of the persecution of geneticists in Mendel's homeland. Before World War II, and shortly after the war, he taught animal breeding and genetics at the Agricultural University as well as genetics at the Faculty of Sciences at the Masaryk University in Brno. Universities in Czechoslovakia were closed during the war. In 1948, after the Communist takeover in Czechoslovakia, genetics was stigmatized as "reactionary science," and Kříženecký was dismissed from his teaching post. In 1958, he was placed under arrest for 18 months as a victim of Lysenkoism. Later, in 1963, after Lysenko was discredited in the USSR, Kříženecký was entrusted with establishing the Mendel Museum in Brno. The sad biographical note on Kříženecký is described in: V. Orel, 'Jaroslav Kříženecký (1896–1964), Tragic Victim of Lysenkoism in Czechoslovakia', *The Quarterly Review of Biology*, 67 (1992), 487–494.

For the Mendel centennial, Kříženecký published papers and books drawing attention to the essence of Mendel's theory, which—according to both Kříženecký and Stern—was not sufficiently understood at that time. The books by Kříženecký are: J. Kříženecký, *Fundamenta Genetica. The Revised Edition of Mendel's Classic Paper with a Collection of 27 Original Papers Published During the Rediscovery Era*, Prague: Academia, 1965; J. Kříženecký, *Gregor Johann Mendel 1822–1884, Texte und Quellen zu seinem Wirken und Leben*, Halle: Leopoldina Akademie, 1965.

8. C. Stern (in Stern and Sherwood, footnote 2), p. v. Stern published a series of papers relating to the Mendel Centennial in addition to the Mendel source book cited in footnote 2.
9. R. C. Olby, 'Constitutional and Hereditary Disorders'. In: W. F. Bynum and R. Por-

ter (eds.) *Companion Encyclopedia to the History of Medicine*, New York: Routledge, 1993: 412–437. The quotation in the text is from p. 421.

10. A. Weismann, 'Ueber die Vererbung'. This is a lecture delivered to the University of Freiburg published in 1883 and republished in: A. Weismann, *Aufsätze über die Vererbung und Verwandte Biologische Frangen*, Jena: Fisher, 1892: 73–121.

11. F. B. Churchill, 'Sex and the Single Organism: Biological Theories on Sexuality in Mid-Nineteenth Century'. *Studies in the History of Biology*, Baltimore: John Hopkins University Press, 1979: 139–177; F. B. Churchill, 'From Heredity Theory to Vererbung. The Transmission Problem'. *ISIS*, 78 (1987), 337–364.

12. E. Mayr, *The Growth of Biological Thought: Diversity, Evolution and Inheritance*, Cambridge, Massachusetts: Harvard University Press, 1982.

13. A. Weismann, 'Die Bedeutung der Sexuellen Fortpflanzung für die Selections-Theorie'. *Verhandlungen der 58. Naturforschender-Versammlung der deutschen Naturforscher* in 1886. Republished in 1892 (footnote 10), p. 361. [English translation in A. Weismann, *Essays Upon Heredity and Kindred Biological Problems*, E. B. Poulton, S. Schönland, A. E. Shipley (eds.) Oxford: Clarendon, 1891.]

14. A. Weismann (footnote 13), p. 361.

15. A. Weismann, (footnote 10), 'Vorwort'.

16. H. Riesler, 'August Weismanns Leben und Wirken nach Dokumenten aus Seinem Nachlass'. In: K. Sandler (ed.) August Weismann (1834–1914) und die Theoretische Biologie des 19 Jahrhunderts Urkunden, Berichte und Analysen, *Freiburger Universitätsblätter*, 24 (1985), 23–42.

17. A. Danailov, 'Zwischen Organismus und Mechanizismus. Der Beitrag August Weismanns zur Theoretischen Biologie'. *Uroboros*, 1 (1991), 201–21.

18. A. Weismann, *Voträge über die Deszendenztheorie*. Jena, Fisher, 1913.

19. H. Spencer, *Principles of Biology*, London: Williams and Nortage, 1864.

20. Ch. Darwin, *The Variation of Animals and Plants under Domestication*, London: Murray, 1868.

21. C. Nägeli, *Mechanisch-Physiologische Theorie der Abstammungslehre*, München and Leipzig, 1884, 273–274.

22. H. de Vries, *Intracellulare Pangenesis*. Jena: Fisher, 1889.

23. O. Meijer, 'The Essence of Mendel's Discovery'. In: V. Orel and A. Matalová (eds.) *Gregor Mendel and the Foundation of Genetics*. Brno: Mendelianum, Brno, 1983: 189–232.

24. R. Leuckart, 'Zeugung'. In R. Eagner (ed.) *Handwörterbuch der Physiologie mit Rücksicht auf physiologische Pathologie*, V. 4, Braunschweig, 1853: 707–1000. Weismann quoted Leuckart's monograph in his work *Die Continuität des Keimplasmas als Grundlage einer Theorie der Vererbung*, published in Jena in 1885 and republished in 1892 (footnote 9), p. 275.

25. Risler (footnote 15).

26. R. Wagner (ed.) *Handwörterbuch der Physiologie mit Rücksicht auf Physiologische Pathologie*, Braunschweig, I–IV, 1853.

27. R. Wagner, 'Nachtrag zu dem vorstehendem Artikel Zeugung' (footnote 26) IV: 1001–1018.

28. R. Wagner, 'Nachtrag zum Nachtrag des Artikels Zeugung' (footnote 26) IV: 1018a–1018d.

29. J. Janko and V. Orel, 'The Cell in Purkyně's Concept of Procreation'. *Folia Mendeliana*, 24–25 (1989/90), 49–57.

30. V. Orel, 'Selection Practice and Theory of Heredity in Moravia before Mendel'. *Folia Mendeliana*, 12 (1977), 179–221.

31. The manifesto deals in generalities but treats heredity as a process distinct from

generation: "a. Animals of healthy and robust constitution are propagated and their characteristics are inherited; b. Traits of ancestors which differ from the traits in their offspring appear again in successive generations; c. Animals which possess the identical and appropriate traits can have offspring with divergent traits. these are varieties, freaks of nature, and unsuitable for propagation, if heredity is the aim; d. The application of inbreeding depends on scrupulous selection of stock animals. Only those animals with strong expression of the required traits do well in heredity." The quotation is from E. Festetics, 'Weitere Erklärung des Herrn Grafen Emmerich Festetics über Inzucht'. *Neuigkeiten und Verhandlungen*, Prague 22 (1819), 169, translated in V. Orel, 'Genetic Laws Published in Brno in 1819'. *Proceedings of the Greenwood Genetics Center,* South Carolina, 8 (1989), 81–82. See also V. Orel, "Empirical Genetic Laws Published in Brno before Mendel Was Born," *The Journal of Heredity* 89 (1998): 79–82.

32. J. K. Nestler, 'Ueber den Einfluss der Zeugung auf die Eigenschaften der Nachommen'. Mittheilungen der k. k. Mährisch-schlesischen Gesellschaft zur Beförderung des Ackerbaues, der Natur- und der Landeskunde in Brünn (1829), 34: 165–169, 35: 273–279, 36: 281–286, 37: 289–300, 38: 300–303, 40: 318–320.

33. V. Orel, J. Janko, and A. Geus, 'The Enigma of Generation in Connection with Heredity in the Teaching of J. E. Purkyně (1787–1869)'. *Folia Mendeliana,* 22 (1987), 7–33.

34. J. M. Ehrenfels, 'Schriftlicher Nachtrag zu den Verhandlungen des Schafzüchter - Versammlung in Brünn in 1836'. *Mittheilungen der k. k. Mährisch-schlesischen Gesellschaft zur Beförderung des Ackerbaues, der Natur- und Landeskunde in Brünn*, 1 (1837), 2–4.

35. E. Bartenstein, Teindl, J. Hirsch, and J. G. Lauer, 'Protokol über die Verhandlungen bei der Schazüchter-Versammlung in Brünn in 1837'. *Mittheilungen der k. k. Mährischschlesischen Gesellschaft zur Beförderung des Ackerbaues, der Natur- und Landeskunde in Brünn*, 29 (1837), 225–231.

36. J. K. Nestler, 'Ueber die Vererbung in der Schafzucht'. Mittheilungen der k. k. Mährisch-schlesischen Gesellschaft zur Beförderung des Ackerbaues, der Natur- und Landeskunde in Brünn, 36 (1837), 281.

37. J. E. Purkinje, 'Erzeugung (generaltio, genesis, procreation)'. *Encyclopädisches Wörterbuch der medizinischen Wissenschaften,* Berlin XI (1834), 515–48; J. Sekerák, 'Source of Purkyně's Idea of Involution and Evolution'. *Folia Mendeliana,* 24/25 (1989/1990), 23–24.

38. V. Orel, J. Janko, and A. Geus (footnote 33).

39. J. E. Purkyně, *'Buňka Zvířecí'* (English: *Animal Cell*). Prague: Riegrův, Slovník Naučný, Prague (1860), 998–1089.

40. J. Kříženecký, 'Wie J. E. Purkyně im Mährischen Karst Philosophierte—Purkyně und Gregor Mendel'. *Folia Mendeliana,* 12 (1987), 69–80.

41. V. Orel, 'Mendel's Elder Friar and Teacher, Matthew Klácel (1808–1882)'. *The Quarterly Review of Biology,* 47 (1972), 435–436.

42. A. Matalová, 'A Monument to F. M. Klácel (1808–1882) in the Vicinity of the Mendel Statue in Brno'. *Folia Mendeliana,* 14 (1979), 251–263.

43. G. Mendel, *Autobiography*, manuscript in Mendelianum (sign. 24), published in original German in the source book by J. Kříženecký (footnote 7), pp. 74–77. English translation published by A. Iltis 'Gregor Mendel's Autobiography'. *The Journal of Heredity*, 45 (1954), 231–234.

44. V. Orel (footnote 30).

45. V. Orel, 'Heredity in the Teaching Programme of Professor J. K. Nestler (783–1841) of Olomouc'. *Acta Universitatis Palackianae Olomoucensis,* 59 (1978), 79–98.

46. I. Sandler and L. Sandler, 'A Conceptual Ambiguity that Contributed to the Neglect of Mendel's Paper, *History and Philosophy of Life Sciences,* 7 (1985), 3–70.

47. J. Huxley, *Evolution: The Modern Synthesis,* London: Allen and Unwin, 1942; E. Mayr and W. B. Provine, *The Evolutionary Synthesis: Perspectives on the Unification of Biology,* Cambridge, Massachusetts: Harvard University Press, 1980.

48. F. A. E. Crew, *The Foundation of Genetics,* Pergamon, Oxford, 1966; L C. Dunn, *A Short History of Genetics,* McGraw-Hill, New York, 1965; A. H. Sturtevant, *A History of Genetics,* Harper and Row, New York, 1965.

49. J. Sapp, 'The Nine Lives of Gregor Mendel'. In: H. E. Le Grand (ed.) *Experimental Inquiries,* Amsterdam, Kluwer, 1990: 137–166.

50. J. Sapp (footnote 49), p. 164.

51. J. H. Bennett (ed.) *Experiments in Plant Hybridisation—Gregor Mendel,* Edinburgh: Oliver-Boyd, 1965. The reference is to the editor's preface.

52. R. A. Fisher, 'Has Mendel's Work Been Rediscovered?', *117.*

53. R. A. Fisher, 'Introductory Notes on Mendel's Paper', (footnote 51), pp. 1–6.

54. J. Huxley, 'Genetics, Evolution and Human Destiny,' In: L. C. Dunn (ed.) *Genetics in the 20th Century—Essays on the Progress of Genetics During the First 50 Years,* New York: McMillan, 1951.

55. C. D. Darlington, 'Mendel and the Determinants', (footnote 54), pp. 315–332.

56. C. D. Darlington, *Genetics and Man,* Harmondwot: Pelican, 1966.

57. C. Zirkle, 'Some Oddities in the Delayed Discovery of Mendelism'. *Journal of Heredity,* 55 (1964), 65–72; G. de Beer, 'Other Men's Shoulders'. *Annals of Science,* 20 (1965), 303–322.

58. R. C. Olby, *Origins of Mendelism,* London: Constable, 1966. Second enlarged edition Chicago: University of Chicago Press, 1985.

59. R. C. Olby (footnote 3).

60. A. Brannigan, *The Social Basis of Scientific Discovery,* Cambridge: Cambridge University Press, 1981.

61. A. F. Corcos and F. F. Monaghan, 'Mendel's Work and Its Rediscovery'. *Plant Science,* 9 (1990), 197–212.

62. R. Falk and S. Sarkar 'The Real Objective of Mendel's Paper: A Response to Monaghan and Corcos'. *Biology and Philosophy,* 6 (1991), 447–451.

63. F. V. Monaghan and A. F. Corcos, "The Real Objective of Mendel's Paper: A Response to Falk and Sarkar's Criticism," *Biology and Philosophy,* 8 (1993), 95–98.

64. V. Orel, "Empirical Genetic Laws" (footnote 31).

65. A. F. Corcos and F. V. Monaghan, *Gregor Mendel's Experiments on Plant Hybrids,* New Brunswick, Rutgers University Press, 1993, p. xvi.

66. A. F. Corcos and F. V. Monaghan (footnote 65), p. xvi.

67. I. Sandler and L. Sandler (footnote 46).

68. M. D. Grmek, 'A Plea for Freeing the History of Scientific Discovery from Myth'. In: M. D. Grmek, R. S. Cohen, and G. Cimino (eds.), *On Scientific Discovery,* Dordrecht: Reidel, 1981: 9–42.

69. J. Kříženecký in Mendel source book (footnote 7), p. 72.

70. Grmek (footnote 68).

71. V. Orel, 'Mendel and New Scientific Ideas at the Vienna University'. *Folia Mendeliana,* 7 (1972), 27–36.

72. We stress the phrase "taken at face value." F. Di Trocchio, in his paper 'Mendel's Experiments: A Reinterpretation', *Journal of the History of Biology,* 24 (1991), 485–519, has proposed that Mendel's experiments were not carried out as reported in the Pisum paper. Di Trocchio proposes that a series of complex polyhybrid crosses were carried out and that Mendel reported the marginal totals for different single traits and pairs of traits as if they had been done as monohybrid and dihybrid crosses.

73. D. L. Hartl and V. Orel (footnote 5).
74. H. Kalmus, 'The Scholastic Origin of Mendel's Concept'. *History of Science*, 21 (1983), 61–83.
75. C. Popper 'Schöpferische Selbstkritik in Wissenschaft und Kunst'. Lecture at the opening of the Music Festival in Salzburg in 1979. A copy of the manuscript was kindly sent to the Mendelianum in Brno.
76. J. A. Komenský (Comenius), Obecná Porada o Nápravě Věci Lidských, (English: *General Consultation on an Improvement of All Things Human*), Prahe: Svoboda, I (1992) p. 374.
77. J. Sapp (footnote 49), p. 162.
78. G. Mendel, *78*.
79. Mendel's letters to C. Nägeli, published by C. Correns in: 'Gregor Mendel's Briefe and Carl Nägeli, 1863–1873'. *Abhandlungen der Mathematisch-Physikalischen Klasse der Königlichen Sächsischen Gesellschaft der Wissenschaften*, Leipzig, 29 (1905), 189–265. An English translation was published in the book by Stern and Sherwood (footnote 2), pp. 56–102, from which Mendel quotations in the present paper are taken.
80. F. V. Monoghan and A. F. Corcos, 'The Real Objective of Mendel's Paper'. *Biology and Philosophy*, 5 (1990), 267–292.
81. G. de Beer, 'Other Men's Shoulders'. *Annals of Science*, 20 (1965), 303–323.
82. I. Sandler and L. Sandler (footnote 46).
83. I. Sandler and L. Sandler (footnote 46).
84. R. Wagner (footnote 27).
85. R. Geschwind, 'Die Hybridation der Forstgehölze'. *Oesterreichische Monatsschrift für Forstwesen* 1864: 399–417; Orel, V. 'History of Plant Hybridization According to Mendel's Contemporary Rudolf Geschwind'. *History and Philosophy of the Life Sciences*, 8 (1986) 251–263.
86. Hornschuh, 'Ueber Ausarten der Pflanzen'. *Flora*, 14 (1848), 17–28, 33–44, 50–64, and 66–86.
87. F. Gärtner, *Versuche und Beobachtunger über die Bastarderzeugung im Pflanzenreiche*, Stuttgart: Herring, 1849.
88. Olby (footnote 3).
89. P. J. Bowler, *The Mendelian Revolution. The Emergence of Hereditarian Concepts in Modern Science and Society*, Baltimore: The Johns Hopkins University Press, 1989.
90. M. Campbell, *A Century Since Mendel*, Adelaide, Illert, 1985.
91. Ch. Darwin, *Die Wirkungen der Kreuz- und Selbst-Befruchtun im Pflanzenreich*, Stuttgart, 1877. Mendel studied this German translation.
92. G. Mendel, *78*.
93. G. Mendel, *79*.
94. V. Orel, 'Response to Mendel's Pisum Experiments in Brno Since 1865'. *Folia Mendeliana*, 8 (1973), 199–211.
95. J. A. Peters (ed.), *Classic Papers in Genetics*, New York: Prentice Hall, 1959.
96. Darwin (footnote 20).
97. C. Nägeli, 'Die Bastardbildung im Pflanzenreiche,' Sitzungsberichte der Königl, bayer, Akademie der Wissenschaften, math. phys. Klasse, München, 2 (1865), 395–443.
98. G. Mendel (footnote 79), p. 57.
99. G. Mendel (footnote 79), p. 63.
100. C. Nägeli, 'Die Theorie der Bastardbildung,' Sitzungsberichte der Königl, bayer, Akademie der Wissenschaften, math. phys. Klasse, München 1 (1866) 93–127.
101. G. Mendel, *89*.

102. N. Pringsheim, 'Über die Befruchtung und Keimung der Algen,' *Preuss, Monatsberichte der Konigl, Preuss, Akademie der Wissenschaften, Berlin* 3 (1855), 1–33.
103. G. Mendel, *100*.
104. F. Gärtner (footnote 87).
105. G. Mendel, *110*.
106. A. Weismann (footnote 18).
107. G. Mendel, *110*.
108. G. Mendel, *111*.
109. G. Mendel, *111*.
110. G. Mendel, *111*.
111. G. Mendel, *111*.
112. G. Mendel, *100*.
113. Footnote 6.
114. D. L. Hartl and V. Orel (footnote 5).
115. R. C. Olby (footnote 3).
116. F. Monaghan and A. Corcos (footnote 63).
117. R. C. Olby, 'Invited Editorial Comment,' to the Paper of F. Weiling, 'Historical Study: Johann Gregor Mendel 1822–1884'. *American Journal of Medical Genetics*, 40 (1991), 1–25.
118. W. Bateson, *Mendel's Principle of Heredity*, Cambridge: Cambridge University Press, 1909.
119. L. A. Callender, 'Gregor Mendel: An Opponent of Descent with Modification,' *History of Science,* 26 (1988), 41–75.
120. R. A. Fisher, *120*.
121. O. Meijer, 'The Essence of Mendel's Discovery'. In: V. Orel and A. Matalová (eds.) *Gregor Mendel and the Foundation of Genetics*, Brno: Mendelianum 1983: 123–178.
122. J. Heimans, *De Elementen der Genetica*, Amsterdam. Cited by Heimans (see footnote 124 below).
123. J. Heimans, 'Hugo de Vries and the Gene Concept'. *The American Naturalist,* 96 (1962), 887. Cited by Heimans (see footnote 124 below).
124. J. Heimans, 'Ein Notitzblatt aus dem Nachlass Gregor Mendels mit Analysen Eines Seiner Krezungsversuche,' *Folia Mendeliana*, 4 (1969), 5–35.
125. R. C. Olby (footnote 3).
126. O. Meijer (footnote 121).
127. G. Mendel, *111*.
128. H. Kalmus (footnote 74).
129. G. Mendel, *111*.
130. G. Mendel (footnote 79), p. 62.
131. R. C. Olby (footnote 3).
132. J. Jindra, 'A Contribution to the Logical Nature of Mendel's Ideas,' *Folia Mendeliana,* 6 (1971), 69–70.
133. H. Kalmus (footnote 74).
134. O. Meijer (footnote 121).
135. R. C. Olby (footnote 3).
136. G. Mendel, *111*.
137. C. Correns in English translation (footnote 2), p. 130.
138. C. Correns (footnote 2), p. 166.
139. F. Monaghan and A. Corcos (footnote 80).
140. G. Mendel, *96*.

141. F. Monaghan and A. Corcos (footnote 80).

142. J. L. Serre, 'Mendel's Rejection of the Concept of Blending Inheritance'. *Fundamenta Scientiae,* Pergamon Press, 2 (1981), 55–66.

143. G. Canguilhem, *Idéologie et Rationalité dans l'Histore de la Vie,* Pais: Vrin, 1977.

144. B. Norton and E. S. Pearson, "A Note on the Background to, and Refereeing of, R. A. Fisher's Paper '[On] the Correlation Between Relatives on the Supposition of Mendelian Inheritance'. *Notes and Records of the Royal Society* 31 (1976), 151–162. Quoted in the paper by A. W. F. Edwards, 'Are Mendel's Results Really Too Close?' *Biological Review,* 61 (1986), 295–312.

145. R. A. Fisher, *134.*

146. R. A. Fisher, 'Marginal Comments on Mendel's Paper,' Bennett (footnote 51), pp. 52–58.

147. C. Stern, 'Foreword,' In: Stern and Sherwood (footnote 2), p. x.

148. S. Wright, 'Mendel's Ratios,' In: Stern and Sherwood (footnote 2), pp. 173–175.

149. I. Pilgrim, 'The Too-Good-to-be-True Paradox and Gregor Mendel,' *The Journal of Heredity,* 75 (1984), 501–502.

150. I. Pilgrim, 'A Solution to the Too-Good-to-be-True Paradox and Gregor Mendel'. *The Journal of Heredity,* 77 (1986), 218–220.

151. F. Weiling, 'Zur Frage der "Überzufällig Grossen Genauigkeit" der Versuche J. G. Mendels' *Mittheilungen der Österreichischen Gesellschaft für die Geschichte der Naturwissenschaften,* 5 (1985), 1–25.

152. F. Weiling, 'Historical Study: Johann Gregor Mendel 1822–1884'. *American Journal of Medical Genetics,* 40 (1990), 1–25.

153. A. W. F. Edwards, 'Are Mendel's Results Really Too Close?' *141.*

154. A. W. F. Edwards, *144.*

155. A. W. F. Edwards, *161.*

156. Th. Dobzhansky, 'Looking back at Mendel's discovery'. *Science,* 156 (1967), 1588–1589.

157. G. Mendel, *88.*

158. A. W. F. Edwards, *142.*

159. G. Mendel, *93* [for "quite similar," see Stem and Sterwood, p. 20].

160. R. A. Fisher, *130.*

161. G. Mendel, *93.*

162. A. W. F. Edwards, *152.*

163. M. Gardner, 'Great Fakes of Science,' *Esquire,* 10 (1977), 88–92.

164. S. Blixt. 'Why Didn't Gregor Mendel Find Linkage?' *Nature,* 256 (1975), 206.

165. E. Novitski and S. Blixt, 'Mendel, Linkage and Synteny,' *Bioscience,* 28 (1979), 34–5.

166. W. W. Piegorsch, 'The Questions of Fit in the Gregor Mendel Controversy,' *History of Science,* 24 (1986), 173–82.

167. F. Di Trocchio, 'Mendel's Experiments: A Reinterpretation', *Journal of the History of Biology,* 24 (1991), 485–519.

168. R. A. Fisher, *122.*

169. G. Mendel (footnote 79), p. 57.

170. F. Weiling, 'Biographie: Johann Gregor Mendel." *Med. Genetik,* 1 (1993), 35–51 and 2 (1993), 208–222.

171. G. Pontecorvo, 'Template and stepwise processes in heredity,' *Proceedings of the Royal Society of London, B,* (1966) 167–169.

172. R. A. Fisher, *139.*

POSTSCRIPT TO CHAPTER 5

■ Amiable Wright and Suspicious Fisher on Mendel's "Personal Equation"

DANIEL L. HARTL

I used to begin the section on transmission genetics in my undergraduate genetics course by telling the students that I never tired of teaching about Mendel's work because no matter how often I went through his experiments I always seemed to gain some new insight that I had not previously understood. Naively, I thought that my experience would motivate students jaded from their encounter with Mendel in high school to appreciate the true depth and profound implications of his discoveries. But what I learned instead was that the students were less impressed by the elegance of Mendel's experiments than by the thought that their college professor was so stupid that even after teaching Mendelism for so many years he still didn't get it.

There are, however, many aspects of Mendel's experiments that are still not fully understood (or in any case not fully understood by me). Among these are important details about exactly how the experiments were carried out. These are not abstruse academic details, but bear on what Sewall Wright called "one of the most disconcerting items that a historian of genetics has to deal with" (Wright 1966, 173). The disconcerting item in question is the alleged excessively good fit of some of Mendel's data to an expected value that R. A. Fisher (1936) claimed to be incorrect. Fisher called this his "abominable discovery" and described his reaction to the discovery as "shocked" (qtd. in Box 1978). The results reported from these experiments constitute the primary evidence in Fisher's allegation that "the data of most, if not all, of the experiments have been falsified so as to agree closely with Mendel's expectations" (Fisher 1936, 000).

It was against this background that I was happy to receive Allan Franklin's invitation to participate in this project reexamining the allegation of data falsification. From my perspective, one of the most insightful and persuasive evaluations of the issue was that of Wright (1966), whose sympathies were clearly with Mendel:

I do not think that Fisher allows enough for the cumulative effect on chi-square of a slight subconscious tendency to favor the expected results in making tallies. Mendel was the first to count segregants at all. It is rather too much to expect that he would be aware of the precautions now known to be necessary for completely objective data.... Checking of counts that one does not like, but not of others, can lead to systematic bias toward agreement. I doubt whether there are many geneticists even now whose data, if extensive, would stand up wholly satisfactorily under the chi-square test. (Wright 1966, 173–74).

Wright emphasizes this point with reference to an experiment reported in Pearl (1940). The experiment was carried out to illustrate Pearl's assertion that "It should be recognized as a general principle, and kept always in mind, in measuring and recording, that every individual has bias or 'personal equation' in his observing and measuring. There is no way to completely eliminate its effects. The most that can be done is to minimize them" (Pearl 1940, 87–88).

In the experiment Pearl recounts, 15 trained observers were asked to examine 532 kernels on exactly the same ear of corn. They were instructed to classify and count the kernels in each of four phenotypic classes (white and starchy, yellow and sweet, yellow and starchy, or white and sweet), in a cross in which the expected proportions on the ear were in the ratio $9:3:3:1$. The standard deviation among the counts in each category was almost the same (averaging about 14), and variation in the counts of the yellow and sweet versus white and starchy classes is shown in figure 1. These classes are expected to be equally frequent, and for each of the observers the number of yellow and sweet kernels is plotted on the left and the number of white and starchy kernels on the right. Among the 15 observers the total counts in these classes ranged from 131 to 195. There are obviously very large differences even among the trained observers, and the results for three of the observers (marked by the asterisks) are highly significantly different from the overall means.

In attributing the overall excessive goodness of fit to Mendel's "personal equation," Wright concludes:

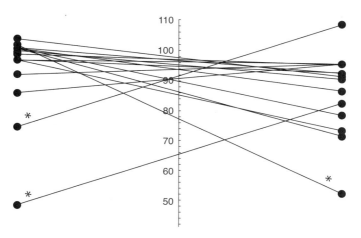

FIG. 1. Variation in counts of yellow and sweet (left) versus white and starchy (right) kernels on the same ear of corn among 15 trained observers. The three marked with asterisks deviate highly significantly from the overall means. Data from Pearl (1940).

In the 91% of the data concerned with one-factor ratios and their testing, the amount of bias toward expectation is slight and on a rough estimate would need less than 2 such misentries in 1000 to reduce chi-square below expectation as much as observed. The more complicated two- and three-factor tests and especially the tests of gametic ratios involve more serious bias, amounting to some 3 percent possibly directed misentries in the last case and about 1.7 percent collectively. In tallying those small experiments, Mendel must have been conscious of how the rows were running, especially where the expected ratio was 1:1:1:1 in the tests of gametic ratios (probability of worse fit is about 0.002 in contrast to 0.05 in the worst of the other cases) and I am afraid that it must be concluded that he made occasional subconscious errors in favor of expectation, especially in this case. Taking everything into account, I am confident, however, that there was no deliberate effort at falsification. (Wright 1966, 175)

Wright also dismissed the possibility of fraud based on Mendel's reporting some very extreme deviates:

Mendel, however, reported the ratios of 43:2 and 14:15 as the extreme F_2's on F_1 plants round × angular, and 32:1 and 20:19 as the extreme F_2's in F_1 yellow × green. These are so difficult to account for among 511 F_1 plants as to suggest some complication, but in any case they would hardly have been reported by one bent on fraud. (Wright 1966, 174)

On the other hand, Wright's account did little to offset Fisher's allegation. It was perhaps too sympathetic to Mendel, and in any case did not refute what Fisher regarded as his primary evidence of fraud, the abominable discovery that some of the data deviate significantly from the correct expectation as derived by Fisher.

The abominable discovery relates to two series of experiments in which Mendel examined the progeny resulting from self-fertilization of individuals with a dominant phenotype in order to ascertain whether each parental plant was heterozygous or homozygous dominant. Mendel clearly expected the ratio to be 2:1 and interprets the results of the first series of experiments as implying: "*The average ratio of 2 to 1 appears, therefore, as fixed with certainty*" (*89*; Mendel's emphasis). In the first series of these experiments, Mendel clearly states that in each set of F_2 progeny, "100 plants were selected which displayed the dominant character in the first generation, and . . . ten seeds of each were cultivated" (*88*). Fisher correctly observed that heterozygous parents will occasionally give rise to 10 progeny, all with the dominant phenotype, by chance alone. In particular, the chance that 10 progeny from a heterozygous parent will all have the dominant phenotype is $(3/4)^{10} = 5.6\%$. These will be misclassified as homozygous dominant parents, and the expected ratio of parents classified as heterozygous to those classified as homozygous is thereby altered from 2:1 to 1.696:1. Mendel reports results of 64:36 for gray-brown seed-coat color, 71:29 for inflated pods, 60:40 for green pods (Mendel evidently did not like this result and repeated it, reporting 65:35 on the retest), 67:33 for axial flowers, and 72:28 for long stem. Fisher's chi-square test for goodness of fit to 1.696:1 for these data yielded a *P* value of 0.069. This is not significant, but Fisher writes: "A deviation as fortunate as Mendel's is to be expected once in twenty-nine trials" (Fisher 1936, *127*).

The figure of once in twenty-nine trials actually corresponds to a *P*-value of 0.034, which is a statistically significant deviation from Fisher's expectation. But as we have seen, the true *P*-value is 0.069, which corresponds to one in 14.5 trials. Fisher's interpretation takes into consideration not only the magnitude of the deviation from 1.696:1 but also its direction.

There are problems with Fisher's interpretation beyond his description of the *P*-value. They arise from the issue of how Mendel might have carried out plantings that resulted in exactly 10 progeny from each of 100 parents in each of six experiments. Fisher (1936) speculated that he must have planted more than 10 in order to have a few in reserve to make up for losses. Novitski (2004) suggested that he planted 10 but classified as heterozygous any parent with one or more homozygous recessive progeny, even if fewer than 10 progeny survived. This counting rule results in an undercounting of homozygous parents, and with the 98% probability of survival that Novitski assumed, the bias is not only opposite in sign to Fisher's correction but also comparable in magnitude. Unfortunately, Novitski seems to have misread Mendel's paper, as the reported probabilities of survival have a 95% confidence interval of 93–95% (Hartl and Fairbanks 2007). With these values, Novitski's counting rule is untenable as it yields expected ratios of parents classified as heterozygous to those classified as homozygous ranging from 2.80:1 to 3.45:1.

Suppose, on the other hand, that Mendel did cultivate extra plants to make sure that he would have 10 survivors. For the likely survivorships, Mendel would have had to plant 14 seeds to have a greater than 99% chance that at least would 10 survive. With this strategy, the average number of survivors would be 13. Are we to believe that he often cultivated more than 10 survivors but disciplined himself to examine only 10? And if he examined more than 10, why didn't he say so?

I think it more likely that Mendel did exactly what he said he did: He cultivated 10 progeny plants. The way any experienced gardener would do this is plant the seeds in the greenhouse and then transplant 10 seedlings to the field, or plant several seeds per hill in the garden and after germination thin the seedlings back to one per hill. This strategy would result in exactly 10 progeny plants per parent for those traits that can be scored only in mature plants. However, two of the traits can be scored directly in seedlings, namely, seed-coat color and stem length (Fairbanks and Rytting 2001). Mendel himself notes that gray-brown seed-coat color is a trait invariably accompanied by reddish axillary pigmentation, and he also notes that the seedlings of dwarf plants with short stems need to be transplanted or will otherwise be overgrown by their taller siblings. Hence, for these two traits in the first series of progeny tests, Mendel almost certainly scored more than 10 plants in the seedling stage. There would have been no need for transplantation or for thinning. For the remaining traits, Mendel reports 263 parental plants classified as heterozygous and 137 classified as homozygous. Compared with Fisher's expectation of 251.6 and 148.4, the P-value is $P = 0.24$. There is not even a hint that the data fit Fisher's expectation too poorly. Indeed, for the two runs of the experiment involving green versus yellow pods. Mendel reports a ratio of 125:75 against Fisher's expectation of 125.8:74.2 ($P = 0.90$).

Three conclusions therefore emerge in regard to the first series of progeny tests. First, there is no statistical reason to assert that Mendel's overall results deviate significantly from Fisher's expectation. Second, two of the traits can be scored in seedlings, seed-coat color (through its proxy axillary pigmentation) and stem length. If Mendel planted enough seeds to ensure at least 10 seedlings, he would have obtained about 13 for each of these traits and would undoubtedly have scored them all. Third, discounting seed-coat color and stem length, the fit to Fisher's expectation is actually very good and in the case of green versus yellow pods almost perfect.

The main statistical support for Fisher's contention of data falsification comes from a second series of experiments in which Mendel tested progeny from a trihybrid cross, which Fisher calls the "trifactorial experiment." Two of the traits were seed shape and seed color, which can be scored directly in the seeds themselves, and gray-brown versus white seed-coat color. Fisher assumed that seed-coat color was scored based on the color of the flowers on mature plants, and interprets Mendel's statement that "this experiment was made in precisely the same way as the previous one" as meaning that exactly 10 progeny plants per tested parental plant were examined. For these experiments Mendel reports a ratio of 321:152, whereas Fisher's expected values are 297.6:175.4, which yields a significant P-value of 0.026.

Fisher is probably wrong in both assumptions. First, it seems likely that seed-coat color was scored in the seedlings through its pleiotropic effect on axillary pigmentation, which implies that more than 10 progeny were scored per parental plant. In fact, any significant discrepancy from Fisher's expectation disappears if Mendel had scored as few as 11 progeny per parent (Hartl and Fairbanks 2007). Second, Mendel's statement that "this experiment was made in precisely the same way as the previous one" does not obviously refer to the first series of progeny tests, but rather seems to refer to the experiment discussed immediately beforehand, which is a bifactorial experiment involving seed shape and seed color. The first series of progeny tests is actually discussed a full two sections earlier in Mendel's paper.

No one can know for sure what went on in Fisher's mind, but I think his suspicion was originally aroused by the paucity of extreme deviates in Mendel's data. Rather than attribute the good fit to an unconscious bias in Mendel's "personal equation," as the always amiable Wright (1966) chose to do, Fisher went looking for the smoking gun of falsification. Motivated by his belief that "Fictitious data can seldom survive a careful scrutiny" (Fisher 1936, *131*), he found what he regarded as suggestive evidence of outright falsification in the first series of progeny tests. Although these results were not significantly different from his own expectation, in the trifactorial experiment he finally found what he may have been looking for—a seemingly clear, statistically significant indication of what he regarded as "the product of some process of sophistication" (*131*).

Fisher was a great admirer of Mendel, however, and graciously attributed the alleged data falsification to "a possibility among others that Mendel was deceived by some assistant who knew too well what was expected" (Fisher 1936, *134*). I am a great admirer of Fisher, and regard his meticulous reconstruction and analysis of Mendel's experiments as a tour de force. But as to the alleged data falsification, there is strong evidence on the other side. Mendel was an expert gardener (Iltis 1932) and a clever experimentalist. In choosing the traits for the trifactorial experiment, he chose two that could be scored in the seeds themselves and one that could be scored in the seedlings. This gave him the opportunity to score all three traits within a single growing season (Fairbanks and Rytting 2001). In assuming that the trait actually scored was flower color, and in possibly misinterpreting Mendel's statement that "this experiment was made in precisely the same way as the previous one" as meaning that exactly 10 progeny plants were examined for each parental plant as in the first series of progeny tests, Fisher may have succumbed to a temptation that befalls many a zealous prosecutor who overlooks exculpatory evidence.

NOTE

Many thanks to Allan Franklin and Daniel J. Fairbanks for their helpful comments on the manuscript.

REFERENCES

Box, J. F. 1978. *R. A. Fisher: The Life of a Scientist*. New York: Wiley.
Fairbanks, D. J., and B. Rytting. 2001. "Mendelian Controversies: A Botanical and Historical Review." *Amer. J. Botany* 88:737–52.
Fisher, R. A. 1936. "Has Mendel's Work Been Rediscovered?" *Annals of Science* 1:115–37.
Hartl, D. L., and D. J. Fairbanks. 2007. "Mud Sticks: On the Alleged Falsification of Mendel's Data." *Genetics* 175:975–79.
Iltis, H. 1932. *Life of Mendel*. New York: Hafner.
Novitski, E. 2004. "On Fisher's Criticism of Mendel's Results with the Garden Pea." *Genetics* 166:1133–36.
Pearl, R. 1940. *Introduction to Medical Biometry and Statistics*. Philadelphia: Saunders.
Wright, S. 1966. Mendel's Ratios. In *The Origins of Genetics*, ed. C. Stern and E. R. Sherwood, 173–75. San Francisco: Freeman.

CHAPTER 6

■ P's in a Pod
Some Recipes for Cooking Mendel's Data

TEDDY SEIDENFELD

1. Introduction

The history, first anonymity and subsequent fame of Gregor Mendel's 1866 research report, "Experiments in Plant-Hybridization," is a familiar story.[1] The tale of neglect until 1900 and then great excitement upon the simultaneous rediscoveries by Carl Correns (in Germany), by Hugo de Vries (in Holland), and by Erich von Tschermak (in Austria), is genetics folklore. Yet, in 1936 R. A. Fisher asked the question, "Has Mendel's Work Been Rediscovered?" What is the point of Fisher's query?

Of course, Fisher was not disputing the authenticity of the document which, in 1900, so stirred the biometrics community.[2] Concerning Fisher's own judgment, there can be no serious question about the importance he vested in Mendelian theory, and the respect he had for Mendel's genius. As a third year undergraduate at Cambridge in 1911, Fisher saw the advantages in reconciling Biometricians (such as K. Pearson and W. Weldon) aligned with Darwinian theory, and the opposing Mendelians (such as W. Bateson). Both camps thought their theories were contrary arguing, e.g., that Mendelian laws are inconsistent with the "continuous evolution" of Darwinism.[3] Fisher's important [1918] paper gives the mathematical details (with due credit to U. Yule) of how the two schools can be joined.[4] In his 1930 book *The Genetical Theory of Natural Selection*, he cemented the bond. Last, in Fisher's [1955] introduction to (a reprinting of Bateson's

1909 version of an English translation of) Mendel's 1866 paper, he praises Mendel for keen methodological insights on central issues facing plant geneticists, without even so much as a hint of the sensational claims raised in his 1936 paper. (Did Fisher grow tired of the controversy he had stirred up twenty years earlier?) Fisher did not write "Has Mendel's Work Been Rediscovered?" either to question Mendel's integrity or to challenge his rightful place among those at the center of modern genetics.

Fisher intended his provocative 1936 paper to chide the 20th century scientific and statistical researcher community for its inability to read with adequate scrutiny.

Mendel's contemporaries may be blamed for failing to recognize his discovery, perhaps through resting too great a confidence on comprehensive compilations. It is equally clear, however, that since 1900, in spite of the immense publicity it has received, his work has not often been examined with sufficient care to prevent its many extraordinary features being overlooked, and the opinions of its author being misrepresented. Each generation, perhaps found in Mendel's paper only what it expected to find; . . . Only a succession of publications, the progressive building up of a *corpus* of scientific work, and the continuous iteration of all new opinions seem sufficient to bring a new discovery into general recognition. (1936, *139*)

The distortions of Mendel's opinions, as Fisher found them, were illustrated by, e.g., Bateson's attempts both to use Mendelian theory against the Darwinians and to represent Mendel as inclined that way too.[5] But what are the "extraordinary" features of Mendel's publication that Fisher uncovers? The fourth section in Fisher's paper, titled "The Nature of Mendel's Discovery," begins with these words,

The reconstruction has been undertaken in order to test the plausibility of the view that Mendel's statements as to the course and procedure of his experimentation are to be taken as an entirely literal account, or whether, on the other hand, there is evidence that data have been assembled from various sources, or the same data rediscussed from different standpoints in different sections of his account. There can, I believe, now be no doubt whatever that his report is to be taken entirely literally, and that his experiments were carried out in just the way and much in the order that they are recounted. The detailed reconstruction of his programme on this assumption leads to no discrepancy whatever. A serious and almost inexplicable discrepancy has, however, appeared, in that in one series of results the numbers observed agree excellently with the two to one ratio, which Mendel himself expected, but differ significantly from what should have been expected had his theory been corrected to allow for the small size of his test progenies. To suppose that Mendel recognized this theoretical complication, and adjusted the frequencies supposedly observed to allow for it, would be to contravene the weight of the evidence supplied in detail by his paper as a whole. Although no explanation can be expected to be satisfactory, it remains a possibility among others that Mendel was deceived by some assistant who knew too well what was expected. This possibility is supported by independent evidence that the data of most, if not all, of the experiments have been falsified so as to agree closely with Mendel's expectations. (134)

What is the basis for Fisher's startling proposal that "most, if not all of" Mendel's data were cooked?[6] "Has Mendel's Work Been Rediscovered?" is a tour de force of statistical analysis using χ^2 Goodness-of-Fit tests. According to Fisher, the two claims pertaining to "sophistication" of data are these:

(i) Mendel conducted tests to separate hybrid from pure-bred plants. Concerning five plant traits (not to be confused with seed traits), plants were to be classified as either pure-dominant or hybrid in a single trait for which they showed the dominant feature. Mendel tested each such plant by looking at 10 of its offspring to see whether the recessive trait was displayed in any. A plant was declared to be pure-bred in case all its offspring showed the dominant trait. This procedure yields about a 5.6% ($\approx .75^{10}$) misclassification of hybrids as pure-breds, against which Mendel's reported observed frequencies are statistically significant. Moreover, in these data, the departures of the observed frequencies in Mendel's data from their "corrected" expectations are all (suspiciously) in the direction of the "uncorrected" Mendelian theory, where a 2:1 hybrid-to-dominant ratio is expected.

(ii) Apart from the misclassification of hybrids, a thorough examination of the entirety of Mendel's published results show an exceptional agreement between observed and expected frequencies which cannot be attributed to luck. The P-value from the composite χ^2 is in excess of .999. In short, Mendel's data are too good to be true! (The χ^2 analysis for these two assertions are conveniently summarized by Fisher's Tables III and V, reproduced here in the Appendix.)

Since 1936, there have been numerous attempts at explaining away the anomalous features in Mendel's evidence that were spotlighted so intensely by Fisher. The well-known statistician A. W. F. Edwards (1986) provides a thorough review of much of the literature. He gives careful reconsideration to Fisher's analysis. Edwards concludes, after adding his own analysis of the distribution of the individual (1 df) χ^2 values, in agreement with Fisher's two central findings (above), that outcomes discrepant with Mendelian expectations have been trimmed. Even current scientifically informed popular literature accepts Fisher's judgment on the point.[7] On the other side of the debate, the botanist F. Weiling (1989) continues his rebuttal (spanning 25 years) to Fisher's criticisms, to the effect that Mendel's data are not cooked. And research botanists continue to cite Mendel's data in prominent publications without hesitation, e.g., Gauch (1993).

My purpose here is several-fold. There are statistical questions to be asked about the logic of "too good to be true" analysis. I want, also, to reexamine some of Fisher's criticisms—to ascertain whether a useful reply to the charge that the data were "sophisticated" can be founded on scientific (genetic) and statistical considerations. After all, it is in the spirit of Fisher's 1936 charge to the reader to reexamine Mendel's original work. As he wrote about it a generation later, "A first-hand study is always instructive, and often, as in this case, full of surprises" (1955, p. 6). I will be satisfied if this essay prompts some to enjoy the pleasure of re-reading both Mendel and Fisher.

2. A Brief Outline of Mendel's Paper

Before discussing Fisher's objections, allow me to sketch the experimental plan of the peas studies. I follow Mendel's account of his attempt to ascertain laws for hybrids.[8]

Mendel's paper is arranged in 11 unnumbered sections. The different trials discussed in these sections are not dated, though some temporal sequences are explicit. In general, it seems that not all trials reported in given section of the paper were conducted in the same year. For example, even when the same Mendelian law is studied for each of the seven pea characteristics, the reader should not assume these were simultaneous trials.[9]

Foremost in the selection of experimental plants were, according to Mendel, two general concerns. As stated in section 2 of his paper:

(1) The plants should possess constant differentiating forms.
(2) There should be little risk of accidental fertilization by foreign pollen.

The second point is obvious. A reason Mendel chose the garden pea for study, then, is because:

a disturbance through foreign pollen cannot easily occur, since the fertilizing organs are closely packed inside the keel and the anther bursts within the bud, so that the stigma becomes covered with pollen even before the flower opens. (*80*)

and:

Among more than 10,000 plants which were carefully examined there were only a very few cases where an indubitable false impregnation had occurred. (*83*)

Regarding the first point, Mendel notes as well that,

In order to discover the relations in which the hybrid forms stand toward each other and also toward their progenitors it appears to be necessary that all members of the series developed in each successive generation be, *without exception*, subjected to observation. (*79–80*)

The significance of this methodological claim is hard to overemphasize, as we shall shortly see!

Beginning with some 34 varieties of peas (*Pisum sativum*), after two-years' trial, Mendel selected 22 for cultivation throughout the years of his experiments, which ran approximately from 1856 through 1864. According to the presentation in the third section of his paper, he used the following seven pea-plant characteristics for intensive study: (the dominant form appears first in each pair)

1. whether ripe seeds are round or wrinkled;
2. whether the ripe seed cotyledon is yellow-orange or green;
3. whether the seed-coat is grey-brown or white—associated with corresponding blossom colors (violet-red or white);
4. whether the ripe pod form is simply inflated or deeply constricted between the seeds;
5. whether the unripe pods are green or bright yellow—with corresponding coloration of the leaf stems;
6. whether flowers distribute along the stem or are terminally bunched at the top;

and

7. whether main stem is (about 6 feet) tall or (about 1 foot) short.

These seven pairs have the feature that the dominant form is almost entirely so in hybrids, as opposed to other characteristics noted by Mendel, e.g., where the hybrid flowering times stand midway between the seed and pollen parents flowering times (*95*).

Of the seven traits, the first two belong to the seed, i.e., are of the next generation, whereas features #3–#7 belong to the (maternal) seed-plant. This division is important for understanding Mendel's experimental design, since his garden averaged about 30 seeds/plant. Thus, for seed shape and seed color, the first two traits, Mendel had sample sizes many times larger than what he could create for the remaining five characteristics, and he had those data in-hand one growing season sooner! It hardly needs saying that, therefore, Mendel had well-grounded expectations for his experiments on single (and even double) factor trials involving the 5 plant characteristics, since he had seen the parallel results (at much large sample sizes) for the 2 pea characteristics a year earlier.

The pea experiments focused on three different aspects of what, today, is called Mendelian theory:

(a) First, Mendel discusses experiments designed to show laws for the heritability of single-factor traits through successive generations bred from hybrids. Each of the seven characteristics (above) was separately studied through several generations. The results appear in Mendel's sections 4–7 *(83–90)*. The grand conclusion of these trials is Mendel's "1:2:1" law for hybrids: their offspring are produced (independently) with probabilities of 1/4, 1/2, and 1/4 (respectively) for recessive, hybrid, and pure-dominant form.

(b) Section 8 *(90–95)* addresses two- and three-factor distributions. The two-factor trials examine laws for the joint heritability of the two seed-characteristics (shape and color). The three-factor trials looked at the simultaneous distribution of seed shape, seed color, and seed-coat color. These experiments led to the conclusion that the traits are distributed independently, each following the "1:2:1" law.[10]

(c) Last among the trials dealing with peas, Mendel's ninth section offers evidence that the genetics for the offspring of hybrids do not depend upon which gamete originates with the pollen and which with the seed parent plant. The laws for offspring are the same when the genetic role of pollen and seed are exchanged. Thus, the ninth section justifies the composite data reported in sections 3–8, where Mendel intentionally grouped data from offspring of hybrids, regardless whether the hybrid was the result of fertilizing a "recessive" seed-plant with "dominant" pollen or the other way around. (See, especially, his remarks on page 84.)

Separately, section 10 of "Experiments in Plant Hybridization" notes some preliminary research on other species, and includes the fertile speculation that some observed plant features are the consequence of numerous "Mendelian" factors *(105)*. Last, there is a very carefully worded concluding section 11, in which Mendel contrasts his findings with some (then) recent work of others investigating hybrids.

3. On the Supposed Error of Misclassification in Mendel's Test of the "2:1" Law

The first of Fisher's two objections rests on the penetrating analysis that using 10 offspring to separate pure-bred dominant plants from hybrids leads to a misclassification of about 5.6% of the hybrids. That is, the probability that all 10 offspring of a hybrid will show the dominant trait is $.75^{10}$, or slightly more than .056. Of course, all the pure breds are correctly classified this way. Thus, if the ratio of hybrid to pure-dominant plants in a given collection is 2:1, Mendel's procedure has an expectation of separating them in the ratio 1.8874:1.1126, rather than 2:1. Where does this affect Mendel's results? According to Fisher, the problem occurs in two settings.

The first occurrence of the misclassification problem is in section 6 (88–89), where Mendel conducts trials to complete his single factor studies of the "1:2:1" law. For each of the seven characteristics, hybrid plants (the "F_1" generation) were created by artificial fertilization of two pure-breeds (as described in section 3), a cross of a pure-dominant with a recessive. The F_1 hybrids produced offspring (the "F_2" generation) by self-fertilization. These F_2 children of the F_1 hybrids displayed the "3:1" law (of dominant to recessive observed types), as described in section 4 of Mendel's paper. The further division of the "3"-group into a "2:1" ratio of hybrids to pure-dominant plants is the locus of Fisher's concern about misclassification.

χ^2s for Mendel's summary of these data are given in a Table appearing on page 133. Regarding the two seed characteristics, form and color (experiments #1 and #2), there is no issue of misclassification since we may assume that each F_2 plant produced about 30 seeds. Hence, for these two trials, with a total of about 1000 F_2 plants to be classified, the (expected) error is less than one plant. That is, the probability of mistaking a hybrid for a pure-bred ($p = .75^{30} < 2 \times 10^{-4}$) is negligible.

Concerning each of the remaining five plant characteristics, Mendel selected 100 F_2 plants to be sorted and cultivated 10 seeds from each.[11] Presumably, an F_2 plant was declared pure-dominant (rather than hybrid) if all its F_3 offspring showed the dominant trait for the characteristic tested. One plant feature (pod color) was tested twice. Thus, there was a total of 600 plants to be sorted between the two kinds: pure-dominant and hybrid (with a total of about 6000 offspring to observe).[12] Of the 600 F_2 plants, 399 were classified as hybrid and 201 were classified as pure-dominant. The χ^2 (1 degree of freedom) for these observed frequencies against the Mendelian "2:1" law (with expectations of 400 hybrid and 200 dominant) is 7.5×10^{-3}; or a P-value of about .93. However, if we follow Fisher's lead and alter the expectations to reflect the 5.6% rate of misclassifying hybrids, the "corrected" χ^2 (1 df) is about 3.3, or a P-value of about .07. That is, with Fisher's correction, the probability of an F_2 plant being classified hybrid is only about .629 rather than .667, yet the data show a frequency of .668.

A similar problem arises in the trifactorial study where plants were to be distinguished as hybrid or pure-dominant for the plant-characteristic of seed-coat color.[13] The "uncorrected" and "corrected" expectations are given on the second row of Fisher's Table III. Combining the two sets of outcomes from the "2:1" and trifactorial experiments, Fisher notes that the "corrected" χ^2 (1 df) is about 8.05; or a P-value of about 4.5×10^{-3}. (The "uncorrected" χ^2 (1 df) is about .093; or a P-value of about .76.) If we calculate the probability of a departure from "corrected" expectations in the direction actually observed (towards the "2:1" law), the P-value is halved. As Fisher writes,

A total deviation of the magnitude observed, and in the right direction, is only to be expected once in 444 trials; there is therefore here a serious discrepancy. (*130*)

What are the possible replies to this analysis? Fisher examines several.

(1) Might it be that the test plants were not a random sample? In selecting the 600 (= 6 × 100) F_2 plants from a population of about 4600, was Mendel disposed to favor hybrids?[14] For example, in the "2:1" monofactorial study, there were more than 900 F_2 plants to choose from in experiment #3 (on seed-coat color), almost 1200 F_2 plants were available from experiment #4, 580 F_2 plants available for experiment #5, about 850 in experiment #6, and over 1050 in experiment #7. A selection bias in favor of hybrids might have increased the proportion of hybrids among the 100 F_2 test-plants chosen for each experiment.

Unfortunately, this rebuttal will not do. Fisher indicates three reasons why, two of which are cogent:

(i) It does not apply to the trifactorial study, where all plants were classified.

(ii) It is implausible that the bias was equally effective for all five characteristics—at best, selection was the result of some gross plant features typical of hybrids, e.g., plant size.

(iii) And (with circular reasoning), if the data were altered, the observed difference between "uncorrected" and "corrected" expectations agrees with the expected number of misclassifications using precisely 10, rather than 9 or 11 F_3 children to sort the F_2 parents.[15]

Like Fisher, I find the coincidences of perfectly offsetting selection biases (the second rebuttal point) more difficult to believe even than the alternative that the data were "cooked"!

(2) Fisher considers, also, the proposal that Mendel might have grown, say, 15 progeny (rather than 10) per F_2 test plant. This would eliminate the

bulk of the misclassification, but it involves a denial of Mendel's own account of his method for the monofactorial study.

(3) Last, since Mendel was aware that low probability events might not appear in limited samples (see, e.g., his remarks on p. *113*, concerning an outcome with 1/8 chance that might fail to be observed in a small sample), why didn't he create the 10 F_3 plants from a backcross; crossing each test F_2 plant against a plant recessive for the trait in question? That would have reduced the misclassification rate for hybrids to less than 2×10^{-4}. The answer is that, unfortunately, the effort needed for 600 artificial fertilizations makes the design infeasible.[16]

What is left to say in Mendel's defense?[17] I propose the following. Mendel did not base his classification of F_2 plants solely on the observations of 10 F_3 children. He carried his experiments to subsequent generations bred from the same hybrids. For example, as Mendel's data support, suppose about 90% of seeds germinate. Assume 10 F_3-seeds per F_2-test-plant were sown. (This is a conservative interpretation of Mendel's words in section 6, 88.[18]) Consider the design that whenever all (that is, on average about 9) F_3 offspring showed the dominant trait then, in the following year, an additional 3 F_4 plants were grown from the seeds of each of the phenotypic dominant F_3 plants. In this sequential procedure, the probability of misclassifying an F_2 hybrid is *less* than 2×10^{-3}. If only 2 F_4 plants were grown, the error in classifying an F_2 hybrid is small, about 5×10^{-3}. Even with only one F_4 observation per F_3 plant, if all 10 F_3 plants were successfully cultivated, the probability of misclassifying an F_2 hybrid still is *less* than .01. Thus, one way around Fisher's first objection is to hypothesize that Mendel used an elementary sequential design.

What is the basis for this speculation? In section 7, titled "The subsequent generations [bred] from the hybrids," coming only one page after describing the data for the "2:1" law, Mendel begins,

The proportions in which the descendants of the hybrids develop and split up in the first and second generations presumably hold good for all subsequent progeny. Experiments 1 and 2 have already been carried through six generations, 3 and 7 through five, and 4, 5, and 6 through four, these experiments being continued from the third generation with a small number of plants, and no departure from the rule has been perceptible. The offspring of the hybrids separated in each generation in ratio of 2:1:1 into hybrids and constant forms. (*89*)

Unfortunately, it is unclear what Mendel means here by a "small number of plants." Might it be that he carried on in the F_4 generation only with a few of the F_3 descendants of the (then) already confirmed F_2-hybrids? Then Fisher's objection is left intact. Or, did Mendel mean that he carried on with some of both kinds, hybrids and pure-breds? Then there is no misclassification bias to correct for.

In any event, misclassifications of the F_2 plants can easily be avoided using as few as 10 F_4 observations whenever all (10) F_3 children of a test plant show the dominant character. That is, by growing ten plants in the F_4 generation, one per suspect F_3 plant, the misclassification rate for F_2 hybrids drops from .056 (reported by Fisher) to less than .01. Thus, according to Fisher's correction to Mendel's design—using 10 F_3 plants per F_2 test plant, or 1000 F_3 plants per experiment—among each 100 F_2's (with probability more than .975) fewer than 47 F_2 plants/experiment would remain unclassified after examining the F_3 generation. That would entail an addition (to whatever else Mendel planned) of no more than 470 plants/experiment in the F_4 generation, or about half the space allocated for the F_3 plants. This sequential design was feasible given the size of Mendel's garden.[19] Mendel is explicit in his paper that he persisted several generations with these experiments. So, a sequential plan to eliminate misclassification of the F_2 hybrids seems to me to be a better reply than Fisher's alternative involving wholesale "sophistication" of the data.

There is another issue which affects Fisher's concern about misclassification, having to do with the assumption that each grouping of 10 F_3 plants from a hybrid F_2 parent constitutes an *independent* sample of 10 (with binomial parameter .75) of displaying the phenotypic dominant trait. This question surfaces, also, in connection with Fisher's assertion that Mendel's data, generally, are too good to be true—to which I turn next.

4. Data Too Good to Be True

4.1 General Remarks

It is a daunting task to address all aspects of Fisher's second challenge—the charge that the balance of Mendel's data conform too well to ("uncorrected") Mendelian expectations. One difficulty I find in responding to Fisher's argument, in-kind, is the absence of an adequate theory of (Fisherian) significance testing. Particularly troubling is the absence of Fisherian theory regulating the alternatives to the null hypothesis. Nonetheless, the central idea behind Fisher's second objection is simple enough to state.

If, for example, one were to flip a supposedly "fair" coin 500 times and observe 250 heads, the coincidence of getting exactly the expected numbers of heads and tails might be noted and then attributed to "luck." However, if in repeated experiments the observed relative frequency of heads distributed about .5 with *much* smaller variance than expected in binomial sampling, that would *seem to be* a good reason to reject the (i.i.d.) Binomial model.

Fisher uses χ^2 tests to quantify this informal reasoning. But even the naive reasoning in the simple case has obvious difficulties. For example, if the data are n flips of the supposedly "fair" coin, how many different "experiments" do they comprise? Does each partition of n constitute a sequence of "experiments" for assessing the Binomial variance?[20]

One point of view allows arbitrary redescriptions of a data set as part of a freewheeling exploration for "significant" outcomes. H. Cramér (1946, §30.2) constructs three tests of seed-data from ten plants that Mendel reports in an illustration of sample variance in his Experiment #2. (Recall, in experiment #2, 258 plant hybrids self-fertilize. The proportions of yellow to green seeds they produce reflect the "3:1" distribution of dominant to recessive phenotypes in the next generation.) Cramér tests the Mendelian "3:1" hypothesis by constructing an array of three χ^2's as follows:

(1) From the ten plants, combined, 355 of 478 seeds were yellow. On the "null" hypothesis of "3:1," that is a χ^2 (1 df) of .137, and a corresponding P-value of about .73.

(2) Take each plant as providing an autonomous experiment, as captured in Cramér's Table 30.21. The sum of the ten χ^2's (each on 1 df) is a χ^2 (10 df) of 7.191, with a corresponding P-value of about .71.

(3) Also, as a test whether the 10 χ^2-values are distributed as i.i.d. χ^2 (1 df) variates, Cramér forms a trinomial partition of χ^2-values, with cell-probabilities of .3, .4, and .3 given the null hypothesis. The 10 χ^2-values of Table 30.2.1 yield cell counts of 2, 6, and 2, respectively. Cramér states the agreement of these data with the trinomial distribution "must be regarded as good."

Given Fisher's (1936) analysis, what I find surprising here is not Cramér's support for Mendel's conclusions, but Cramér's methodology.[21] He writes,

Thus all our tests imply that the data of Table 30.2.1 are consistent with the 3:1 hypothesis. If either test had disclosed a significant deviation, we should have had to reject the hypothesis, at least until further experience had made it plausible that the deviation was due to random fluctuation. (p. 423)

However, repartitioning the evidence to hunt with various χ^2 tests for significant P-values makes little sense without critical control over the space of alternatives to the null hypothesis. All that a P-value provides in this context is an index of rarity, one measure of the data's so-called "discrepancy" with the null hypothesis. But, each outcome of an experiment can be made "rare" when suitably redescribed using one of many (nonequivalent) test-statistics.

Here is a completely trivial example: Index the discrepancy of an outcome inversely to its probability of occurrence under the hypothesis.[22] On n independent flips of a fair coin, the exact sequence of heads and tails observed has a probability of 2^{-n}. Unless the same discrepancy index (the improbability of the sequence) establishes that data are *not* "rare" under a serious rival to the null hypothesis, what is the point of the small P-value for sequence observed, and the large P = value (of 1) for the set of outcomes at least as discrepant as the one observed?

Consider, for example, Fisher's use of χ^2-tests to evaluate Mendel's data about the "2:1" law—the ratio of heterozygous to homozygous dominant offspring among the "3-group" in the "3:1" law, discussed in the previous section. Fisher's Table III (*130*) reports the combined data, pooled from six (of the eight) experiments Mendel conducted on this subject. In these six experiments (of 100 trials each) 399 out of a total of 600 plants had the (relevant) dominant trait (χ^2 = .0075, P = .93 on 1 df). And in Table V (*133*), Fisher uses the sum of these six 1 df χ^2's ($\Sigma\chi^2$ = 4.58, P ≈ .60, on 6 df) and the sum of the remaining two 1 df χ^2's ($\Sigma\chi^2$ = .5983, P ≈ .74, on 2 df) as part of an analysis of all of Mendel's data. He offers the P-value only for the sum of these eight 1 df χ^2's ($\Sigma\chi^2$ = 5.1733, P = .74, on 8 df). In his Table V grouping of Mendel's data, by summing χ^2 ($\Sigma\chi^2$ = 41.6056, 84 df), Fisher arrives at a truly exceptional P-value, P > .9999, data clustering too closely about their expected Mendelian values to be believed!

However, in the spirit of Cramér's analysis, using iterations by examining distributions of distributions of χ^2 or transformations of P-values with χ^2 having other degrees of freedom[23], there exist unlimited varieties of possible meta-analyses of Mendel's data. We are threatened making P-soup! The different meta-analyses offer very different perspectives on the data, as we see next.

The two figures in the Appendix, Figures N1 and N2 (Nobile, 1992) graph the asymptotic joint distribution (n→∞), under a Binomial (θ = 1/3) null hypothesis of P-values for χ^2s with six samples of size 100n, calculated either by:

(1) summing six independent (1 df) χ^2s (P_U), to yield a χ^2 (6 df)
or by
(2) pooling the data first to create a single (1 df) χ^2 (P_Z).

This corresponds to the case, discussed before, of Mendel's six experiments for plant characteristic data in the "2:1" law. Observe, especially, how much of the joint distribution is off the diagonal line, where $P_U = P_Z$, especially near the large ("too good to be true") P-values.[24]

Evidently, the unstated alternative model to the null hypothesis that is used with a traditional test of the simple null hypothesis (i.e., reject $\theta = ⅓$ when the P-value is low) is not the same in the one kind of meta-analysis as in the other. For example, the simple binomial model (varying θ) works fine with the "pooled" 1-df χ^2 test, whereas the sum of χ^2-test (6 df) is made sense of by considering a background model in which each of the six experiments might have its own separate binomial value. That is, in this case the alternative models are nested.

In order to respond to Fisher's argument, that the data are too good, we have to provide a family of alternative hypotheses, rivals to the "null" hypothesis that Mendel's data follow the Mendelian laws, to which we may turn when the "null" hypothesis is deemed to be overly discrepant with the observed data. Also, we are obliged to make precise the standards by which "discrepancy" is judged, to insure that there exist some alternatives which are not discrepant with the data whenever the "null" hypothesis is. The cogency of Fisher's argument rests, then, on the plausibility of these tacit alternative hypotheses, so introduced.

In the informal example above, where a sequence of observed relative frequencies agrees "too well" with Binomial expectations for independent experiments, consider a rival statistical model:

(a) with the same first moment as in the Binomial "null" distribution,

but

(b) with a smaller second moment.

Then, whenever the "null" hypothesis is suspect because the data are "too good to be true," such an alternative hypothesis fits the first two moments of the sequence of relative frequencies better than the "null" does.

The χ^2 distribution (on k degrees of freedom) is the distribution for the sum

$$\chi^2 = \sum_{i=1}^{k} x_i^2$$

where the x_i are i.i.d. standardized, normal $N(0,1)$ variates. More generally, if the x_i are i.i.d. normal $N(\mu,\sigma^2)$ variates, then

$$\sum_{i=1}^{k} \frac{(x_i - \mu)^2}{\sigma^2}$$

is distributed as χ^2 on k degrees of freedom. Suppose the "null" hypothesis h_0 is that binary variables, $y_i \in \{cell_1, cell_2\}$, (i = 1, ..., n) form an i.i.d. Binomial sequence, with parameter θ_0. Then the cumulative cell_1 count, n_1, is approximately normally distributed $N[n\theta_0, n(1-\theta_0)\theta_0]$. Let n_2 be the cumulative cell_2 count; so $n_1 + n_2 = n$. Hence, the familiar chi-square Goodness of Fit test-statistic,

$$\chi^2 = \frac{(n_1-n\theta_0)^2}{n\theta_0} + \frac{(n_2-n(1-\theta_0))^2}{n(1-\theta_0)}$$

$$= \frac{(n_1-n\theta_0)^2}{n(1-\theta_0)\,\theta_0}$$

is distributed, approximately, as χ^2 (on 1 df), in accord with (*).

If, as Fisher alleges, exceptionally small values of χ^2 tell against the null-hypothesis just as exceptionally large values do, then we may ask which alternatives to h_0 have reduced discrepancy with the data, when discrepancy is indicated by such extreme (small or large) χ^2's. To repeat, unless there is some alternative hypothesis which is not discrepant, the so-called "discrepancy" cannot be the basis for discrediting just the "null."

With k-cells ($k \geq 2$), for large values of the χ^2 test-statistic, shifting to an alternative hypothesis within the Multinomial family, e.g., using the m.l.e. or the minimum χ^2 estimator, reduces the χ^2 test-statistic from the unacceptably large magnitude. As is very well known, provided one of the Multinomial hypothesis is correct, asymptotically (with increasing sample size), such estimators are consistent and even efficient. Hence, using the magnitude of χ^2 as the index of discrepancy, the usual test, satisfies the condition that some alternative hypothesis exists with low discrepancy.

However, Fisher asserts quite generally that (very) small values of the χ^2 test-statistic also indicate discrepancy with an hypothesis.

> The term Goodness of Fit has caused some to fall into the fallacy of believing that the higher the value of P the more satisfactory is the hypothesis verified. Values over .999 have sometimes been reported which, if the hypothesis were true, would only occur once in a thousand trials. . . . In these cases the hypothesis considered is as definitely disproved as if P had been .001. (1925, §20, pp. 80–81)

Then we face a dilemma, at least with 1 df tests. If hypotheses with small values of χ^2 (1 df) are discrepant too, the Binomial minimum χ^2 or the m.l.e. will *not* survive as alternatives to a Binomial "null" hypothesis when large values of χ^2 are observed. An instance occurs in Fisher's analysis of Mendel's data for the "2:1" law; the data discussed in the previous section.

Regarding the 600 F_2 plants which Mendel sorted as either pure-bred dominants or hybrids, his counts were 399 hybrids and 201 dominant types. This corresponds to a suspiciously low χ^2 of 7.5×10^{-3} on the "null" hypothesis that the Binomial ratio is 2:1. (That is a P-value of .93 on 1 degree of freedom for the "uncorrected" Mendelian "2:1" law.) However, also Fisher cites the relatively large χ^2 value 3.3 (a P-value of about .07) for the "corrected" Mendelian hypothesis—corresponding to the 1.89:1.11 ratio which incorporates the 5.6% misclassification rate for hybrids—as another reason to believe that the data were "cooked." The problem here is that a hypothesis corresponding to the Binomial m.l.e. (or one sufficiently close to it) *must* yield small χ^2 values on 1 df; whereas, with increasing sample sizes, Binomial hypotheses with cell expectations different from the (limiting) sample frequencies *must* yield large χ^2 values. Then the Binomial model is damned because its expectations fit too well or if not well enough!

Thus, we shouldn't use Fisher's two-tailed reasoning with χ^2 values on 1 df. Of course, Fisher's criticism of Mendel uses a χ^2 value on 84 df, so he may (consistently) work both tails of the distribution: When χ^2 is too high appeal to Binomial alternatives with different first moments (but retain the Binomial second moment), and when χ^2 is too small, appeal to alternatives with the same first moment as the "null," but with a reduced second moment. As we see next, Fisher's argument using χ^2 tests makes explicit use of one variant of such an alternative hypothesis.

Specifically, a recipe for an alternative model that conforms to the two statistical requirements (above) is ready to hand. Propose that the data are "cooked," i.e., that observed relative frequencies which depart from Binomial expectations by more than some critical amount have been censored, or they have been altered to fall within the allowed range. Using small values of χ^2 to ascertain whether the observed frequencies in Mendel's data cluster too tightly about their Mendelian expectations, Fisher reports having

> had the shocking experience lately of coming to the conclusion that the data given in Mendel's paper must be practically all faked.[25] (Bennett, 1983, p.199)

Consider one group of experiments in Mendel's study, from §5 of his paper, titled [F$_2$] "The First Generation [Bred] from the Hybrids," to illustrate this point. These trials constitute the first phase, testing the "3:1" ratio, in the investigation leading up to the "1:2:1" law for offspring of hybrids. Mendel allowed the F$_1$ hybrids to self-fertilize and tabulated the ratio of phenotypic dominant to recessives in the next generation. The data for the "3:1" law are displayed in Table A.

TABLE A: Mendel's 7 experiments testing the "3 : 1" law in the F_2 generation

	Dominant form	Recessive form	Total
Experiment 1 (seed shape)	5,474 round	1,850 wrinkled	7,324 seeds
Experiment 2 (seed color)	6,022 yellow	2,001 green	8,023 seeds
Experiment 3 (seed-coat color)	705 grey-brown	224 white	929 plants
Experiment 4 (pod shape)	882 inflated	299 constricted	1,181 plants
Experiment 5 (pod color)	428 green	152 yellow	580 plants
Experiment 6 (flower position)	651 axial	207 terminal	858 plants
Experiment 7 (stem length)	787 long	277 short	1,064 plants

Fisher's Table V reports a χ^2 for these data of 2.1389, P = .95 (on 7 degrees of freedom). That value is obtained by summing the seven separate χ^2's (each an "experiment" with 1 degree of freedom) and using the statistical fact the sum of χ^2's is again χ^2 (on the sum of degrees of freedom). Thus, considering these seven experiments under the "3:1" law, the odds are 19:1 that, in a repetition of these trials, the new χ^2 (on 7 degrees of freedom) will exceed the magnitude (2.1389) observed in Mendel's data.

Fisher's Table V offers a cumulative χ^2 analysis for **all** of Mendel's data, constituting 84 degrees of freedom.[26] The upshot is a P-value in excess of .9999. That is, given Mendelian expectations, in a repetition of all of Mendel's experiments, the odds are better than 10,000:1 that a new χ^2 will exceed the value achieved by Mendel's data. Next, I try to make more precise when small values of χ^2 might plausibly indicate data are "cooked."

4.2 An Alternative Model?

Apart from manipulating the data, are there sensible, rival accounts that lead to a reduced second moment for Mendel's data? The Mendelian model posits that distinct plants are the result of probabilistically independent fertilizations. A random pollen grain fertilizes a random egg cell. Regarding egg cells, biology supports this hypothesis. However, it is plausible that pollen cells form on the anther in a restricted pattern. Specifically, during meiosis, a hybrid germ cell becomes a tetrad of 2 dominant and 2 recessive pollen cells. It is plausible to suppose that these move to the surface of the anther roughly as a tetrad, maintaining their proximity. If this is correct, the dominant and recessive pollen form something approximating a checkerboard pattern on the anther.

When the anther bursts in an open flower there is mixing. The tetrad configuration appears irrelevant to the Mendelian model. However, in the garden pea, self-fertilization occurs when the keel is very tightly packed. Recall, Mendel selected the pea plant for that feature—to minimize foreign pollen. Also, Mendel's garden averaged only about 30 peas/plant (over 511 plants) in experiments 1 and 2, compared to the 106.5 peas/plant in Bateson-Kilby's (1905) study (over 283 plants) of the same 3:1 law, or the 217.6 peas/plant in Darbishire's (1909) 1908 large-scale experiment (over 482 plants) for testing the 3:1 law. Evidently, Mendel's garden was not Eden for pea plants. Given their yields, the stems were not likely to have been subject to high water pressure from their roots, which (I understand) determines when and how vigorously the anthers burst inside the keel.

Two questions are obvious: First, is there evidence to confirm or to refute the speculative genetics that Mendel's peas are not independently distributed within self-fertilizing pods? Second, does it matter to Fisher's analysis if the model of pea genetics is not quite Mendelian but, instead, reflects this alternative distribution of pollen cells? How much of Fisher's .9999 P-value can be explained away with some subtle correlation among the pollen?

To the best of my knowledge, the issue whether peas in a pod are probabilistically independent has not been rigorously tested by field trials, at least not for the varieties of peas Mendel grew and under similar circumstances.[27] There are some intriguing numbers in Mendel's data, however, to suggest that not all is i.i.d., as we see next.

Suppose that Mendel tried to waste as few plants as possible. When counting peas, he counted all the peas in a pod. Concerning experiments with the first generation [F_2] bred from the hybrids, Mendel writes,

> Expt. 1. Form of seed.—From 253 hybrids 7,324 seeds were obtained in the second trial year. Among them were 5,474 round or roundish ones and 1,850 angular wrinkled ones. Therefrom the ratio 2.96 to 1 deduced.
>
> Expt. 2. Color of albumen.—258 plants yielded 8,023 seeds, 6,022 yellow, and 2,001 green; their ratio, therefore, is as 3.01 to 1.
>
> In these two experiments each pod yielded usually both kinds of seed. In well-developed pods which contained on the average six to nine seeds, it often happened that all the seeds were round (Expt. 1) or all yellow (Expt. 2); on the other hand there were never observed more than five wrinkled or five green ones in one pod. It appears to make no difference whether the pods are developed early or later in the hybrid or whether they spring from the main axis or from a lateral one. In some few plants only a few seeds developed in the first formed pods, and these possessed exclusively one of the two char-

acters, but in the subsequently developed pods the normal proportions were maintained nevertheless.

So, Mendel reports a total of 15,347 seeds taken from 511 plants. Is it "significant" that there were never more than 5 recessives in a single pod? The answer depends upon the distribution of peas in a pod. For example, with an average of 6.65 peas/pod distributed as:

20.% pods; 22.% pods; 38.% pods; 12.5% pods; and 7.5% pods
at 5 peas;   6 peas;   7 peas;   8 peas;   9 peas

on the Mendelian model, the chance of not seeing either 6 or 7 recessives in a single pod (among about 2300 pods) is approximately .015. If we increase the average peas/pod to 6.925 with

15% pods; 20% pods; 30% pods; 27.5% pods; and 7.5% pods
at 5 peas; 6 peas; 7 peas; 8 peas;   9 peas

the chance of observing no more than 5 recessives in any of the (about 2215) pods is approximately .0055. Thus, on the simple Mendelian model, gauged by "significance," his data are at least mildly surprising for their absence of large numbers of recessives in any pod.[28]

Reconsider the second question: Does it matter to Fisher's analysis if the model of pea genetics is not quite Mendelian but, instead, reflects this alternative distribution of pollen cells? In response, re-examine Fisher's Table V. We see that if we want to reduce the overall expected 1-sided P-level for such a table to about .98, then we require a total χ^2 (84 df) of about 59.07. That corresponds to a model with about 70% of the Mendelian (Binomial) variance, but with the same first moment. Then we would expect an observed χ^2 (summed on 84 df) of about 59. Likewise, if we want to reduce the expected overall 1-sided (84 df) P-level to the more extreme .99 level, then we require a total χ^2 (84 df) of about 56.15, i.e., a model with about ⅔ of the Mendelian variance. Even a rival model with only 75% of the Mendelian one carries an expected χ^2 (84 df) with a corresponding one-sided P-value in excess of .95. However, there is more to explain about the distribution the component χ^2 from Mendel's data than just the 84 df sum, as we shall discover.

What follows next is a simplistic model of self-fertilization for peas, based on an 80% seed survival, that has the same (Binomial) first moment and 74.1% of the variance of the Mendelian model, corresponding to a one-sided χ^2 P-value of about .96 on (Fisher's) 84 df. (For comparison, at only 70% seed survival, this model has about 77.8% the variance of the Mendelian model, which corresponds to an 84 df one-sided χ^2 P-value of about .93) The alternative model uses a 'hypergeometric' distribution for selection of the surviving pollen, following-up on the speculative account of how self-fertilizing hybrid peas depart from the i.i.d. "fair" (binomial) chance that either kind of pollen (dominant versus recessive) fertilizes eggs within the same pod. Here, of course, it is a departure from "independence, i.e., it is the first 'i.' of the i.i.d. Mendelian model that is removed. Call it the "Correlated Pollen" [CP] model. (Perhaps this is what Weiling intends with his appeal to a hypergeometric model of pea selection?)

The Correlated Pollen Model

Suppose that within the pea-flower for hybrids, 10 egg cells form according an i.i.d. "fair" (binomial) distribution. However, approximating the speculated, checkerboard pattern that pollen have on the anther, suppose that exactly 5 of every 10 pollen cells arriving at the egg cells are dominant. Last, assume that, with equal probability, 2 of these 10 zygotes spontaneously abort, leaving 8 peas/pod. The result is a model where pollen cells are negatively correlated within a pod.

4.3. The "3 : 1" Law

Table B displays the two sets of probabilities for the number N (out of 8 peas in a pod) of phenotypic dominants in the "3 : 1" law for self-fertilizing hybrids. The top row gives the ordinary Mendelian probabilities, whereas the bottom row gives the probabilities for Correlated Pollen model, with 74.1% of the Mendelian variance but the same first moment.

TABLE B: Comparison between two models of the "3 : 1" law for 8 peas in a pod

N is the number (out of 8) of dominant peas

	$N=0$	$N=1$	$N=2$	$N=3$	$N=4$
Mendelian	1.50×10^{-5}	13.66×10^{-4}	3.85×10^{-3}	2.31×10^{-2}	8.65×10^{-2}
Correlated	0.00	0.00	0.00	6.94×10^{-3}	6.94×10^{-2}

	$N=5$	$N=6$	$N=7$	$N=8$
Mendelian	2.08×10^{-1}	3.11×10^{-1}	2.67×10^{-1}	1.00×10^{-1}
Correlated	2.36×10^{-1}	3.61×10^{-1}	2.57×10^{-1}	6.94×10^{-2}

A more useful comparison is the distribution of the number N of dominant peas out of 32, corresponding to 4 pods/plant. That is, with 4 pods of 8 peas/plant, the probabilities for outcomes of N phenotypic dominants out of 32 is given in Table C.

TABLE C: χ^2, P-value, and probabilities for N (out of 32) dominant peas in the 3 : 1 law

χ^2	P-value	N	Mendelian model	Correlated Pollen model
10.667	.001	32	1.00×10^{-4}	2.30×10^{-5}
8.167	.004	31	1.07×10^{-3}	3.44×10^{-4}
6.000	.014	30	5.54×10^{-3}	2.39×10^{-3}
4.167	.041	29	1.85×10^{-2}	1.04×10^{-2}
2.667	.105	28	4.46×10^{-2}	3.16×10^{-2}
1.500	.221	27	8.32×10^{-2}	7.14×10^{-2}
0.667	.410	26	1.25×10^{-1}	1.24×10^{-1}
0.167	.670	25*	1.55×10^{-1}	1.70×10^{-1}
0.000	1.000	24*	1.61×10^{-1}	1.87×10^{-1}
0.167	.670	23*	1.43×10^{-1}	1.65×10^{-1}
0.667	.410	22*	1.10×10^{-1}	1.19×10^{-1}
1.500	.221	21	7.32×10^{-2}	6.89×10^{-2}
2.667	.105	20	4.27×10^{-2}	3.23×10^{-2}
4.167	.041	19	2.19×10^{-2}	1.16×10^{-2}
6.000	.014	18	9.90×10^{-3}	3.63×10^{-3}
8.167	.004	17	3.96×10^{-3}	9.44×10^{-4}
10.667	.001	16	1.40×10^{-3}	1.49×10^{-4}
			$\Sigma P > 9.99 \times 10^{-1}$	$\Sigma P > 9.99 \times 10^{-1}$

* Denotes those outcomes where the probability is greater under the Correlated model than under the Mendelian model.

The expected sum of χ^2 (1,000 df) from 1,000 plants (rounded to integer numbers of plants, above), each with 32 peas/plant, is 1,000 for the Mendelian model and only 734 for the Correlated Pollen model, reflecting the reduction in variance for the latter. Of course, under the Correlated Pollen model, no pod ever shows more than 5 recessives, since (by design!) there are at most 5 recessive pollen grains fertilizing each pod.

In his report of Experiments 1 and 2, Mendel illustrates his data with a sequence of ten plants for each of the two pea characteristics, shape and color. These 20 plants produced 915 seeds, an average of 45.75 seeds/plant. That is fully 50% greater than the overall average of 29.39 peas/plant (14,432 seeds) from the other 491 plants in Experiments 1 and 2. Thus, these illustrative plants were somewhat hardier than the average plant in Mendel's garden. Nonetheless, the sum of these 20 χ^2 is only about 12.49, corresponding to a one-sided P value of about .90.

Contrast the usual (i.i.d. binomial) Mendelian and the Correlated Pollen models, each with 4 pods of 8 peas/plant resulting in 9 possible values of χ^2 as illustrated in Table C. Grouped by χ^2 values, Table D gives the expected numbers of plants out of 20, together with the integer values obtained by rounding these to whole plants. Below each column is the expected χ^2 (20 df) and P-value that results. Again, Mendel's 20 plants have $\Sigma\chi^2$ = 12.49 on 20 df and P = .90.

TABLE D

χ^2	Expected numbers and integer numbers of plants (out of 20)			
	Mendelian model		Correlated Pollen model	
0.000	3.22	3.00	3.74	4.00
.167	5.96	6.00	6.72	7.00
.667	4.70	5.00	4.84	5.00
1.500	3.12	3.00	2.80	3.00
2.667	1.76	2.00	1.28	1.00
4.167	0.80	1.00	0.44	0.00
6.000	0.32	0.00	0.12	0.00
8.167	0.10	0.00	0.01	0.00
10.667	0.00	0.00	0.00	0.00
Expected χ^2 (20 df)	19.57	18.33	14.67	11.67
and P-values	.50	.56	.79	.93

Unfortunately, Mendel does not provide us with additional data at the plant-level for his experiments relating to the "3:1" law apart from his report about 4 extreme cases.[29] Instead, he gives pooled data for each of the 7 experiments of the "3:1" law taken from the remaining 2 pea-characteristics experiments and the 5 plant characteristics experiments. Also, there are 2 tests of the "3:1" law contained in the two-factor experiment and 3 tests in the three-factor experiment. These data have χ^2 that sum to 3.98 (12 df), with a P-value of about .98. Thus, these 32 tests of the

"3:1" law yield χ^2s that sum to 16.47 (32 df), and a corresponding P-value of about .98 also.

It is mere speculation whether, in each case, these experiments of the "3:1" law are the results of Mendel cultivating all the peas from a set of pods, rather than sampling a few peas from different pods. If the former was the situation, i.e., if Mendel wasted as few peas as possible from each pod, the Correlated Pollen model might apply. That model supports an expected P-value of about .96, compared to the observed P-value of .98.

What about other studies of the "3:1" law that were conducted in early years of this century? Both Bateson-Kilby's (1905) study (with more than 280 plants), and Darbishire's (1909) large-scale experiment (with over 450 plants) give plant-by-plant counts and show a cumulative χ^2 that is within 6% of the Mendelian model (see note 26). That is, these studies reflect very little variance reduction; in contrast with the Correlated Pollen model, which calls for about a 25% reduction in Binomial variance. Moreover, the scatter of P-values from these two experiments approximate the uniform U[0,1] distribution, in accord with the Mendelian model. Group the P-values from the 1-df χ^2 by deciles. The "null" hypothesis is that the P-values are uniformly distributed: that is, the "null" probability is .10 for each decile of P-values. Tested this way, Darbishire's 1908 data (482 plants) has a χ^2 of 13.23 (9 df), for a P-value of about .15. Bateson & Kilby's data (283 plants) has a χ^2 of 4.03 (9 df), for a P-value of about .91.

The plants in these two studies have profuse yields: respectively, 7 and 3 times the average yield per plant compared with Mendel's garden. In these two studies, moreover, there is no correlation between plant yield and P-value. For the Bateson-Kilby data, a linear regression of P-value on plant yield fits with an intercept of .516, a slope of -2.74×10^{-4}, and a correlation of about .06. With Darbishire's 1908 data set, the linear regression of P-value on plant yield has an intercept of .510, a slope of 4.06×10^{-5}, and a vanishing correlation of .02. In short, there is no indication of any departure from the Mendelian model in either of these two studies. Certainly, there is no warrant for a Correlated Pollen model here, nor does consideration of just the plants with small yields in these studies offer any evidence for a Correlated Pollen model.

However, in Darbishire's 1907 trials of the "3:1" law his plant yields were a better approximation of Mendel's. From 87 plants he reports 3904 peas, or 44.87 peas/plant. Discount 20 plants, whose yields were less than 24 peas/plant.[30] The net result is a data set of 67 plants (each with at least 24 peas/plant) that produced 2703 yellow and 974 green peas: a total of 3677 peas for an average of nearly 59 peas/plant. This is roughly on the same order of Mendel's 20 plants illustrating the "3:1" law, whose average yield was about 45 peas/plant. Overall, these data give a χ^2 (1 df) of 4.35, with a P-value of about .04—"significant" at the .05 level. Nonetheless, even with such a "significantly" large number of recessives, the 67 χ^2s (1 df) for Darbishire's 1907 study total only 59.86, or about 88% of the Mendelian variance: for a 1-sided P-value of about .62. The distribution of the 67 P-values is "significantly" far from uniform, as one might anticipate from the overabundance of recessives: Grouped by deciles the $\chi^2 = 19.42$ (9 df), with a one-sided P-value of .022. But that thinking misses an unusual feature of the data.

The noticeable point in the departure from the uniform distribution of P-values in Darbishire's 1907 data-set is not due to an excessive number of plants (out of 67) that have individual P-values that are too small. Figures 1 and 2 display the histograms of P-values by deciles in Darbishire's 1907 data-set (N=67) and in Mendel's tests (N=32) of the "3:1" law. In fact, in Darbishire's data, the lower third of P-values is under-represented (15/67 = 22.4%) against a uniform distribution. (The upper third of P-values is over-represented (29/67 = 43.3%) and the middle third shows about as expected (23/67 = 34.3%) for a uniform distribution.) The shortfall in low P-values in Darbishire's results duplicate a similar short-left-tail histogram of P-values from Mendel's (32 df) data on the 3:1 law. (More on this feature of plotted P-values later on!)

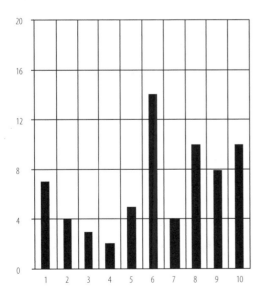

FIG. 1.   67 P-values for Darbishire's 1907 data

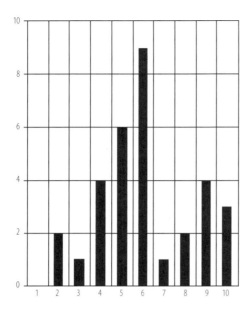

FIG. 2. 32 P-values for Mendel's test of the "3:1" law

In summary, regarding the "3:1" law, Mendel reports the absence of pods with more than 5 recessives for either pea-characteristic (color or shape). This is at least mildly surprising under the Mendelian model (a P-value less than .02). An examination of the 32 separate tests of "3:1" law present in Mendel's data produces a histogram of P-values (Figure 2) with a variance reduction also in accord with the speculative Correlated Pollen model. On the other hand, neither of the large-scale studies by Darbishire or Bateson & Kilby give any support whatsoever to such a speculative model. But these studies were under conditions noticeably different from Mendel's, at least in terms of plant yield. Whether the Correlated Pollen model is negatively associated with, e.g., soil moisture (a factor that positively affects the number of pollen competing to fertilize the eggs, I believe)—a factor also positively associated with yield—is a question for field trials to decide. One (smaller) study, by Darbishire, that approximates Mendel's plant yields also has a reduced sum-of-χ^2, despite having a "significantly" different first moment from the Mendelian model.

4.4. The "2:1" Law

Next, let us consider the evidence Mendel offers for his "2:1" law that offspring of self-fertilizing hybrids have a 2:1 ratio of hybrid to pure dominants. We have already considered one aspect of these data in connection with Fisher's (first) objection, relating to the alleged misclassification of the hybrids as pure-bred dominants. But that is not what we are concerned with here. The question now is whether the Correlated Pollen model is a serious rival to Fisher's alternative of "faked" data, based on the data being "too good to be true"?

Unfortunately, unlike his discussion of the "3:1" law, Mendel gives us only pooled counts, not any counts by outcomes of individual plants (let alone by pod) for his experiments on the "2:1" law. There are 15 such tests. Eight come from experiments on the phenotypic dominant plants that were used to test the "3:1" law. Each of the two pea-characteristics and four of the five plant characteristics were tested once to determine how the "3" divide up between hybrid and pure dominants. There were two trials testing pod-color. Four tests are available from the bi-factorial trials, which yielded 499 peas (out of 529) of the relevant sorts.[31] To maintain rough parity with the sample sizes from the other 12 tests, in the tri-factorial study (involving altogether 639 test plants of which 632 are relevant to tests of the "2:1" law), I have pooled the counts using the coarse categories, of the individual characteristics, producing only 3 more tests.[32] A table of the resulting χ^2s is just below, showing the "too good to be true" totals.

Mendel's data on the "2:1" law

Experiment	χ^2	P-value
1 (seed shape)	.17348	.67
2 (seed color)	.42486	.51
3 (seed coat)	.32083	.57
4 (pod shape)	.50005	.48
5a (pod color)	2.00022	.16
5b (pod color)	.12502	.74
6 (flower position)	.00500	.94
7 (plant height)	1.28000	.26
8 (bi-fac. pea shape)	.10631	.75
9 (bi-fac. pea color)	.08140	.78
10 (bi-fac. pea shape)	.75000	.38
11 (bi-fac. pea color)	.04412	.83
12 (tri-fac. pea color)	.00938	.92
13 (tri-fac. pea shape)	.33129	.57
14 (tri-fac. seed coat)	.30550	.58
	$\Sigma\chi^2 = 6.45746$	.97 (15 df)

The Mendelian model, with independence between peas, tells a simple story about distribution of hybrids within a pod. From a self-fertilizing hybrid, the (conditional) chance that phenotypic dominant pea is hybrid is 2/3. Moreover, this chance is independent of the number of phenotypic dominants in a pod. The Correlated Pollen model provides a more complicated account.

In a pod of 8 peas, both models give the same (marginal) distribution for the number of hybrids.[33] However, because of the correlation between pollen in the C-P model, the conditional probability of a pea being hybrid, given that it is phenotypic dominant, increases with the number, n, of such phenotypic dominant peas there are in a pod of 8. Let 'H' stand for the event that a pea ("randomly" chosen from the n) is hybrid rather than pure-dominant. In the C-P model, $p(H \mid n) = (.5, n = 3), (.5625, n = 4), (.61765; n = 5), (.66346; n = 6), (.70077; n = 7)$, and $(.73125; n = 8)$. Under this model, the probability distribution of pods, by the number n ($3 = n = 8$) of phenotypic dominant peas out of 8 in a pod, is in the ratios

16 : 160 : 544 : 832 : 592 : 160
for n = 3   4   5   6   7   8 phenotypic dominant peas.

This yields an expected "ratio" of hybrids to pure-dominants of only .65879, rather than the 2/3rds value of the Mendelian model.

However, this is not the relevant average to use for a contrast with Mendel's data. Mendel collected peas from pods to reach a number of test plants in creating his samples of the 2:1 law. That is, he collected about 500 seeds for each of his first two experiments (about 10% of the total available from the pervious year's two experiments on the 3:1 law) and he used 100 seeds for each test of the plant characteristics. I assume that he used all the phenotypic dominants in a pod and randomly sampled pods until he reached his quota of test seeds.

Among 100 randomly selected pods from the C-P distribution, we expect (to the nearest integer)

              1   7   24   36   26   7 pods
with n =   3   4   5   6   7   8 phenotypic dominant peas.

As there are n phenotypic peas/pod, this induces a distribution of peas, so that

           0.5%   4.6%   19.8%   35.7%   30.1%   9.3% of the peas come from pods
with n =   3   4   5   6   7   8 phenotypic dominant peas.

Rounded to whole numbers of (100) peas, this gives a C-P expected ratio (Hybrids to Pure-Dominants) of .6665—compared to the Mendel value of .6667. In the C-P model, the variance for this ratio is 3.29×10^{-2}. Under the same sampling rule, the Mendelian variance for this ratio is 4.39×10^{-2}. Hence, the C-P model has (approximately) 75% of the Mendelian variance for tests of the "2:1" law. The expected sum-of χ^2 P-values, then, is about .96, which figure matches the value obtained from Mendel's 15 tests of the "2:1" law.

4.5 Mendel's Data from Artificial Fertilizations

The C-P model is a speculative proposal about the details of self-fertilizing peas which allows a re-analysis of Mendel's data relating to the "3:1" and "2:1" laws. However, fully 15 of Fisher's 84 degrees of freedom in his Table V reflect Mendel's experiments on gametic ratios. These come from section 9 of Mendel's paper, "The Reproductive Cells of the Hybrids." There, he shows that it is irrelevant whether hybrids are formed by fertilization with a dominant pollen and a recessive egg or vice versa.

Five experiments, each with 3 df, sum to a combined χ^2 of only 3.6730, with a P-value of .9987. These data reflect approximately 550 plants, all the results of *artificial* fertilizations. For each test there were 4 cells, which had an equal expected count under the Mendelian hypothesis. Four of the tests involved the two pea-characteristics, and used approximately 100 test-plants each, with expectations of approximately 25 plants/cell. The fifth test used two plant characteristics (flower color and plant height), and had expectations of about 40 plants per cell. Twelve plants, total, were used to create the peas for the first four tests. Mendel writes that he made 45 fertilizations leading to the 166 plants of the fifth test.

Whatever merit there is to using the C-P model for re-examining the other data, it is of no relevance with these. They are the results of artificial fertilizations, all, where Mendel "dusted" the pollen onto each stigma for the flowers. I have no account of why these ratios cluster so closely around their expected values, each of which is ¼ of its sample size.

In these experiments, classification of a plant was (in the case of pea-characteristics) based on whether all of a plant's peas were of one (dominant) kind or whether it was hybrid, showing both dominant and recessive traits. Similarly, in the studies on plant characteristics, entire plants were classified, not merely individual peas. I find it hard to understand how misclassifications occurred.

5. What Model of "Cheating"?

The distribution of Mendel's Ps is the focus of A. W. F. Edwards's (1986) important article. He concludes (*161*) that the data were adjusted (rather than censored) to avoid extreme segregations in the record. I think there is rather good reason to agree with Edwards in that, at least, Mendel's "cheating" was not either the result of censoring extreme values, nor a more aggressive strategy of faking outcomes that cluster at the expected values.

Rather than reproducing Edwards's chart of Fisher's (1 df) signed-χs consider, instead, a histogram of the (78) P-values from χ^2.[34] Under the Mendelian hypothesis, these are uniformly distributed on the unit interval. The sum of the 78 (1 df) χ^2 = 41.49, with a P-value of about .9997, just as in Fisher's Table V. If we examine this distribution by deciles, the resulting (9 df) χ^2 is 19.18, with a P-value of about .023. The discrepancy is maximum for the first decile, reflecting the absence of a "left-tail" in Mendel's data, as we are well aware. However, the uppermost decile, rather than showing an excess over its expected value (of 7.8) has only 4 values. Even the upper-two deciles fail to exceed their expectations. In short, there is no evidence here of aggressive cheating, where outcomes at the highest decile of P-values are over-represented.

The final two histograms show the P-values from the data as Mendel organized it in his paper. Following his presentation, by my count, there are 48 experiments or separate steps in the tests that he devises.[35] I have plotted these by their P-values, which are uniformly distributed under the Mendelian model regardless of the number of degrees of freedom involved in each test. Compare this histogram with that for Mendel's data in the "3:1" law. The bulge near the median in the histogram is "significant" by χ^2. Contrast the middle two deciles [.4, .6] with the 8 deciles in the complementary set. Under the Mendelian hypothesis these have probability .2 and .8. But the (1 df) χ^2 = 5.33 for this hypothesis, with a P-value of .021.

What model of cheating, then, can the reader propose that replaces extremely discrepant outcomes with ones clustered about the median of χ^2s?[36] I challenge the reader to try to adjust binomial data from sample sizes in Mendel's experiments, so that the following three features appear in the resulting distribution of P-values from the (1 df) χ^2s:

1. There is a significant reduction in the left-tail of the Ps.
2. There is no significant departure from uniformity in the right tail of the Ps.
3. There is a significant concentration of the Ps about their median, i.e., about .50.

To be fair to Mendel, this exercise should be attempted without the aid of χ^2 Tables, which distribution, the reader recalls, K. Pearson discovered only in 1900!

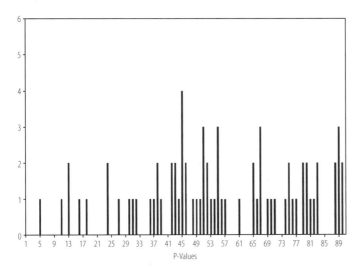

FIG. 3.   A. W. F. Edwards's partition of Mendel's data into 1-df chi-squares (N = 78)

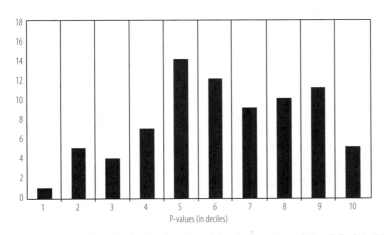

FIG. 4.   P-values, by deciles, in Edwards's 1-df partition of Mendel's data (N = 78)

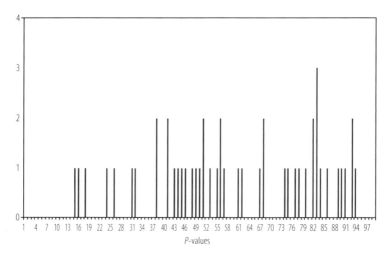

FIG. 5. Partition of Mendel's data into chi-squares by experiment ($N = 48$)

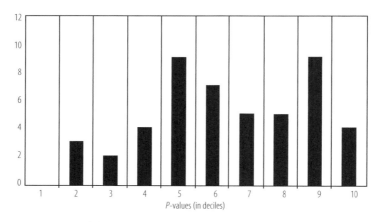

FIG. 6. Mendel's data (by experiment) ($N = 48$)

6. Conclusions

I have here reviewed each of Fisher's two principal objections to the data in Mendel's classic paper.

There is an easy reply to Fisher's charge that Mendel "cheated" in producing data for the "2 : 1" law. Data, in fact, which are significant against the hypothesis of a corrected ratio (of hybrids to pure-dominants) that reflects the 5.6% misclassification error Fisher uncovered in Mendel's protocol. The reply is that Mendel's protocol included (as he wrote) extending the testing through successive generations bred from the same hybrids. This sequential aspect of Mendel's design provides a simple rebuttal to Fisher's claim of misclassification.

This paper offers no easy nor complete answer to Fisher's second objection that, overall, Mendel's data conform too well to their theoretical expectations. There is insufficient variability in Mendel's counts under the Mendelian model. That is a certainty. But how is it to be explained, if not by positing reduced variance by cheating?

Regarding the "3:1" and "2:1" laws of segregation for self-fertilizing hybrids, I propose the Correlated Pollen model, an alternative to the usual Mendelian (i.i.d.) distribution of peas in a pod. The C-P model has the same first moment as the Mendelian model with about 75% of the Mendelian variance. This speculative model is enough to recover the P-values in Mendel's data for each of the two, main Mendelian laws. Also, the C-P model fits much better than the Mendelian model the values that Mendel reports for the maximum number of recessives in pods, from more than 2 thousand pods. It does slightly worse than the (i.i.d.) Mendelian model in fitting the extreme value (per plant) that Mendel reports for the "3:1" law from 511 plants.

The C-P model gets no support whatsoever from the two large-scale studies by Kilby & Bateson, and Darbishire on the "3:1" law, both conducted in the early 1900s. However, these studies do not duplicate the small yields (per plant) in Mendel's garden. I offer reason to think that the C-P model, if it applies at all, does not fit luscious plants. One of Darbishire's experiments that duplicated Mendel's lower yields shows an anomalous, small sum-of χ^2s, as in Mendel's data.

Even with all this effort, I have no insight to offer about the extremely low sum of χ^2s (15 df) from the 5 experiments on gametic ratios. These tests were intelligently designed to show that, regarding the Mendelian laws, a hybrid is a hybrid regardless which parent carried the dominant trait. The C-P model fails to apply to these data because all the test plants were the result of artificial fertilizations.

Where do we stand, more than sixty years after Fisher's shocking allegations against the authenticity of data in Mendel's paper? The allegation of misclassification (of hybrids) admits such a straightforward reply that I no longer find merit in that aspect of Fisher's criticism. But, unless some alternative model with reduced variance, like the C-P model, can be justified, I see little hope of explaining away the Ps that are "too good to be true."

The C-P model can be subjected to simple field trials, providing that classic strains can be cultivated under circumstances similar to those found in Mendel's garden. The difference with the Mendelian model is evident for the "2:1" law and makes testing straightforward. The C-P model (but not the Mendelian model) introduces a probabilistic dependence between the ratio of pure-breds to hybrids and the number of phenotypic dominants in a pod.

Regardless the outcome, no matter how peas self-fertilize, I urge the reader to study Mendel's classic paper and Fisher's provocative article. Mendel's work is a standard of clarity and a delight for its intelligent, sequential designs. Fisher, as always, is a brilliant statistician and imposing geneticist. As with many of his other writings, coming to an understanding of how he argues is the key, regardless what the reader thinks, in the end, of his conclusion.

Appendix

TABLE III—Comparison of numbers reported with uncorrected and corrected expectations

	Number of plants tested	Number of non-segregating progenies observed	Number expected		Deviation	
			Without correction	Corrected	Without correction	Corrected
1st group of experiments	600	201	200.0	222.5	+1.0	−21.5
Trifactorial experiment	473	152	157.7	175.4	−5.7	−23.4
Total	1073	353	357.7	397.9	−4.7	−44.9

TABLE V—Deviations expected and observed in all experiments

		Expectation	χ^2	Probability of exceeding deviations observed
3 : 1 ratios	Seed characters	2	0.2779	
	Plant characters	5	1.8610	
		— 7	— 2.1389	.95
2 : 1 ratios	Seed characters	2	0.5983	
	Plant characters	6	4.5750	
		— 8	— 5.1733	.74
Bifactorial experiment		8	2.8110	.94
Gametic ratios		15	3.6730	.9987
Trifactorial experiment		26	15.3224	.95
Total		64	29.1186	.99987
Illustrations of plant variation		20	12.4870	.90
Total		84	41.6056	.99993

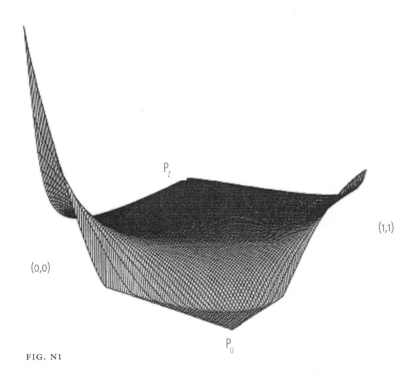

FIG. N1

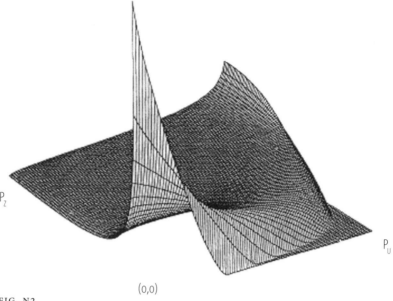

FIG. N2   (0,0)

NOTES

This paper was originally published as Seidenfeld, Teddy. 1998. "P's in a Pod: Some Recipes for Cooking Mendel's Data." PhilSci Archive, University of Pittsburgh Library System. http://philsci-archive.pitt.edu/archive/00000156/.

I have struggled with the Fisher-Mendel dispute for several years and owe thanks to many for helping me to find whatever is of value here. During Fall term 1997, where the current version was mostly completed, I was the guest of the STICERD Center and the Philosophy of Science Center at the London School of Economics. I have benefited considerably from reactions to talks I have given on this subject at: the Popper Seminar of LSE; the Department of Biometrics of Cornell University; the Harrisburg Chapter of the American Statistical Associations; and the Philosophy of Science Seminar of the University of Pittsburgh.

1. Mendel presented his work to the Natural History Society of Brünn (now Brno, Czechoslovakia) the year before, in 1865.

2. W. B. Provine (1971, chapter 3) gives an excellent account of the impact of the rediscovery on then ongoing disputes about the continuity of evolution. Particularly noteworthy were the 1902–1903 debates between W. F. R. Weldon and W. Bateson, whether Darwinian continuous evolution and the biometric blending of heritable traits were consistent with the Mendelian laws of dominance and segregation. (See, too, Karl Pearson's [1908] overview of Weldon's theory.)

3. In a preface to that undergraduate presentation Fisher concludes,

> I have almost entirely devoted myself to the two lines of modern research which are of particular interest in Eugenics, that is to Biometrics and Mendelism; and perhaps experts and professionals

will forgive the absence of more complicated details in both branches, if I explain that my object has been to give a fair view of the merits of the two methods, whose advocates have shown so little appreciation of the other school. (Bennett, 1983, p. 51)

4. The central theme in Fisher's 1918 work is to recover, e.g., a (normal) distribution of observed magnitudes and the observed correlations among related individuals using a large number of (independently) segregating Mendelian factors. Fisher's argument is a direct rebuttal to K. Pearson's 1909 Royal Society paper. In fact, Mendel anticipated the use of multiple factors in his speculation, derived from his abbreviated experiments, on seed and flower colors in *Phaselous* (Mendel, *105*). Ironically, K. Pearson's and R. Punnett's non-supportive referee reports of Fisher's 1916 submission to *Biometrika* (of his 1918 paper) focused on the question whether Mendelian theory could responsibly use more than 2 or 3 factors per observable trait (Norton and E. S. Pearson, 1976, pp. 153–155). Put in current terms, they questioned the empirical basis for evaluating Mendelian hypotheses with so many unmeasured parameters.

5. See Bateson's remark, from his 1909 biographical notice on Mendel, quoted by Fisher (1936, *119*).

6. Fisher's concern, even suspicion, about the close fit between Mendelian theory and Mendel's data dates, at least, from his 1911 (undergraduate!) presentation to the Cambridge University Eugenics Society. There, at age 21, he writes (without reference),

> It is interesting that Mendel's original results all fell within the limits of probable error; if his experiments were repeated the odds against getting such good results is about 16 to one. It may have been just luck; or it may be that the worthy German abbot, in his ignorance of probable error, un-consciously placed doubtful plants on the side which favored his hypothesis. (Bennett, 1983, p.57)

However it was Weldon (1901, p. 233), not Fisher, who first published a detailed analysis of the fit of Mendel's data to Mendelian theory, including the 16:1 probable error odds that such a good fit would not be duplicated. Regarding the data for the 3:1 law, he writes,

> These results then accord so remarkably with Mendel's summary of them that if they were repeated a second time, under similar conditions and on a similar scale, the chance that the agreement between observation and hypothesis would be worse than that actually observed is about 16 to 1.

What is, I find, much more remarkable than the young Fisher's loose scholarship in a 1911 unpublished undergraduate paper, is his total neglect of Weldon's analysis, 25 years later, in 1936. For, not only did Weldon scrutinize Mendel's data by analyzing how many times results exceeded their probable error, thereby providing the 16:1 odds quoted above. Also, Weldon used the then new method of χ^2 Goodness-of-Fit tests to analyze some of Mendel's more complicated experiments. (Parenthetically, Pearson published his famous paper introducing χ^2 in 1900, but Elderton's tables appeared only in 1901, in the same inaugural issue of *Biometrika* as Weldon's paper. Thus, Weldon's may well be the first application of χ^2 as a "Goodness of Fit" test; ironically, used to argue that the fit is suspiciously too good.)

Regarding the 3-factor experiments, for testing the simultaneous distribution of three-factor inheritance (with 3 categories—two for homozygotes and one for hybrids) Weldon (1901, p. 235) notes that, using Pearson's χ^2,

> Applying the method of Pearson (No. 25) the chance that a system will exhibit deviations as great as or greater than these from the result indicated by Mendel's hypothesis is about 0.95 (see Elderton, this Journal, *ante*, p. 161), or if the experiment were repeated a hundred times, we should expect to get a worse result about 95 times, or the odds against a result as good as this or better are

20:1. Fisher's (1936, *24*) Table V χ^2 analysis of Mendel's trifactorial experiment on 26 (= 33 −1) df leads to the same conclusion. Thus, Weldon anticipated Fisher not only in his finding that Mendel's counts showed surprisingly little departure from their expected values, but also even in his early application of small values of χ^2 for analyzing the extreme goodness of this fit.

Weldon follows his remark, above, with a repetition of the well known fact (and one reported by Mendel himself) that the Mendelian laws fail to apply to all the observable traits in plants, even for peas. Then Weldon concludes his critical discussion of the scope of Mendel's laws by saying,

> In trying to summarize the evidence on which my opinion rests, I have no wish to belittle the importance of Mendel's achievements. I wish simply to call attention to a series of facts which seem to me to suggest fruitful lines of enquiry.

The reader is urged to contrast this with Fisher's parallel conclusion in 1936 (*139*), quoted in the main text, above. Just how much of Weldon's 1901 essay did Fisher unconsciously internalize?

7. For example, even S. Mawer (1997, pp. 237–239) concedes that "Mendel cheated." I am not sympathetic, however, with Mawer's excuses for Mendel that both Burt and Lysenko also cheated with their data but, contrary to Mendel, created false (and dangerous) theories, nor that Darwin was simply wrong in his speculations about heritability.

8. Throughout, I shall refer to the translation appearing in Bateson's 1909 volume. This is the version used by Fisher in his 1936 paper. (Pagination refers to this volume and is noted in italics.)

9. Fisher (1936, Table VI) proposes a chronology of the experiments. Based on a subsequent review of the historical record, he revised Table VI by reducing by one each year indicated. (See Bennett, 1965, p. 59.)

10. There are seven chromosomes for the garden pea. S. Blixt (1974, pp. 187–188) indicates that the 7 characteristics studied by Mendel are governed by alleles located on 4 different chromosomes. Using Mendel's ordering, as reported above, we find these 7 characteristics are associated (respectively) with chromosomes #7, #1, #1, #4, #5, #4, and #4. The reason the three-factor trial supports "independent" heritability, despite the coincidence of seed color and seed coat genes on chromosome #1, is because these alleles are at remote sites. Likewise for the comparison of flower position with either pod shape or plant height. According to Lamprecht (1968a), as reported by Blixt, only pod shape and plant height might show linkage in trials the size Mendel had; but it is not known that Mendel investigated joint frequencies for this pair.

11. There is some discussion in the literature whether Mendel *planted* 10 seeds, or *cultivated* 10 plants. I will here not cite the various authors who take opposite sides on this matter. The point, however, is rather simple. If only 10 seeds were planted, of which about 8–9 would be expected to grow, then the misclassification problem is worsened. If Mendel planted more than 10 seeds, in order to be confident of having at least 10 mature plants to observe, then the misclassification problem is diminished but also his writing is misleading for not indicating how many test plants were observed. Fisher considers the latter alternative (*130*) and dismisses it.

12. Experiment #5a (for pod color) yielded a 60:40 division of 100 plants. Mendel retested this trait with another batch of 100 (called experiment #5b in the Appendix). The retest showed a 65:35 division. If the retest was in a subsequent year, as Fisher suggests (*127*), where did the second hundred plants come from? In this special case, did Mendel manage to grow a second crop in the same year? (The pea-plant requires only about 90 days to reach maturity.) Or, as I suspect, did he simply combine the results of two years' experimentation on the plant characteristics in this one grouping of his data? In that case, it fur-

ther endorses my speculation (below) that Mendel reported on this part of his study using the benefit of hindsight, after all the (subsequent years') data were in.

13. Recall, the misclassification issue does not affect the sorting of plants with regard the two seed-traits, for which each test plant yields perhaps 30 seeds, as explained in the main text. The bifactorial trials, therefore, were not involved, nor was there a problem for two of the three traits in the trifactorial study.

14. In a footnote (*139–140n5*) Fisher quotes the biologist, J. Rasmussen, that in picking from batches of whole, dry plants, a selection bias in favor of hybrids is quite plausible.

15. Fisher's third rebuttal point is without merit that I can see. The "adjustment" had to agree with the "correction" based on 10, rather than some other number of F_3 test plants. That is because Fisher calculated the "correction" (of 5.6% unclassified hybrids) based on Mendel's announced number (10) of test plants. Whatever number N of test plants Mendel had announced, that would have served as the basis for the "correction." So, of course there is no surprise that the "adjustment" Fisher calls for agrees with N = 10.

16. Mendel writes, "Artificial fertilization is certainly a somewhat tedious process, . . ." (*80*). I estimate that all of Mendel's (reported) data on peas required fewer than 500 artificial fertilizations.

17. F. Weiling (1971, p. 76) suggests that only 8 of 10 seeds might germinate. I do not see how this explains away the misclassification problem. If only 8 F_3 plants grow, then the resulting misclassification worsens. In another paper (1989, p. 136), in response to Edwards (1986), Weiling appeals a hypergeometric distribution—which has lower variance than the binomial. After conditioning on some expected values (instead of calculating a distribution), he manages to reduce the misclassification rate to about .038. (Recall, Fisher gave a misclassification rate for hybrids of .056).

Among several difficulties I have with Weiling's statistics, I do not understand the basis for his use of the hypergeometric distribution. It is true, as he writes, that the process of choosing 10 of 30 particular seeds from a plant (as Mendel is posited to have done to make the 10 F_3 offspring per F_2-parent) follows a hypergeometric distribution, with smaller variance than the i.i.d. Binomial distribution. However, under Mendelian theory, these 30 seeds follow the i.i.d. Binomial distribution. Hence, the net (marginal) distribution for the 10 seeds, chosen from the 30, is again i.i.d. Binomial, not hypergeometric, contrary to what Weiling asserts. The challenge, taken up below, is to justify the claim that the 30 seeds are not an i.i.d. sample from the Binomial distribution.

18. It is interesting to note that Mendel uses the same word for all the test plants used in the 5 experiments for classifying the F_2 generation. Regarding the 100 F_2 in experiment #3, for example, he writes, "The offspring of 36 plants yielded exclusively grey-brown seed-coats, while of the offspring of 64 plants, some had grey-brown and some had white."

The German word for 'offspring' Mendel uses here is 'Nachkommen' which (according to my colleague, Wilfried Sieg) is exactly the right term to refer to future generations as well the immediate progeny. However, Mendel uses 'Nachkommen' also at the beginning of the same section (§6) in a context where it is evident he intends to refer to the next (F_3) generation only. Thus, I do not see how to settle the issue, whether Mendel employed a sequential design, merely by this choice of word in describing the protocol.

19. Fisher (*140n6*) attributes to Rasmussen an estimate of space for 4000–5000 plants.

20. Recall, the number of ways of partitioning n elements into m non-empty subsets, $S_n^{(m)}$, is given by Sterling numbers of the second kind:

$$S_n^{(m)} = \frac{1}{m!} \sum_{k=0}^{m} (-1)^{m-k} \binom{m}{k} k^n.$$

Thus, there are $\sum_{m=1}^{n} S_n^{(m)}$ alternative representations of the data as a sequence of "experiments."

21. The parallel three tests relating to the 10 plants Mendel selected from Experiment #1, concerning the "3:1" ratio of round to wrinkled seeds, lead to the same conclusion: (1) Ten plants yielded 437 seeds of which 336 showed the dominant trait, for a χ^2 (1 df) of .831, and a corresponding P-value of about .36. (2) The sum of the ten, 1-df χ^2 is 5.299 with a corresponding P-value of .87. (3) Last, regarding Cramér's test that these ten values are trinomial with cell probabilities .3, .4, and .3, the respective cell-counts are 3, 6, and 1 for a χ^2 (2-df) of 2.33 and P-value of .33. None of these three is "significant." Nor do these analyses change when the combined (20) data are similarly analyzed in three tests.

22. Fisher often refers to such a discrepancy index as an "exact" test, e.g., Fisher, 1925, §21.02. Of course, he does so when the outcomes are suitably described, e.g., through a minimal sufficient statistic with respect to an unstated, larger model that contains the null as a special case. For example, cell frequencies are the right test statistic for an exact test of independence in contingency tables, where Multinomial sampling serving as the unstated background model.

23. See, for example, Fisher's advice in §21.1 of (1925) where, for example, he recommends meta-analysis by converting P-values back into χ^2 values on 2 df before summing. This rule he gives because $-2\ln(P)$ is the corresponding χ^2 values on 2 df. Supplied with a table of logarithms, anyone can do meta-analysis!

24. Nobile (1992, Table 5) recomputes an analogue of Fisher's Table V for the whole of Mendel's data, aggregating the results of Mendel's six experiments wherever these appear by summing the corresponding χ^2 statistics as in P_z. The effect is a χ^2 of 19.0 on 40 df, with a P-value of .998.

25. This was written by Fisher to E. B. Ford, in a letter dated 2 January 1936. Fisher confesses that his suspicions about Mendel's data were aroused by the problem associated with the misclassification of hybrids, discussed (above) in section 3.

26. As has been noted by others, e.g., Edwards (*147*) there are two, statistically minor points concerning Table V.

(1) The last row of Fisher's table, "Illustrations of plant variation," constitutes 20 degrees of freedom, includes a double-counting of 20 plants out of 511 plants (or a doubling counting of 915 seeds out of 15,347 seeds) from the first two of Mendel's seven experiments on the "3:1" law. When the data from these 511 plants are separated from the other results of the first two experiments, the χ^2's change as follows: In experiment 1, the new χ^2 (1 df) is .57504 (to replace .26288). In experiment 2, the new χ^2 (1 df) is .04811 (to replace .01500). The net change for the sum is a new χ^2 (2 df) of .62315, to replace .27788. (This corresponds to a new P value of about .73, to replace a P value of .87.) Concerning Fisher's Table V summary of the "3:1" ratios, the new χ^2 (7 df) is 2.48413, or a new P-value of about .93 instead of Fisher's .95. Since this correction adds only .3452 to the overall χ^2 (84 df), no doubt Fisher would persist in his claim that the overall P-value, .999, remains too good to be true.

(2) Also, Fisher's Table V includes all the data pertaining to the supposed misclassified hybrids used in testing the "2:1" law—data both from the five monofactorial and from the single trifactorial studies. These constitute another 15 degrees of freedom. However, Fisher uses the "uncorrected," Mendelian "2:1" expectations in calculating the cumulative χ^2 on 84 degrees of freedom. If we accept Fisher's objection that there was a 5.6% misclassification rate for hybrids, then Fisher incorrectly uses the uncorrected "2:1" expectations to

calculate χ^2. It seems to me that, to be fair to Mendel, Fisher should have calculated in one of two ways instead:

Either (i) Fisher might have taken his first objection to heart and used the "corrected" expectations regarding the plant characteristic data for the "2:1" law. Considering the six, 1 df trials (n = 100 each). With a "corrected" expectation, the cumulative χ^2 for these increase from 4.58 to 7.36. Regarding the data for the relevant nine 1 df cells in the trifactorial study, the χ^2 for these increase from 6.29 to 10.68. The net gain of 7.17 between these two raises Fisher's Table V χ^2 on 84 df to 48.78, with a corresponding P of about .9975. [Of course, I reject this defense of Mendel, as I argue that there is no need to used a "corrected" expectation for his "2:1" law data.]

Or (ii) Fisher should have excluded these 15 degrees of freedom altogether and calculated a cumulative χ^2 on 69 degrees of freedom, just as Edwards does (155).

27. Edwards (1986, *148*) argues that peas are excellent randomizers, contrary to the view offered by Weiling (1966) that they carry sub-binomial variance, for example. Edwards's reasoned position rests on a reexamination of the large data sets in Bateson & Kilby's (1905) and Darbishire's (1908, 1909) studies—all experiments that (having among other goals) re-test the Mendelian "3:1" law in self-fertilizing pea hybrids. Edwards (*148*) reports Weiling combining these data in the fashion of Fisher's meta-analysis. That yields a total χ^2 = 1008.8 on 1,062 df, corresponding to about 95% (= 1,008.8/1,062.0 %) of the variance under ordinary (Mendelian) i.i.d. Binomial sampling. So Edwards finds no basis here for Weiling's claim that sub-binomial variability explains away Fisher's analysis with an extremely low χ^2 = 41.6 on 84 df, corresponding to less than 50% of the Mendelian variance, and I agree.

The one-sided P-value even for this meta-analysis is about .87, however. In that sense, these data are not quite as "unremarkable" as Edwards makes out, though (of course) they do not even remotely approximate the "too good to be true" χ^2 values Fisher gives for Mendel's data.

Another first-rate investigation of these two studies is found in Stephens' (1994) work. He, too, concludes (§6.4) that these large-scale studies (each) fit the traditional Mendelian model, but not so with Mendel's own data.

My analysis of these studies lead to the following χ^2 values, taking each plant as its own "experiment," i.e., with 1 df/plant. I have ignored those plants (as indicated) where either the expected number of recessives fell below 6 peas/plant, i.e., where fewer than 24 peas grew on a single plant, or where the authors report that a plant had shed "many" peas before counting. By deleting plants with fewer than 24 peas, in effect, I have increased the average χ^2 value (per d.f.), as the binomial distribution is truncated at the origin. For example, with 23 peas, even if all 23 show the dominant trait, the 1 df χ^2 = 5.99 with a P-value of .014.

Bateson & Kilby (1905) χ^2 = 289.17 on 283 df (with 38 plants not counted);
Darbishire (1909) 1906 data: χ^2 = 7.44 on 13 df;
Darbishire 1907 data χ^2 = 59.35 on 67 df (with 20 plants not counted);

and

Darbishire 1908 data χ^2 = 440.48 on 482 df (with 4 plants not counted).

This gives a total χ^2 of 796.44 on 845 df, corresponding to about 94% of the Mendelian variance, with a one-sided P-value of about .88. Again, my ("conservative") analysis carries fewer degrees of freedom where I have not counted plants, as indicated above.

28. Weldon (1901, p. 230) arrives at the opposite conclusion. Weldon considers data

from Mendel's second experiment, on seed color, and then he calculates the odds that a pod (with 6, 7, 8, or 9 seeds) contains all recessive (only green) peas. This doubly dilutes the force of Mendel's observation of no more than five recessive peas, either for shape or color, in a single pod.

29. Mendel (86) does give extreme plant counts in both directions for each of the two pea-characteristics. For Experiment 1 (pea shape) he reports that one plant yielded 43 round peas (which is dominant) against 2 that were angular, and another plant had 14 round and 15 angular peas. Regarding Experiment 2 (pea color) a plant had 32 yellow peas (dominant) and 1 green one. Fourth, a plant had 20 yellow and 19 green peas. Respectively, the 1 df χ^2s for these four cases are: 10.14 (a P-value of about 1.5×10^{-3}); 11.05 (a P-value of about 9.0×10^{-4}); 8.50 (a P-value of about 3.5×10^{-3}) and 11.70 (a P-value of about 6.3×10^{-4}). The latter being the extreme of the four.

However, taking into account the discreteness of the Binomial distribution, the exact Binomial counts for these "tails" are less extreme than the χ^2 values. Respectively, the Binomial probability of an outcome as or more extreme than Mendel reports are, respectively, only: 1.54×10^{-3}; 1.90×10^{-3}; 3.63×10^{-3} and 1.15×10^{-3}. For example, the Binomial probability that all 258 trials in Experiment 2 were less discrepant than the fourth case is about .742. The corresponding result for Experiment 1 is about .677. More importantly, the probability under the Mendelian model that all 511 trials are less discrepant outcomes than the fourth case (above) is about .554. By contrast, under the Correlated Pollen model, the probability that all χ^2s from 511 trials (for plants of 32 peas each, i.e., 4 pods of 8 peas each) do not exceed 10.67 is about .916. That approximation gives the Mendelian model a factor of about 5:1 over the Correlated Pollen model, for the maximum discrepancy in 511 trials. That is, the best such outcome would be that the probability is .50 that the maximum would be exceeded again in 511 trials. But $(1-.554)/(1-.916)$ is about 5.3. Again we face the question of how to index "discrepancy."

30. These total 227 peas: 166 yellow-dominant, and 61 green-recessive. To repeat, there are two reasons for "cleaning" the data this way. First, the rule-of-thumb for χ^2 calls for expected values of at least 6 per cell; hence, calling for plants with at least 24 peas to test the "3:1" law. The second reason is that the Correlated Pollen model yields increasing variance as fewer peas per pod are counted. In the extreme, with only 1 pea/pod counted, the Correlated Pollen model is the Mendelian model.

31. I have tested each character twice for the "2:1" by partitioning the bifactorial data on the phenotype of the other character.

32. The question is how fine to partition the tri-factorial experiment into separate tests of the "2:1" law? At the extreme, there are 9 genotype configurations for the other two characteristics, opening the door to 27 tests (with 26 df), achieved by partitioning a test for each trait on each such genotypic configuration of the other two.

To partition by the phenotype of the other two traits still yields 12 tests, as the following table illustrates. However, such a partition leads to tests with samples that differ by an order of magnitude, e.g., N = 269 in Exp. 1 and N = 27 in Exp. 4.

Mendel's trifactorial data on the "2 : 1" law partitioned into 12 tests using the phenotypes of the other two traits

Experiment	χ^2	P-value	N
1 pea shape	.00186	.97	269
	AA vs Aa among [BBCC + BbCC + BBCc + BbCc]		
2 pea shape	.74419	.39	86
	AA vs Aa among [bbCC + bbCc]		
3 pea shape	.02041	.89	98
	AA vs Aa among [BBcc + Bbcc]		
4 pea shape	.66667	.41	27
	AA vs Aa among bbcc		
5 pea color	.98327	.32	269
6 pea color	.68750	.41	88
7 pea color	.02041	.89	98
8 pea color	.23529	.62	34
9 seed-coat	.22491	.63	269
10 seed-coat	.42186	.52	96
11 seed-coat	.27841	.60	88
12 seed-coat	.00000	1.00	30
$\Sigma\chi^2 =$ 4.285 (12 df)		.98	

A: Round pea; a: wrinkled pea   B: Yellow albumen; b: green albumen   C: Grey seed-coat; c: white seed-coat

If these 12 degrees of freedom replace the 3 entries in the text that represent the tri-factorial data, (i.e., numbers 12, 13, and 14) in the table, the upshot is a $\chi^2 = 10.096$ (24 df), and a P-value of .99.

33. Under both models, with a self-fertilizing hybrid the chance of a hybrid pea is 1/2, and separate peas are independent for this genotype. In the Mendelian model, it is evident that the chance that n out of 8 peas are hybrid is just the Binomial chance of n "heads" out of 8 flips of a fair coin. For the C-P model reason as follows: exactly one egg-type matches with each pollen type to form a hybrid. But egg-types are i.i.d. Binomial, with chance 1/2 for each type. Hence, as with the Mendelian model, the chance that n out of 8 peas in a pod is hybrid is given by the Binomial distribution for i.i.d. data.

34. Edwards deletes the 6 "questionable" values from the (n = 100) tests of the "2:1" law that purportedly involves misclassification of hybrids.

35. Specifically, these 48 P-values comprise:

22 1 df tests of the 3:1 law with two pea-characteristics—I adjusted Mendel's two grand totals to avoid double-counting the twenty illustrative plants.;

5 1 df tests of the 3:1 law with plant characteristics;
8 1 df tests of the 2:1 law;
2 3 df tests in the bifactorial study;
2 1 df tests of the 2:1 law in the bifactorial study;

1 3 df test in the trifactorial study (according to the number of hybrid characters, 0–3);
1 5 df test in the trifactorial study of those hybrid in two traits;
1 11 df test within the trifactorial study of those hybrid in one trait;
1 7 df test within the trifactorial study of those hybrid in no traits
5 3 df tests in the gametic study.

Note that under the Mendelian null-hypothesis these P-values are uniformly distributed regardless the number of degrees of freedom associated with each.

36. The problem iterates in the literature. Novitski's (1995) reanalyzes Mendel's 5 experiments involving the "Reproductive Cells of the Hybrid," where Mendel tests the hypothesis that hybrids are the same regardless which parental gamete carries dominant trait. Novitski introduces a new factor of approximately 1:29 to the already low 1:700 significance level as reported by Fisher for these 15 (= 5 × 3) df, to conclude that the observed significance level for these data are at the exceptionally small level of approximately 1:20,000. The new factor of 1:29 he arrives at by numerical simulation of the conditional distribution for the variance of five 3-df χ^2s, given the sum of these five χ^2s (= 3.67) calculated from Mendel's data. That is, given the sum (3.67), under the Mendelian null hypothesis, the odds are about 30:1 for 5 χ^2s with larger variance than is found in Mendel's data. Novitski (p. 65) finds that this "places some additional weight on the conclusion of the majority of others who have looked at Mendel's results that his data, in the context in they were reported in his paper, are highly improbable." Alas, Novitski does not consider whether the added factor, of the reduced (conditional) variance in these 5 χ^2s, in fact makes it more or less plausible that Mendel's data were faked.

REFERENCES

Bateson, W. and Kilby, H. (1905) "Experimental Studies in the Physiology of Heredity: Peas," *Royal Soc. Reports to the Evolution Committee* **2**, 55–80.

Bateson, W. (1909) *Mendel's Principles of Heredity*. Cambridge: University Press; (1990) Classics of Medicine Library, Div. of Gryphon Editions, Inc.: Birmingham, Alabama.

Bennett, J. H., ed. (1983) *Natural Selection, Heredity, and Eugenics*. Oxford: Clarendon Press

Blixt, S. (1974) "The Pea" in R. C. King (ed.) *Handbook of Genetics*, vol. 2. New York: Plenum Press, chapter 9.

Cramér, H. (1946) *Mathematical Methods of Statistics*. Princeton: Princeton Univ. Press.

Darbishire, A. D. (1908) "On the Results of Crossing Round with Wrinkled Peas, with Especial Reference to their Starch-grains," *J. Royal Soc. of London B, Proceedings* **80**, 122–135.

Darbishire, A. D. (1909) "An Experimental Estimation of the Theory of Ancestral Contributions in Heredity," *J. Royal Soc. of London B, Proceedings* **81**, 61–79.

Elderton, W. Palin (1901) "Tables for Testing the Goodness of Fit of Theory to Observation," *Biometrika* **1**, 155–163.

Edwards, A. W. F. (1986) "Are Mendel's Results Really Too Close?" *Bio. Rev.* **61**, 295–312.

Gauch, H. G., Jr. (1993) "Prediction, Parsimony and Noise," *American Scientist* **81**, 468–478.

Fisher, R. A (1911) "Mendelism and Biometry," in Bennett, J. H (1983), 52–58.

Fisher, R. A. (1918) "The Correlation between Relatives on the Supposition of Mendelian Inheritance," 399–433.

Fisher, R. A. (1925) *Statistical Methods for Research Workers*. (14th ed., 1973) New York: Hafner Press.

Fisher, R. A. (1936) "Has Mendel's Work Been Rediscovered?" *Annals of Science* **1**, 115–137.

Fisher, R. A. (1955) "Introductory Notes on Mendel's Paper," in J. H. Bennett, ed. (1965) *Experiments in Plant Hybridization*. London: Oliver & Boyd, pp. 1–6.

Mawer, S. (1997) *Mendel's Dwarf*. London: Doubleday.

Mendel, G. (1865) "Experiments in Plant-Hybridization," *Verh. naturf. Ver. in Brunn, Ab.* **4**; translated by the Royal Horticultural Soc. and appearing in Bateson (1909), 318–361.

Mendel, G. (1869) "On Hieracium-Hybrids Obtained by Artificial Fertilization," *Verh. naturf. Ver. in Brunn, Ab.* **8**; trans. Royal Horticultural Soc. and appearing in Bateson (1909), 362–368.

Nobile, A. (1992) "A Note on Goodness-of-fit χ^2 Statistics," Tech. Report 550, Dept. of Statistics, Carnegie Mellon University, Pittsburgh.

Norton, B. and Pearson, E. S. (1976) "A Note on the Background to, and Refereeing of, R. A. Fisher's 1918 Paper "On the Correlation between Relatives on the Supposition of Mendelian Inheritance," *Royal Society of London, Notes and Records* **31**, 151–162.

Novitski, C. E. (1995) "Another Look at Some of Mendel's Results," *J. Heredity* **86**, 62–66.

Orel, V. (1968) "Will the Story on 'Too Good' Results of Mendel's Data Continue?" *BioScience* **18**, 776–778.

Pearson, K. (1900) "On the Criterion that a Given System of Deviations from the Probable in the Case of a Correlated System of Variables Is Such that It Can Be Reasonably Supposed to Have Arisen from Random Sampling," *Phil Magazine* **50**, 157–175.

Pearson, K. (1908) "On a Mathematical Theory of Determinantal Inheritance, from Suggestions and Notes of the Late W. F. R. Weldon," *Biometrika* **6**, 80–93.

Piegorsch, W. W. (1983) "The Questions of Fit in the Gregor Mendel Controversy," *Commun. Statist.-Theor. Meth.* **12**, 2289–2304.

Piegorsch, W. W. (1990) "Fisher's Contributions to Genetics and Heredity, with Special Emphasis on the Gregor Mendel Controversy." *Biometrics* **46** (4): 915–24.javascript: PopUpMenu2_Set(Menu2085640);

Provine, W. B. (1971) *The Origins of Theoretical Population Genetics*. Chicago: U. Chicago Press.

Stephens, M. (1994) "The Results of Gregor Mendel: An Analysis, and Comparison with the Results of Other Researchers." Thesis submitted for the Diploma in Mathematical Statistics, University of Cambridge. (A copy may be obtained from the author, who may be reached via his home page: www.stats.ox.ac.uk/~stephens/index.html. I thank Dr. Stephens for supplying me with a personal copy.)

Weiling, F. (1971) "Mendel's 'Too Good' Data in Pisum-Experiments," *F. Mendeliana* **6**, 75–77.

Weiling, F. (1986) "What About R. A. Fisher's Statement of the 'Too Good' Data of J. G. Mendel's Pisum Paper?" *J. Heredity* **77**, 281–283.

Weiling, F. (1989) "Which Points Are Incorrect in R. A. Fisher's Statistical Conclusion: Mendel's Experimental Data Agree Too Closely with His Expectations?" *Angew. Botanik* **63**, 129–143.

Weldon, W. F. R. (1901) "Mendel's Laws of Alternative Inheritance in Peas," *Biometrika* **1**, 228–253.

Weldon, W. F. R. (1902) "On the Ambiguity of Mendel's Categories," *Biometrika* **2**, 44–55.

Weldon, W. F. R. (1902) "Mr. Bateson's Revisions of Mendel's Theory of Heredity," *Biometrika* **2**, 286–298.

POSTSCRIPT TO CHAPTER 6

■ A Brief Account of a Trial Conducted at Pillsbury Labs, 2000–2001

TEDDY SEIDENFELD

In fall 2000 and spring 2001, Dr. Rebecca J. McGee, senior researcher at the Pillsbury-Green Giant Agricultural Research Department in Le Sueur, Minnesota, graciously undertook some trials involving classic pea lines with an eye on testing several aspects of the "Correlated Pollen" model. In particular, some plants were stressed with poor soil and little irrigation, and others were pampered, in order to determine whether that affected the distribution of pea characteristics per pod.

A pilot sample with 25 offspring, using a cross between *Caractacus* and *Champion of England*, was initiated to test feasibility of the stressed-plant arrangement. These 25 stressed plants produced a total of 282 seeds that were classified into four categories according to pea shape and color. The Mendelian ratios 9:3:3:1 yield expectations of, respectively, 158.625, 52.875, 52.875, 17.625 peas. The observed counts (see below) were, respectively, 156, 54, 54, 18. This produces an exceptionally good fit: a χ^2 of approximately 0.01 on 3 d.f., with a *P*-value in excess of .99. Of course, merely reproducing Mendelian data "too good to be true" by χ^2 does not constitute a test of the special features that constitute the rival "Correlated Pollen" model. For that, the experiment was replicated with 18 other crosses, whose identities and conditions are reported in the accompanying two charts.

Autumn 2000 crosses: F_1

Cross	Female	Female cotyledon color	Female seed shape	Male	Male cotyledon color	Male seed shape	# F_1 seeds	F_1 cotyledon color	F_1 seed shape	Pred F_1 Ht
00474	Alpha	Yellow	Dimpled	Early Badger	Green	Wrinkled	6	Yellow	Smooth	Short
00475	Caractacus	Yellow	Smooth	Champ England	Green	Wrinkled	50	Yellow	Smooth	Short
00476	Caractacus	Yellow	Smooth	Little Gem	Green	Wrinkled	25	Yellow	Smooth	Short
00477	Caractacus	Yellow	Smooth	Notts Excelsior	Green	Wrinkled	27	Yellow	Smooth	Tall
00478	Early Bird	Yellow	Smooth	Early Blue	Green	Wrinkled	47	Yellow	Smooth	Tall
00479	Early Bird	Yellow	Smooth	Kentish Invicta	Green	Dimpled	34	Yellow	Smooth	Tall
00480	Early Blue	Green	Dimpled	Champ England	Green	Wrinkled	80	Green	Smooth	Tall
00481	Early Blue	Green	Dimpled	JI 1573	Yellow	Smooth	32	Yellow	Smooth	Tall
00482	Kentish Invicta	Green	Smooth	Champ England	Green	Wrinkled	63	Green	Smooth	Tall
00483	Kentish Invicta	Green	Smooth	Little Gem	Green	Wrinkled	13	Green	Smooth	Tall
00484	Little Gem	Green	Wrinkled	WBH 1485 JI 15	Green	Dimpled	60	Green	Dimpled	Tall
00485	Ne Plus Ultra	Green	Wrinkled	Improved Stratagem	Green	Wrinkled	1	Green	Wrinkled	Tall
00486	JI 1573	Yellow	Smooth	Tall Telephone	Green	Wrinkled	19	Yellow	Smooth	Tall
00487	Roi des Gourmands	Yellow	Smooth	Kentish Invicta	Green	Smooth	9	Yellow	Smooth	Tall
00488	Roi des Gourmands	Yellow	Smooth	Little Gem	Green	Wrinkled	21	Yellow	Smooth	Tall
00489	Sel Duke of Alb JJ304	Green	Wrinkled	WBH 1485 JI 15	Green	Dimpled	83	Green	Dimpled	Tall
00490	Sel Duke of Alb JJ924	Green	Wrinkled	Yorkshire Hero	Green	Wrinkled	48	Green	Wrinkled	Tall
00491	Witham Wonder	Green	Wrinkled	Early Blue	Green	Dimpled	40	Green	Wrinkled	Tall
00492	Yorkshire Hero	Green	Wrinkled	WBH 1485 JI 15	Green	Dimpled	30	Green	Smooth	Tall

WBH 1485 JI 15 has black hilum

Autumn 2000 crosses: Crosses made

Cross	Female	Male	# plants—stressed	# plants—pampered	seg R	seg I	seg Le
00474	Alpha	Early Badger	6		n	y	y
00475	Caractacus	Champion of England	25	24	y	y	y
00476	Caractacus	Little Gem	23		y	y	y
00477	Caractacus	Notts Excelsior	30		y	y	y
00478	Early Bird	Early Blue	21	18	n	y	y
00479	Early Bird	Kentish Invicta	19		n	y	n
00480	Early Blue	Champion of England	33	18	y	n	n
00481	Early Blue	JI 1573	22		n	y	n
00482	Kentish Invicta	Champion of England	36	17	y	n	y
00483	Kentish Invicta	Little Gem	7		y	n	y
00484	Little Gem	WBH 1485 JI 15	16		y	n	y
00485	Ne Plus Ultra	Improved Stratagem	1		n	n	y
00486	JI 1573	Tall Telephone	8		y	y	y
00487	Roi des Gourmands	Kentish Invicta	9		n	y	n
00488	Roi des Gourmands	Little Gem	17		y	y	y
00489	Sel Duke of Alb JI304	WBH 1485 JI 15	20		y	n	n
00490	Sel Duke of Alb JI924	Yorkshire Hero	19		n	n	y
00491	Witham Wonder	Early Blue	16		y	n	n
00492	Yorkshire Hero	WBH 1485 JI 15	25		y	n	y

The following summary of the Le Sueur data reflects statistical analysis by Daniel Heinz and Erich Huang, graduate students in the statistics department of Carnegie Mellon University. With respect to the Correlated Pollen model, there was generally no difference observed between stressed and pampered plants. Moreover, the counts on the pea-color trait conformed generally to the ordinary Mendelian model, with no significant relationship observed between the variance of the trait and the number of peas-per-pod. However, the counts for pea-shape showed a significant relationship between the variance for that trait and number of peas-per-pod. This finding is tempered by the fact that the counts for pea-shapes departed from their Mendelian expectations. That is, the counts for pea-shapes have a suspicious first moment as well.

In short, I judge the trial conducted at the Pillsbury Labs shows no relevant effects on the distribution of pea traits per pod based on the level of stress of the plant. And I find the data inconclusive as to whether any of the classic lines shows a significant relationship between the variance for the pea-shape and the number of peas per pod. This is worth retesting, in my opinion, with a focus on particular crosses, e.g., *Caractacus* × *Champion of England* deserves special attention!

Data from the pilot sample of 25 stressed plants

Cross	Plant	Pod	R_I_	R_ii	rrI_	rrii	Unclassifiable
00475	1	1	3	2		1	
00475	1	2	2		2		
00475	1	3	3			1	
00475	2	1	2	1	1		
00475	2	2	1				
00475	3	1	2	2		1	
00475	3	2	6				
00475	3	3	2	1	1		
00475	3	4	1	1	1		
00475	3	5	1				
00475	4	1	1	1	1	1	
00475	4	2	3	2			
00475	4	3	2		2		
00475	4	4	2			1	
00475	5	1	2	2	1		
00475	5	2	5				
00475	5	3	2	1		1	
00475	5	4	3				
00475	6	1	4	1	1		
00475	6	2	3	1			
00475	6	3	1	1	1		
00475	7	1	6				
00475	7	2	1	1	3		
00475	7	3		1		1	
00475	8	1	2	1	1		
00475	8	2	3	2			

(table continues)

(Continued)

Cross	Plant	Pod	R_I_	R_ii	rrI_	rrii	Unclassifiable
00475	8	3	1		2		
00475	8	4	1		2		
00475	9	1	2	1	1		
00475	9	2	2				
00475	10	1	1	3	2		
00475	10	2	3	1			
00475	10	3	2	1			
00475	11	1	3	1			
00475	11	2	4		1		
00475	11	3	1	1	1		
00475	11	4	3				
00475	12	1	3		1	1	
00475	12	2	2		2		
00475	12	3	3				
00475	12	4		1			
00475	13	1	2	1			
00475	13	2	2				
00475	14	1	3	1			
00475	14	2	2	2			
00475	15	1	4		1	1	
00475	15	2	2	1			
00475	15	3	2	1			
00475	15	4	1				
00475	16	1	2	1	1		
00475	16	2	3		2		
00475	17	1	2	2	1	1	
00475	17	2	2	1	1		
00475	17	3	1	1			
00475	18	1	1		2	1	
00475	18	2	1	2			
00475	18	3	1		2		
00475	18	4	1				
00475	19	1	3	2			

(table continues)

(Continued)

Cross	Plant	Pod	R_I_	R_ii	rrI_	rrii	Unclassifiable
00475	19	2		2	1	1	
00475	19	3	2	1	1		
00475	20	1	2		3		
00475	20	2	2	1	2		
00475	20	3					1 **rrii** or **R_ii**
00475	21	1	2		1	1	
00475	22	1	3		1	1	
00475	22	2	2		1	1	
00475	22	3	1		1	1	
00475	23	1	1	2	2	1	
00475	23	2					1 **R_I_** or **rrI_**; 2 **R_ii** or **rrii**
00475	23	3	2				
00475	24	1	2	2			
00475	24	2	3				
00475	25	1	1	2	1		
00475	25	2	3		1		
00475	25	3	1		1	1	
Totals			156	54	54	18	

Caractacus × Champion of England

R_: Smooth seed
I_: Yellow cotyledon
Le_: Long internodes (tall plant); All F1 plants of 00475 are **Le_**.
A_: Pink flowers

rr: Wrinkled seed
ii: Green cotyledon
lele: Short internodes (short plant)
aa: White flowers; All F1 plants of 00475 are **aa**.

CHAPTER 7

■ Mendelian Controversies
A Botanical and Historical Review

DANIEL J. FAIRBANKS AND BRYCE RYTTING

The science of genetics traces its origin to Gregor Mendel's classic experiments with the garden pea (*Pisum sativum* L.) and common bean (*Phaseolus vulgaris* L.). Mendel presented his findings to the Brünn Natural History Society in two lectures in the spring of 1865 and then published the lectures in the following year as a single paper under the title "Versuche über Pflanzen-Hybriden" (Experiments on Plant Hybrids), hereafter referred to as "Versuche" (Mendel, 1866). "Versuche" contains data from eight years of experimentation, statistical analysis of those data, and mathematical models of the fundamental laws of inheritance. Although his paper would eventually become the foundation for the science of genetics, Mendel did not live to see that day. He died in relative obscurity in 1884, 16 years before his work became widely known. This is not to say that "Versuche" was entirely ignored: it was cited at least 15 times between 1865 and 1899 (Olby, 1985). However, no one recognized its relevance to the science of inheritance until 1900 when three European botanists, Hugo de Vries, Carl Correns, and Erich von Tschermak, independently observed the same phenomena and arrived at the same interpretation as Mendel. In the first years of the 20th century, genetics established itself as a core discipline in biology and Mendel's work finally began to receive widespread recognition.

The year 2000 marks a century since the rediscovery of Mendel's work and the birth of genetics. During that century, Mendel's name became indispensable to science. The fundamental laws of inheritance are now known as Mendel's laws and the science on which they are based is called Mendelian genetics. However, because Mendel's importance was unrecognized during his lifetime, little original information about his scientific work was preserved. Most unfortunately, his scientific records were apparently burned around the time of his death (Olby, 1985; Orel, 1996).

In part because of the paucity of original documents, controversy plagued discussions of Mendel's work throughout the 20th century. Some authors praise Mendel as a brilliant scientist whose work was ahead of its time, others are critical of his methods, and a few claim he was a fraud. There is substantial disagreement about his objectives, the accuracy of his presentation, the statistical validity of his data, and the relationship of his work to evolutionary theories of his day. In the following pages we address five of the most contentiously debated issues by looking at the historical record through the lens of current botanical science: (1) Are Mendel's data too good to be true? (2) Is Mendel's description of his experiments fictitious? (3) Did Mendel articulate the laws of inheritance attributed to him? (4) Did Mendel detect but not mention linkage? (5) Did Mendel support or oppose Darwin?

We begin with a brief overview of Mendel's data.

A Brief Review of Mendel's Data

Mendel chose *Pisum* for his work after preliminary experiments with several plant species and an examination of botanical literature on plant hybridization, particularly C. F. Gärtner's (1849) *Versuche und Beobachtungen über die Bastarderzeugung im Pflanzenreiche* (Experiments and Observations on Hybrid Production in the Plant Kingdom). For his hybridization experiments, Mendel selected 22 pea varieties that he had confirmed through two years of testing to be true-breeding. He reported data from hybridization experiments on seven traits that differed among the varieties. In the following list of these traits, we include some information that is not in Mendel's paper but is pertinent for our later discussions, such as the modern designations of the genes Mendel studied and the chromosomes on which they reside.

1) Seed shape. In mature seeds, the dominant phenotype is a smooth or slightly indented round seed, and the recessive phenotype is a wrinkled angular seed. The varieties with wrinkled angular seeds were classified at the time as *Pisum quadratum*. The gene Mendel studied that governs this trait is r on chromosome 7.

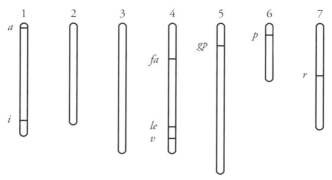

FIG. 1. Chromosomal locations of the genes that Mendel studied

2) Cotyledon color. In mature seeds, the dominant phenotype is yellow cotyledon color, and the recessive phenotype is green cotyledon color. The gene Mendel studied that governs this trait is i on chromosome 1.

3) Seed coat color. The dominant phenotype is a colored-opaque seed coat, and the recessive phenotype is a colorless-transparent seed coat. The gene Mendel studied that governs this trait is a on chromosome 1. Mendel noted that in his experiments, variation for seed coat color was always associated with variations for flower color and axillary pigmentation. He always found colored seed coats, colored flowers, and anthocyanin pigmentation at the axils of the stipules on the same plants, and colorless seed coats, white flowers, with no axillary pigmentation on the same plants in both parents and progeny. This complete association of phenotypes that Mendel observed was a case of pleiotropy, which Mendel observed because he studied alleles of the a gene. Pea researchers in the early 1900s reported epistatic interactions of these traits in a few experiments with other genes (White, 1917). However, variation for seed-coat color, flower color, and axillary pigmentation in most pea varieties is due to variation for alleles of the a gene, and consequently the pleiotropic association of these three traits is complete in most experiments.

4) Pod shape. The dominant phenotype is inflated pods, due to a parchment layer inside the pod, and the recessive phenotype is constricted pods, due to the absence of the parchment layer. The constricted-pod varieties were classified at the time as *Pisum saccharatum*, currently known as edible-pod sugar peas. The gene Mendel studied that governs this trait is either v on chromosome 4, or p on chromosome 6.

5) Pod color. The dominant phenotype is green unripe pods with green venation in the leaves, and the recessive phenotype is yellow unripe pods

with yellow venation in the leaves. The gene Mendel studied that governs this trait is *gp* on chromosome 5.

6) Flower position. The dominant phenotype is flowers borne at upper axillary positions along the plant, and the recessive phenotype is terminal flowers with stem fasciation near the plant apex. Varieties with terminal flowers were classified at the time as *Pisum umbellatum*. The gene Mendel studied that governs this trait is *fa* on chromosome 4.

7) Stem length. The dominant phenotype is a long stem between the internodes, causing a tall plant, and the recessive phenotype is short internodes, causing a dwarf plant. Some of Mendel's varieties had semi-dwarf phenotypes; however, he conducted experiments on stem length only with tall and dwarf parents. The gene Mendel studied that governs this trait is *le* on chromosome 4.

The first two of these traits are considered seed traits because they are observed in the seed cotyledons, which consist of embryonic tissue. Because each seed embryo is genetically a different individual, seed-trait phenotypes may differ among the seeds on a single heterozygous plant. The remaining five traits are considered plant traits because they are observed in whole plants. The seed coat consists of maternal rather than embryonic tissue, so its phenotype is the same on every seed coat of a single plant. Thus, it is considered a plant trait along with its pleiotropic counterparts, flower color, and axillary pigmentation.

The data from "Versuche" are summarized in Table 1. Mendel reported the results of the F_1 and F_2 generations of seven monohybrid experiments, one for each trait, and observed near 3:1 phenotypic ratios in the F_2 generation of each experiment (Experiments 1–7 in Table 1). He allowed some of the F_2 plants with the dominant phenotype to self-pollinate and obtained F_3 progeny from which he determined the genotypes of those F_2 plants. In each case, the ratio of homozygotes to heterozygotes among the F_2 individuals was close to 2:1 (Experiments 8–15 in Table 1).

He then conducted dihybrid or trihybrid experiments for all combinations of the traits but reported data for only one dihybrid experiment (for seed shape and cotyledon color; Experiments 16a and 16b in Table 1), and one trihybrid experiment (for seed shape, cotyledon color, and seed coat color; Experiment 17 in Table 1). In both experiments, he planted the F_2 seeds and allowed the F_2 plants to self-fertilize, then determined the genotypes of the F_2 individuals from the phenotypes of their F_3 progeny. The phenotypic and genotypic ratios for both the dihybrid and trihybrid experiments were consistent with the laws of segregation and independent assortment.

TABLE 1. A summary of Mendel's experiments on Pisum hybrids

Experiments	Results				Expected ratio
	F_2 generation of monohybrid experiments				
1. Seed shape[a]	Round	5474	Angular	1850	3 : 1
2. Cotyledon color[a]	Yellow	6022	Green	2001	3 : 1
3. Seed coat color	Colored	705	White	224	3 : 1
4. Pod shape	Inflated	882	Constricted	299	3 : 1
5. Pod color	Green	428	Yellow	152	3 : 1
6. Flower position	Axial	651	Terminal	207	3 : 1
7. Stem length	Long	787	Short	277	3 : 1
Totals	Dominant	14949	Recessive	5010	3 : 1
	F_3 progeny tests of F_2 individuals from monohybrid experiments				
8. Seed shape	Heterozygous	372	Homozygous	193	2 : 1
9. Cotyledon color	Heterozygous	353	Homozygous	166	2 : 1
Totals for seed traits	Heterozygous	725	Homozygous	359	2 : 1
10. Seed coat color	Heterozygous	64	Homozygous	36	2 : 1
11. Pod shape	Heterozygous	71	Homozygous	29	2 : 1
12. Pod color	Heterozygous	60	Homozygous	40	2 : 1
13. Flower position	Heterozygous	67	Homozygous	33	2 : 1
14. Stem length	Heterozygous	72	Homozygous	28	2 : 1
15. Pod color (repeat)	Heterozygous	65	Homozygous	35	2 : 1
Totals for plant traits	Heterozygous	399	Homozygous	201	2 : 1
	F_2 generation of dihybrid experiment for seed shape (A, a) and cotyledon color (B, b)				
16a. Phenotypic	Round, yellow	315	Angular, yellow	101	
	Round, green	108	Angular, green	32	9 : 3 : 3 : 1
16b. Genotypic[b]	AB	38	ABb	65	
	AaB	60	AaBb	138	
	aB	28	aBb	68	
	Ab	35	AaB	67	
	ab	30			1 : 2 : 1 : 2 : 4 : 2 : 1 : 2 : 1
	F_2 generation of trihybrid experiment seed shape (A, a), cotyledon color (B, b), and seed coat color (C, c)				
17. Genotypic[b]	ABC	8	ABCc	22	ABbCc	45
	ABc	14	AbCc	17	aBbCc	36

(table continues)

(continued)

Experiments	Results				Expected ratio		
	AbC	9	aBCc	25	AaBCc	38	
	Abc	11	abCc	20	AabCc	40	
	aBC	8	ABbC	15	AaBbC	49	
	aBc	10	ABbc	18	AaBbc	48	
	abC	10	aBbC	19			
	abc	7	aBbc	24	AaBbCc	78	
			AaBC	14			
			AaBc	18			
			AabC	20			
			Aabc	14			
			1:2:1:2:4:2:1:2:1:2:4:2:4:8:4:2:4:2:1:2:1:2:4:2:1:2:1				

Progeny genotypes[b] of reciprocal dihybrid testcross experiments for seed shape (A, a) and cotyledon color (B, b) (female parent listed first)

18. AaBb × AB	AB	20	ABb	23	AaB	25	AaBb	22	1:1:1:1
19. AB × AaBb	AB	25	ABb	19	AaB	22	AaBb	21	1:1:1:1
20. AaBb × ab	AaBb	31	Aab	26	aBb	27	ab	26	1:1:1:1
21. ab × AaBb	AaBb	24	Aab	25	aBb	22	ab	27	1:1:1:1

Progeny genotypes[b] of dihybrid testcross experiments for flower color (A, a) and stem length (B, b) (female parent listed first)

22. Aab × aBb	AaBb	47	Aab	40	aBb	38	ab	41	1:1:1:1

[a] Mendel also reported data for 10 individual plants from each of these experiments to illustrate the variation among plants. Because these data are a subset of experiments 1 and 2, we have not included the data for individual plants in this table.

[b] Throughout this table, we have used Mendel's genotypic designations, which are A = homozygous for A, Aa = heterozygous for A and a, a = homozygous for a, etc. Experiments 16a and 16b carry the same number because the data are from the same plants. The total number of individuals in experiment 16b is less than in experiment 16a because some F_3 seeds failed to germinate and others produced plants that failed to bear seeds.

He also conducted a series of testcrosses and reported data for four dihybrid testcrosses for seed shape and cotyledon color and one dihybrid testcross for flower color and stem length (Experiments 18–22 in Table 1). In the dihybrid testcrosses for seed shape and cotyledon color, he hybridized F_1 plants that were doubly heterozygous with the homozygous parental genotypes in all four possible combinations of reciprocal crosses. He carried each experiment through the number of generations required to determine the genotypes of the testcross progeny. In all four experiments, he observed near 1:1:1:1 ratios in the testcross progeny. He also observed a near 1:1:1:1 ratio in the progeny of the dihybrid testcross for flower color and stem length.

Mendel reported general non-numeric results for several experiments with the common bean (*Phaseolus vulgaris*) and stated that in most cases the results were similar to those obtained with *Pisum*. He also reported the numeric results of one preliminary experiment for flower color in *Phaseolus*. He hybridized a variety that had colored flowers with one that had white flowers, and in the F_2 generation he observed 30 plants with colored flowers and one with white flowers. He interpreted these results as adhering to either a 15:1 ratio for two factors, or a 63:1 ratio for three factors.

Having reviewed Mendel's data, we now address five of the most important controversies about his work.

Are Mendel's Data Too Good to Be True?

In 1902, two years after Mendel's work was rediscovered, W. F. R. Weldon suspected that Mendel's results were very close to expected values and tested this suspicion with Pearson's newly developed χ^2 test. He concluded that Mendel's observed ratios were astonishingly close to his expectations (Weldon, 1902). Weldon's analysis created little controversy and was quickly forgotten (Magnello, 1998). The next statistician to question the proximity of Mendel's results to expected values was Ronald A. Fisher (1936) who published a now-famous paper in which he closely examined Mendel's paper and reconstructed the thought process of the experiments. Fisher's analysis is careful and thorough and reveals his admiration for Mendel's work. However, his paper is best known for its conclusion, the same one that Weldon had arrived at 32 years earlier, that Mendel's results were consistently so close to expected ratios that the validity of those results must be questioned. Fisher's work spawned a series of papers dealing with this issue. Citations of these papers can be found in several reviews (Edwards, 1986; Piegorsch, 1986; Di Trocchio, 1991; Weiling, 1991; Nissani, 1994; Orel, 1996). Unfortunately, all this effort has failed to yield a definitive solution: according to Nissani (1994, p. 182), "the subject remains every bit as controversial today as it was in 1936."

Like Weldon's analysis, Fisher's was based on consistently low χ^2 values produced when he subjected Mendel's data to χ^2 tests. We present the χ^2 values and their associated probabilities with the appropriate degrees of freedom for each of Mendel's independent *Pisum* experiments in Table 2. As Edwards (1986) noted, about half of all independent experiments should yield χ^2 values with probabilities < 0.5 and half with probabilities

> 0.5. Of Mendel's 22 experiments, only four yield χ^2 values with probabilities < 0.5, and six yield χ^2 values with probabilities > 0.90, indicative of the bias toward expectation in Mendel's data (Table 2).

As Fisher (1936) and several subsequent authors (Sturtevant, 1965; Edwards, 1986) have pointed out, Mendel's data are most suspicious where they closely approach an incorrect expected ratio. After presenting the F_2 segregation ratios of his seven monohybrid experiments, Mendel proposed that two-thirds of the F_2 individuals with the dominant phenotype

TABLE 2. Results of chi-square tests for Mendel's *Pisum sativum* experiments

Experiment from table 1	Degrees of freedom	χ^2	Probability
1.	1	0.2629	0.6081
2.	1	0.0150	0.9025
3.	1	0.3907	0.5319
4.	1	0.0635	0.8010
5.	1	0.4506	0.5021
6.	1	0.3497	0.5543
7.	1	0.6065	0.4361
8.	1	0.1735	0.6771
9.	1	0.4249	0.5145
10.	1	0.3200	0.5716
11.	1	0.8450	0.3580
12.	1	2.0000	0.1573
13.	1	0.0050	0.9436
14.	1	1.2800	0.2579
15.	1	0.1250	0.7237
16a.	3	0.4700	0.9254
16b.	8	2.8110	0.9457
17.	26	15.3224	0.9511
18.	3	0.5778	0.9015
19.	3	0.8621	0.8346
20.	3	0.6182	0.8923
21.	3	0.5306	0.9121
22.	3	1.0843	0.7809

should be hybrids (heterozygotes) and the remaining third should be constant (homozygotes) for the trait in question, giving a ratio of 2:1. To test this hypothesis, he allowed F_2 plants with the dominant phenotypes to self-fertilize, then observed the phenotypic traits of the F_3 progeny. For the seed traits, cotyledon color and seed shape, this was a relatively easy task because the cotyledons of the seeds on the F_2 plants displayed the F_3 phenotypes at maturity, so there was no need for him to grow F_3 plants to score these traits. The five plant traits presented some difficulty because the F_3 phenotypes could only be scored in the F_3 plants. Because of limited garden space, Mendel chose 100 F_2 plants with the dominant phenotype for each of the five traits and grew ten F_3 descendents from each of these plants. If all ten F_3 descendents had the dominant phenotype, he classified the F_2 plant as constant (homozygous); if the F_3 descendents had both dominant and recessive phenotypes, he classified the F_2 plant as hybrid (heterozygous). One of the experiments (for pod color) yielded results that Mendel felt were too far from the predicted ratio of 2:1, so he repeated the experiment and obtained results that were more acceptable to him. By the conclusion of this set of six experiments (Experiments 10–15 in Tables 1 and 2), Mendel had scored the progeny of 600 F_2 plants, 399 classified as heterozygotes, and 201 classified as homozygotes, a ratio that was extremely close to his predicted 2:1 ratio.

Fisher (1936) explained that although the predicted ratio of 2:1 is genotypically correct, Mendel should have misclassified some heterozygotes as homozygotes:

> In connection with these tests of homozygosity by examining ten offspring formed by self fertilization, it is disconcerting to find that the proportion of plants misclassified by this test is not inappreciable. If each offspring has an independent probability, .75, of displaying the dominant character, the probability that all ten will do so is $.75^{10}$, or 0.0563. Consequently, between 5 and 6 percent of the heterozygous parents will be classified as homozygotes, and the expected ratio of segregating to nonsegregating families is not 2:1, but 1.8874:1.1126, or approximately 377.5:222.5 out of 600. Now among the 600 plants tested by Mendel 201 were classified as homozygous and 399 as heterozygous. Although these numbers agree extremely closely with his expectation of 200:400, yet, when allowance is made for the limited size of the test progenies, the deviation is one to be taken seriously.... We might suppose that sampling errors in this case caused a deviation in the right direction, and of almost exactly the right magnitude, to compensate for the error in theory. A deviation as fortunate as Mendel's is to be expected once in twenty-nine trials. (127)

Later in his paper, Fisher cited these experiments once again as evidence that Mendel's data were questionable:

A serious and almost inexplicable discrepancy has, however, appeared, in that in one series of results the numbers observed agree excellently with the two to one ratio, which Mendel himself expected, but differ significantly from what should have been expected had his theory been corrected to allow for the small size of his test progenies. To suppose that Mendel recognized this theoretical complication, and adjusted the frequencies supposedly observed to allow for it, would be to contravene the weight of evidence supplied in detail by his paper as a whole. Although no explanation can be expected to be satisfactory, it remains a possibility, among others that Mendel was deceived by some assistant who knew all too well what was expected. This possibility is supported by independent evidence that the data of most, if not all, of the experiments have been falsified so as to agree closely with Mendel's expectations. (134)

The last sentence (or part of it) is the most frequently quoted passage from Fisher's paper, and is often quoted out of context. Fisher did not accuse Mendel of fraud, nor did he claim that Mendel's description of his experiments was fictitious, as later historians were to do. Nonetheless, he suspected that an assistant manipulated the data and he was most disturbed by the fact that Mendel's data in the F_2 progeny tests were biased toward a 2:1 ratio rather than the ratio expected when the presumed effect of Mendel's misclassification is taken into account.

A χ^2 test of Mendel's observed values of 399 and 201 and Fisher's expected values of 377.5 and 222.5 yields a χ^2 value of 3.3020 with one degree of freedom, which is not statistically significant because its probability is 0.0692. However, the probability derived from a χ^2 test is the probability of a deviation as great as or greater than the observed deviation in either direction from the expected value. Fisher's calculation of a probability of one in 29 that Mendel would observe the deviation he did assumes a deviation of both the magnitude *and in the direction* he observed. This halves the probability to 0.0346, which corresponds closely to one in 29 trials.

Fisher (1936) stated his assumption that the probability of 0.75 for each individual displaying the dominant phenotype required independence. Weiling (1986, 1989) argued that Mendel sampled ten seeds per plant without replacement in the F_3 progeny tests, and that the sampling, therefore, was not independent. He assumed that the average pea plant in Mendel's experiments had 30 seeds per plant, 23 of which had the dominant phenotype (0.75 × 30 = 22.5, rounded to 23). Based on this assumption, Weiling determined that the average probability of misclassification was 23/30 × 22/29 × 21/28 × 20/27 × 19/26 × 18/25 × 17/24 × 16/23 × 15/22 × 14/21 = 0.0381, instead of 0.0563 as determined by Fisher.

However, although Weiling's estimate is correct for a plant with 30 seeds, 23 of which have the dominant phenotype, it cannot be used to estimate the average probability of misclassification for a population of plants. For any particular number of seeds per plant, the average probability of misclassification must be determined as the sum of the probabilities of misclassification for all possible combinations weighted by the expected frequencies of those combinations according to the binomial distribution. When this is done, the average probability of misclassification is consistently 0.0563. In other words, if Mendel's data are from random seed samples collected from a binomially distributed population, Fisher's estimate of 0.0563 as the probability of misclassification is correct, even when the effect of sampling seeds without replacement is taken into account.

Because Mendel provided data for each of the six experiments with 100 F_3 plants, we can partition the χ^2 test to examine each experiment individually, then calculate a series of χ^2 values which can then be summed and the probability determined with six degrees of freedom. Fisher (1936) partitioned these experiments in his calculation of the χ^2 values in Table 5 of his paper, but did not partition them when he concluded that "[a] deviation as fortunate as Mendel's is to be expected once in twenty-nine trials." As Piegorsch (1983) pointed out, when χ^2 values calculated with Fisher's expectations after correction for misclassification and partitioning of the six experiments, the summed χ^2 value is 7.6582 with a corresponding probability of 0.2642 with six degrees of freedom. When this probability is halved to 0.1321, it corresponds to a probability of about one in 7.6 trials, which is not statistically significant and much less serious than that implied by Fisher's estimate of one in 29 trials.

The proximity of Mendel's F_3 progeny data to an incorrect expectation is not as questionable as it might seem when viewed in a botanical context. Fisher's analysis is based on the assumption that Mendel scored exactly ten F_3 progeny from every F_2 plant in his experiments for plant traits. However, had Mendel sown exactly ten F_3 seeds from each F_2 plant, he would have scored fewer than ten F_3 progeny in some cases because of losses due to germination failure, and his misclassification of heterozygotes as homozygotes would have been even greater than that proposed by Fisher. For example, in experiments with nine plants, Mendel would have misclassified on average 7.51% of the heterozygotes as homozygotes ($0.75^9 = 0.0751$). On the other hand, Mendel probably sowed more than ten seeds in a space to be occupied by ten plants, then thinned the seedlings to ten to ensure that there were ten F_3 progeny from each F_2 plant. Indeed, Mendel's description of his method, "*von jeder 10 Samen angebaut*" is most appropriately translated as "10 seeds were cultivated," rather than "10 seeds were sown."

Had Mendel sown more than ten seeds from each F_2 plant, then he could have scored two of the plant traits in seedlings before thinning. Differences in stem length, as Mendel noted in his paper, can be easily scored in seedlings a few days after germination. Also, as Mendel further noted in his paper, variation for seed-coat color was perfectly correlated with variation for axillary pigmentation in his experiments. Mendel could score F_3 plants for the presence or absence of axillary pigmentation as early as two to three weeks after germination and identify the phenotypes for flower color and seed-coat color that the plants would attain if grown to the flowering stage or to maturity. Indeed, almost half of the deviation from Fisher's expectations in Mendel's F_2 progeny tests comes from these two experiments, in which the 136 heterozygotes reported by Mendel exceed Fisher's expectation by ten plants (nine from the experiment for stem length). If the results of these two experiments are excluded, χ^2 values for Fisher's expectations are not statistically significant for either summed data ($\chi^2 = 1.3795$, 1 df, $P = 0.2402$) or partitioned data ($\chi^2 = 4.0687$, 4 df, $P = 0.3968$).

Fisher raised the same concern about Mendel's trihybrid experiment (Experiment 17 in Table 1). In the trihybrid experiment, Mendel determined the genotypes for all three traits (seed shape, cotyledon color, and seed-coat color) of each F_2 individual in the F_3 progeny. He could determine the seed-shape and cotyledon-color genotypes directly from the F_3 seeds on the F_2 plants (although he had to remove at least part of the opaque seed coats from the seeds on plants with colored seed coats to determine cotyledon color). However, to determine the F_2 genotypes for seed coat color he had to grow F_3 plants. Fisher (1936) speculated that Mendel must have grown ten F_3 progeny from each F_2 plant with colored seed coats, as in the F_3 progeny tests for seed-coat color (Experiment 10 in Tables 1 and 2). However, Mendel's description of his method is vague; he simply referred to it as "further investigations [*Weiteren Untersuchungen*]." Had Mendel scored exactly ten F_3 progeny from each F_2 plant with colored seed coats, the bias in his data toward an incorrect expectation is even greater than in the monohybrid F_3 progeny tests when the summed data are used ($\chi^2 = 4.9617$, 1 df, $P = 0.0259$). When halved, the probability corresponds to about one in 77 trials.

To explain the bias in the trihybrid experiment, Orel and Hartl (1994, *196*) suggested that "if Mendel had cultivated 12 seeds per plant rather than 10, then $\chi^2 = 3.0$, for which $P > 0.05$ and the insinuation of data tampering evaporates." Fisher (1936, *130*) recognized this possibility, stating in reference to the trihybrid experiment that "if we could suppose that larger progenies, say fifteen plants, were grown on this occasion, the greater part of the discrepancy would be removed. However, even using families of 10 plants the number required is more than Mendel had assigned to any previous experiment." Fisher's concern in this passage is Mendel's lack of garden space for growing the large number of plants required for this experiment (4730 F_3 plants if he grew ten F_3 plants from each F_2 plant that had colored seed coats).

Hennig (2000) raised an additional concern about this experiment, questioning why Mendel chose seed-coat color as the third trait for analysis in a trihybrid experiment. It created more work for Mendel than the other plant traits because for many of the seeds he had to remove part of the seed coats to score seed color. Also, "he would know about the first two components of his trihybrid cross (pea shape and color) more than nine months before he found out about the third" (Hennig, 2000, p. 127).

However, we can dismiss Fisher's concern about lack of garden space and Hennig's concern about a nine-month delay in scoring when we consider that Mendel could distinguish the plants in question for presence or absence of axillary pigmentation as seedlings. Because this trait can be scored in seedlings, it is an excellent choice for the third trait in the trihybrid experiment because it creates at most a three-week delay between data collection for the first two traits and the third. Garden space is not as critical because many seedlings can be grown in the space occupied by a single mature plant. Mendel probably harvested the F_3 seeds for this experiment in July or early August. Had he planted the seeds in the garden soon after harvest, he could have scored the seedlings for axillary pigmentation within three weeks, long before cold weather affected the plants. Alternatively, Mendel had a 27.5 × 4.5 m greenhouse available to him (Orel, 1996) and he could have grown as many as 15–20 seedlings per pot for scoring in the greenhouse during the fall and winter months. If he had allotted one pot for the F_3 seedlings from each F_2 plant, he had more than ample space in the greenhouse for the 473 pots required for this experiment.

When these statistical and botanical aspects of Mendel's F_3 progeny tests are considered, there is no reason for us to question his results from these experiments. However, we must still account for the bias that is evident when the data for all of the experiments that he reported are compared as a whole (Table 2). After Fisher, numerous authors have sought reasonable explanations for the bias in Mendel's data that do not imply fraud, but most cannot withstand botanical or historical scrutiny.

Wright (1966) and Beadle (1967) proposed that Mendel might have unconsciously misclassified individuals with questionable phenotypes to favor his expectations. From a botanical point of view, this explanation can account for only a negligible degree of bias. The five plant traits display very distinct phenotypes in multiple positions within each plant so that each plant's phenotype can be readily identified without error. The two seed traits, seed shape and cotyledon color, however, are potentially subject to misclassification of phenotypes. Round seeds often have indentations that under certain circumstances might lead an untrained researcher to misclassify a round seed as being wrinkled. Also, some seeds whose genotype should confer a round-seed phenotype do not fully develop in the pod and may appear wrinkled. However, such seeds have irregular wrinkles, and Mendel used for his seed-shape experiments varieties with seeds that have regular angular wrinkles (in Mendel's words, *kantig runzlig Samen*). Angular wrinkled seeds have a cube-shaped phenotype and for this reason were classified as *Pisum quadratum* in Mendel's day. Under these circumstances, the number of seeds that Mendel misclassified for seed-shape phenotypes was probably negligible.

The trait most likely to be misclassified is cotyledon color because it is subject to some degree of environmental variation. Green seeds may turn yellow when mature plants are left unharvested too long in the sun. Yellow seeds may appear green if they are harvested before reaching full maturity. Mature pea seeds of some varieties may have segments of green and yellow coloration in the cotyledons. Mendel was aware that misclassification of seed color was possible: "in individual seeds of some plants green coloration of the albumin is less developed and can be easily overlooked." However, he dismissed the possibility of misclassification of cotyledon color, stating in reference to this trait that "with a little practice in sorting, however, mistakes are easy to avoid" (87).

Also, because of his experimental design, Mendel only could have misclassified cotyledon color and seed shape in a limited number of individuals. For some of the F_2 individuals in his monohybrid experiments and for all individuals in his dihybrid and trihybrid experiments, Mendel identified not only the phenotypes, but also the genotypes of individual seeds through examination of their self-fertilized progeny. This procedure would have allowed him to correct any initial phenotypic misclassifications for individuals whose genotype had been determined.

Olby (1985) and Beadle (1967) suggested that Mendel might have stopped counting individuals when the numbers were close to the ratios he expected. However, as Campbell (1985) and Orel (1996) pointed out, Mendel explicitly denied this practice when he wrote near the beginning of his paper, "To discover the relationships of hybrid forms to each other and to their parental types it seems necessary to observe *without exception all* members of the series of offspring in each generation" (*80*).

Despite this statement Olby (1985) claimed that the data for Mendel's seed shape experiment indicate that he indeed did not count all of the individuals in this experiment:

> If Mendel stopped recording his seeds before he had exhausted the material, one would expect that his totals would be less than that of an average crop for the population of mother plants grown. This is so. Mendel stated that fully ripe pods contained between 6 and 9 seeds. If we take 6 as the average number, in order to make an allowance for unripe pods, then the 7,324 seeds which Mendel harvested from 253 plants would have come from 1,046 pods, thus giving 4 to 5 pods per plant. (Olby, 1985, p. 211)

Olby considered this estimate of pods per plant to be too small, implying that Mendel did not count all of the seeds in this experiment. Di Trocchio (1991) raised a similar concern:

> From a calculation made by Margaret Campbell [1976], it appears that Mendel obtained an average of 28 to 37 seeds per plant.... If the pea plant actually produced so few seeds, this vegetable would be a rarity in the markets! Instead, we know that each plant produces on average more than 60 pods, and Mendel himself informs us that his plants produced pods that contained an average of 6–9 seeds; he would therefore have obtained at least 400–500 seeds from every plant. (Di Trocchio, 1991, p. 504)

If all the information provided by Mendel on numbers of seeds per plant is taken into account, the average number of seeds per plant is close to 30 (16,590 seeds divided by 550 plants = 30.16 seeds per plant from the monohybrid, dihybrid, and trihybrid experiments for seed shape and cotyledon color). Is 30 seeds per plant too low for 19th century pea cultivation? Fisher (1936, *000*) addressed this question and quoted Dr. J. Rasmussen, a pea geneticist, who wrote to Fisher: "About 30 good seeds per plant is, under Mendel's conditions (dry climate, early ripening, and attacks of *Bruchus pisi*) by no means a low number."

Olby's and Di Trocchio's estimates that the average number of seeds per pod in Mendel's experiments is six to nine are based a passage in "Versuche":

> In these two experiments [Experiments 1 and 2 in Table 1] each pod usually yielded both kinds of seed. In well-developed pods that contained on the average, six to nine seeds, all seeds were fairly often round (Experiment 1) or all yellow (Experiment 2); on the other hand, no more than 5 angular or 5 green ones were ever observed in one pod. (86)

According to Mendel, the average number of seeds per pod was not six to nine for all pods, but rather for "well-developed pods [*gut ausbebildeten Hülsen*]." His point in this passage was not to give the average number of seeds per pod in all of his experiments, but to illustrate variation for phenotypic ratios within well-developed pods that have a relatively large number of seeds. The average number of seeds per pod in all of Mendel's experiments was probably between three and four, which results in an average number of about seven to ten pods per plant.

We searched for a significant body of data collected during the 19th century on numbers of seeds per pod, pods per plant, and seeds per plant in *Pisum sativum*. The earliest data we found were from a field test of 24 garden-pea varieties conducted at the New York Agricultural Experiment Station in Geneva, New York during 1888 (Curtis, 1889). The average number of seeds per pod was 4.47, the average number of pods per plant 10.82, and the average number of seeds per plant 47.18, averages that are only slightly higher than those Mendel observed. This confirms Rasmussen's claim. Of the 24 varieties tested, four produced averages of less than 30 seeds per plant. Therefore, although Olby's and Di Trocchio's skepticism about the average number of seeds per plant in Mendel's data might be valid for modern pea varieties grown under conditions of high fertility, the number of seeds per plant reported by Mendel cannot be considered unreasonably low for pea varieties grown in the 19th century.

Another explanation of the bias in Mendel's data is botanical. Sturtevant (1965), Thoday (1966), and Weiling (1989, 1991) proposed that because pollen grains are produced in tetrads that consist of the four products of a meiotic event, and because the mature pollen grains from a tetrad may remain juxtaposed following dehiscence, there is a possibility that ovules in a self-pollinated pea flower may be fertilized by two or more pollen grains from the same tetrad. This model has often been called the "urn model" because it is analogous to a person sampling items without replacement from an urn. If such an event is common in pea fertilization, it could bias genetic data away from binomial distributions and toward mean ratios. Beadle (1967) determined that such an effect is insufficient to explain the bias in Mendel's data. There is currently no empirical evidence to support the urn model in *Pisum*, but it is one that can be empirically tested because it should produce a significant deviation from a binomial distribution for phenotypes of individual seeds from the same pod (or plants grown from those seeds). We have initiated the necessary experiments but do not yet have the results.

We believe that the most likely explanation of the bias in Mendel's data is also the simplest. If Mendel selected for presentation a subset of his experiments that best represented his theories, χ^2 analyses of those experiments should display a bias. His paper contains multiple references to experiments for which he did not report numerical data, particularly di- and trihybrid experiments. For example, he conducted dihybrid or trihybrid experiments for all combinations of the seven characters he studied. However, he reported data for only one dihybrid and one trihybrid experiment. In his words: "Several more experiments were carried out with a smaller number of experimental plants in which the remaining traits were combined by twos or threes in hybrid fashion; all gave approximately equal results" (*94*). He also conducted dihybrid testcross experiments with all seven traits but reported only those for seed shape and cotyledon color, and flower color and stem length: "Experiments [dihybrid testcrosses] on a small scale were also made on the traits of *pod shape*, *pod color*, and *flower position*, and the results obtained were in full agreement: all combinations possible through union of the different traits appeared when expected and in nearly equal numbers" (*100*). In his second letter to Nägeli (Stern and Sherwood, 1966), Mendel described a true-breeding genotype that he obtained in 1859 (the fourth year of his experiments) from a tetrahybrid experiment for cotyledon color, seed coat color, pod shape, and stem length. He did not report the results of this or any other tetrahybrid experiment in his paper. He described in the concluding remarks of his paper a pentahybrid reciprocal backcross experiment carried through several generations, but he only reported a few selected data from this experiment. He also reported in his paper that he conducted experiments on the timing of flowering, peduncle length, and brownish-red pod color but he likewise did not report the data for these experiments.

Mendel made it very clear that the data reported in his paper are from a subset of experiments that he conducted. In Mendel's second letter to Nägeli, he referred to his paper as "the unchanged reprint of the draft of the lecture mentioned; thus the brevity of the exposition, as is essential for a public lecture" (Stern and Sherwood, 1966, p. 61). Had he included all of his data, the paper would have been much longer. Mendel's choice to present data from a subset of his experiments created a bias that was detected only when 20th century scientists subjected his data to statistical analysis.

Several authors have been quick to label Mendel as a fraud on the basis of Fisher's analysis. As examples, Orel (1996) cited articles by Doyle (1968, "Too many small χ^2's or hanky-panky in the monastery?") and Gardner (1977, "Great fakes of science"), and a book by Broad and Wade (1983, *Betrayers of the Truth*). Orel (1996, p. 207) then stated, "These selected examples show how great scientific achievements can be discredited by dilettantes who claim a combination of two incompatibles: the rigorousness of a meticulous scientist, and falsification of the results." We conclude that, although the bias in Mendel's experiments is evident, there are reasonable statistical and botanical explanations for the bias, and insufficient evidence to indicate that Mendel or anyone else falsified the data.

Is Mendel's Description of His Experiments Fictitious?

Some authors claim that although the data in Mendel's paper may be accurate, his description of the experiments is fictitious. This assertion stems mostly from suppositions about his monohybrid experiments. A monohybrid experiment is one in which two homozygous individuals that differ from each other in only one trait are hybridized. The F_1 progeny are called monohybrids and are heterozygous for only one of the genes under study.

After reporting the results of his monohybrid experiments on each of the seven traits, Mendel wrote, "In the experiments discussed above, plants were used which differed in only one essential trait [*wesentliches Merkmal*]" (*90*). Several authors doubt Mendel's claim and argue that he did not conduct true monohybrid experiments. The first to do so was William Bateson, the most ardent defender of Mendelism in the first decade of the 20th century. In a footnote to the Royal Horticultural Society's English translation of Mendel's paper, Bateson referred to Mendel's claim:

This statement of Mendel's in the light of present knowledge is open to some misconception. Though his work makes it evident that such varieties may exist, it is very unlikely that Mendel could have had seven pairs of varieties such that the members of each pair differed from each other in *only* one considerable character (*wesentliches Merkmal*). (*115n15*)

In the introduction to his paper, Fisher (1936) quoted Bateson's statement that Mendel's experiments might be fictitious and proposed that a reconstruction of Mendel's experiments might determine whether or not they were. After a detailed analysis and proposed reconstruction of the experiments, Fisher concluded that "there can, I believe, be no doubt whatever that his report is to be taken entirely literally, and that his experiments were carried out in just the way and in much the order that they are recounted" (*134*).

Half a century later, Corcos and Monaghan (1984) resurrected Bateson's claim and held it to be of "considerable importance." They entitled their paper "Mendel had no 'true' monohybrids" and concluded that Mendel's "'monohybrid' experiments were performed with varieties [that differed] in several traits but that in each offering he concentrated his attention on only one" (p. 499).

Such claims that Mendel's experiments were fictitious have little foundation when viewed from a botanical perspective. Taken in the context of Mendel's paper, we must interpret his statement that "plants were used which differed in only one essential trait" as meaning that each pair of parental varieties used for the monohybrid experiments differed from each other in only one of the seven traits he studied. Much of the confusion on this issue arises from the fact that *Pisum sativum* is a domesticated species and among the many cultivated varieties there are several different phenotypes. Bateson's claim, which other authors have accepted (Corcos and Monaghan, 1984; Di Trocchio, 1991; Bishop, 1996), is based on the notion that the pea varieties available to Mendel were so highly varied that he could not have paired his pea varieties to create seven monohybrid experiments.

Contrary to this claim, the nature of variation in pea varieties (both old and modern) facilitates, rather than prevents, the construction of monohybrid experiments. Pea varieties fall into three general categories: garden varieties (also called shelling varieties), field varieties, and sugar varieties. Most garden varieties have white seed coats (also white flowers and no axillary pigmentation), inflated pods, green pods, and axillary flowers; they vary for seed shape, cotyledon color, and stem length. White seed coats are desirable for garden varieties because colored seed coats can discolor the cooked peas. Inflated pods are desirable because they facilitate shelling unripe peas from the pod. Because garden varieties typically do not vary for seed-coat color, pod shape, pod color, and flower position, Mendel could easily design monohybrid experiments among them for seed shape, cotyledon color, and stem length. Garden varieties with all possible combinations of the differing phenotypes for these three traits were readily available in Mendel's day, and still are available among modern commercial garden varieties. Field varieties were used mostly as fodder in Mendel's day and they typically display the dominant phenotypes for all seven traits. Many sugar varieties also display the dominant phenotypes for all traits except pod shape. Thus, Mendel's monohybrid experiment for pod shape may have included a field variety and a sugar variety as parents. There are several possibilities for Mendel's monohybrid experiment for seed-coat color. He could have hybridized a field variety with a garden variety that differed only for seed-coat color. Also, sugar varieties with white seed coats were available in his day, so he could have used two sugar varieties that differed only for seed-coat color. Varieties with terminal flowers have always been rare novelties. The most readily available terminal-flowered variety in the 19th century was called the Mummy Pea and it is probably the variety that Mendel used. White (1917) described two Mummy varieties, one with white flowers and one with colored flowers. The colored-flower variety was the one probably available to Mendel, and this variety differs from most field varieties only in flower position. Thus Mendel could have easily designed a monohybrid experiment for flower position. Most pea varieties have green pods. However a few garden and sugar varieties, called gold varieties, have yellow pods. Mendel could have matched any one of the gold varieties with another garden or sugar variety in a monohybrid experiment. A mathematical minimum of eight varieties is required for seven monohybrid experiments. We conclude that Mendel could have easily designed and conducted seven monohybrid experiments with 22 varieties at his disposal.

Di Trocchio (1991) accepted the argument of Bateson (1913) and Corcos and Monaghan (1984) that Mendel's experiments were fictitious. He then took the argument a step further, claiming that instead of conducting monohybrid experiments, Mendel must have hybridized the 22 varieties in all possible combinations, then disaggregated the data into fictitious mono-, di-, and trihybrid experiments in his presentation, for the sake of simplicity. Bishop (1996) used Di Trocchio's (1991) proposed reconstruction of Mendel's experiments as evidence that Mendel began his experiments in 1861 and conducted them over a period of four years. Mendel stated in "Versuche" that he conducted his experiments over a period of eight years, and he clarified the dates as 1856–1863 in his second letter to Nägeli (Stern and Sherwood, 1966). Bishop argued that the dates in Mendel's second letter to Nägeli must have been wrong and that Mendel wrote them because the letter was "obviously a defensive response to the latter's [Nägeli's] criticism" (Bishop, 1966, p. 206). Bishop's claim that Mendel's experiments as reported in his paper were fictitious was part of an attempt to demonstrate that they were inspired by his reading Darwin's (1859) *On the Origin of Species by Means of Natural Selection or the Preservation of Favoured Races in the Struggle for Life* (hereafter referred to as the *Origin*). Bishop surmised that Mendel was inspired by Darwin's work in 1861 (when Mendel may have first heard of Darwin's theory of natural selection in a lecture) and that he thereafter began his experiments to counter Darwin's theory and promote the theory of special creation.

Di Trocchio's (1991) and Bishop's (1996) claim that Mendel hybridized his 22 varieties in all possible combinations runs counter to the experimental design that Mendel described and the logic on which it is based. Mendel first tested each trait individually in monohybrid experiments, then subsequently combined traits in twos and threes to determine whether the patterns of inheritance were independent. When he began his monohybrid experiments, he probably did not know whether or not the inheritance of one trait influenced the inheritance of another. Monohybrid experiments were essential to his experimental design if he intended to study the inheritance of a particular trait in the absence of any possible confounding influences from other differing traits. Once he had determined that the inheritance of each trait in isolation followed the same pattern, he could then study the patterns of inheritance for combinations of two or more traits. Taken literally, Mendel's account describes a well-conceived experimental design that would not have been difficult for him to perform.

Did Mendel Articulate the Laws of Inheritance Attributed to Him?

The two laws of inheritance most often attributed to Mendel are segregation and independent assortment. The law of segregation, stated in modern terms, is the idea that during meiosis two alleles of a single locus, one inherited from each parent, pair with each other, and then segregate from one another into the germ cells so that each germ cell carries only one allele of that locus. Segregation in heterozygous individuals produces in equal proportions two different types of gametes, each with one of the two alleles. The law of independent assortment, stated in modern terms, is the idea that the segregation of alleles of a single locus has no influence on the segregation of alleles at another locus. The result is completely random and uniform combinations of alleles of different loci in the self-fertilized progeny of dihybrid (or multihybrid) individuals.

Several authors question Mendel's articulation of these laws. Olby (1985) attributed segregation of character elements (not necessarily what we now perceive as alleles) to Mendel:

The whole theory rests on one inference which no one else had the thought of making. It was simply the prediction of the number of different forms that would result from the random fertilisation of two kinds of "egg cells" by two kinds of pollen grains. Naudin had postulated the segregation of specific essences in the formation of germ cells; Mendel postulated the segregation of character elements. (Olby, 1985, p. 101)

However, although Olby attributes the laws of inheritance to Mendel, he also concluded that "the laws of inheritance were only of concern to him [Mendel] in so far as they bore on his analysis of the evolutionary role of hybrids," and that "Mendel did not have the conception of pairs of factors or elements determining his pairs of contrasted characters" (Olby, 1979, p. 67). Monaghan and Corcos (1990, p. 268) fully rejected the idea that Mendel articulated the laws of segregation and independent assortment, stating that "he [Mendel] did not explain his results by employing invisible particulate determiners, paired or otherwise," and that "the traditional Mendelian laws of segregation and independent assortment are not given in the paper." They also concluded that "the first Mendelian law, the law of segregation is not present anywhere in Mendel's paper. That it cannot be found has been said many times by quite a few writers" (p. 287). Callender (1988, pp. 41–42) called it "the myth of 'Mendel's Law of Segregation'; a law not to be found in either of Mendel's papers, nor in his scientific correspondence, nor in any statement that can be unambiguously attributed to him." Monaghan and Corcos (1990, 1993) claimed that two of Mendel's rediscoverers, de Vries (1900) and Correns (1900), were the first to articulate the law of segregation, and that Thomas Hunt Morgan (1913) was the first to articulate the law of independent assortment.

The claim that Mendel did not articulate or perceive the law of segregation in his interpretation of his experiment is based in part on the difficulty that modern readers have finding in Mendel's paper statements that resemble the current concept of segregation. Genetic terms, such as *allele*, *locus*, and *chromosome*, had not been coined in Mendel's day, nor was the cellular process of meiosis understood. Therefore, we must look for statements in Mendel's paper indicating that he perceived paired hereditary factors that segregate from one another during the formation of germ cells.

As pointed out by Olby (1985), Hartl and Orel (1992), Orel and Hartl (1994), Weiling (1994), and Fairbanks and Andersen (1999), Mendel referred to segregation of hereditary elements several times in his paper. We (along with Olby, 1985) believe that the clearest statement is in the concluding remarks of "Versuche":

One could perhaps assume that in those hybrids whose offspring are variable a compromise takes place between the differing elements of the germinal and pollen cell great enough to permit the formation of a cell that becomes the basis for the hybrid, but that this balance between antagonistic elements is only temporary and does not extend beyond the lifetime of the hybrid plant. Since no changes in its characteristics can be noticed throughout the vegetative period, we must further conclude that the differing elements succeed in escaping from the enforced association only at the stage at which the reproductive cells develop. In the formation of these cells, all elements present participate in completely free and uniform fashion, and only those that differ separate from each other. In this manner the production of as many kinds of germinal and pollen cells would be possible as there are combinations of potentially formative elements. (*111*)

Those who search for statements on segregation in Mendel's paper often overlook this paragraph, probably because it is near the end of the paper embedded in a discussion in which Mendel attempted to reconcile his observations of predictable variation in the offspring of hybrids with those of other hybridists who reported that some hybrids breed true. Also, some reprints of English translations of Mendel's paper omit the part of his paper that includes this paragraph (for example, see Peters, 1959). This paragraph, however, is a remarkably lucid summary of the law of segregation. Mendel's reference to "potentially formative elements [*bildungsfähigen Elemente*]" implies the existence of invisible particulate determinants of inherited traits. We might well view the term "element [*Element*]", which Mendel used five times in this passage, as the equivalent of the modern term "allele." This view is reinforced by Hennig's (2000) observation that Mendel used the German term *Element* only ten times in his paper, all near the end of the paper in reference to the plant's genotype. Mendel's reference to the "enforced association [*erzwungenen Verbindung*]" of differing elements indicates that he perceived the differing elements as being paired in hybrids (heterozygotes). His statement that differing elements "separate from each other [*sich gegenseitig ausschliessen*]" shows a clear understanding of segregation that is similar to the modern view. He also correctly recognized that segregation takes place "only at the stage at which the reproductive cells develop" (i.e., during meiosis).

This paragraph, however, reveals one aspect of Mendel's perception that differs from the modern concept of segregation. According to the modern concept, alleles at a single locus, whether different as in heterozygotes or the same as in homozygotes, segregate from one another during meiosis. The above passage suggests that Mendel perceived segregation as an anomaly restricted to hybrids (heterozygotes). He called the differing elements "antagonistic elements [*widerstrebenden Elemente*]" whose association in the hybrid is a "compromise [*Vermittlung*]," and wrote that "only those elements that differ separate from one another," statements that, rephrased in modern terms, suggest that only those alleles in the heterozygous condition, and not those in the homozygous condition, are paired and segregate from one another.

In the paragraph that precedes the one we cited, Mendel explained his understanding of this concept:

> When the reproductive cells are of the same kind and like the primordial cell of the mother [i.e., a homozygous cell], development of the new individual is governed by the same law that is valid for the mother plant. When a germinal cell is successfully combined with a dissimilar pollen cell we have to assume that a compromise takes place between those elements of both cells that cause their differences. The resulting mediating cell [heterozygous cell] becomes the basis of the hybrid organism whose development must necessarily proceed in accord with a law different from that of the two parental type. (110)

Mendel's apparent perception of segregation as a phenomenon restricted to heterozygotes sheds light on another aspect of his paper. Although Mendel represented heterozygotes with a two-letter designation (*Aa*), as modern geneticists usually represent them, he consistently represented homozygotes with a single letter (*A* or *a*), rather than the two letters (*AA* or *aa*) used today. For example, Mendel represented the genotypic ratio of the F_2 generation of a monohybrid experiment as *A* + 2*Aa* + *a*, instead of *AA* + 2*Aa* + *aa*. According to the passage above, Mendel may have concluded that like elements (alleles) do not pair with one another and do not segregate in plants that are not hybrids (i.e., are not heterozygotes), and that therefore a single letter was an accurate way to represent such plants. Hartl and Orel (1992, p. 250) defended Mendel's understanding of segregation, reminding us that Mendel was not aware of chromosomes, and when the law of segregation is stated only in terms of different alleles, rather than in terms of chromosomes, "Mendel's view of segregation occurring only in the heterozygotes (i.e. with different alleles) could easily be defended as being completely consistent even with the modern use of the term."

Although many authors have overlooked the passage we cited above, in which Mendel described the law of segregation, few have missed the following often-quoted statement of independent assortment, which Olby (1979) called the climax of Mendel's paper. This statement appears immediately after Mendel's presentation of his di- and trihybrid experiments:

> In addition, several more experiments were carried out with a smaller number of experimental plants in which the remaining traits were combined by twos and threes in hybrid fashion; all gave approximately equal results. Therefore there can be no doubt that for all traits included in the experiment, this statement is valid: *The progeny of hybrids in which several essentially different traits are united represent the terms of a combination series in which the series for each pair of differing traits are combined.* This also shows at the same time that the behavior of each

pair of differing traits in a hybrid association *is independent of all other differences in the two parental plants.* (94)

Mendel's liberal use of italics in this passage indicates that he wished to emphasize his conclusion of the independent inheritance of different traits.

Did Mendel Detect but Not Mention Linkage?

Linkage is defined as a significant deviation from independent assortment due to proximity of genes on the same chromosome. In his statement on independent assortment quoted above, Mendel concluded that for all of the traits he studied "the behavior of each pair of differing traits in a hybrid association is independent of all other differences in the two parental plants" (94). Scientists have often been intrigued by the notion that Mendel studied seven traits in a species that has a haploid number of seven chromosomes, implying that Mendel discovered one gene on each of the seven chromosomes, and for this reason he observed independent assortment. According to some accounts, had he chosen to study just one more trait he would have detected linkage. Dunn (1965), under the assumption that Mendel studied one gene on each of the seven chromosomes, calculated the probability of doing so as $6/7 \times 5/7 \times 4/7 \times 3/7 \times 2/7 \times 1/7 = 0.0061$ ($< 1\%$), again calling Mendel's experimental results into question.

However, based on genetic maps of pea chromosomes, Nilsson (1951), Lamprecht (1968), Blixt (1975), and Novitski and Blixt (1979) showed that Mendel did not study one gene per chromosome. Instead, he studied two genes on chromosome 1 (*a* and *i*), no genes on chromosomes 2 and 3, either two genes (*fa* and *le*) or three genes (*fa*, *le*, and *v*) on chromosome 4, one gene (*gp*) on chromosome 5, possibly one gene (*p*) on chromosome 6, and one gene (*r*) on chromosome 7 (Fig. 1). The gene in doubt is the one that governs pod shape. Recessive alleles of the *v* gene on chromosome 4 and the *p* gene on chromosome 6 both confer constricted (unparchmented) pods when homozygous, and it is not known which of the two Mendel studied.

The idea that two genes on the same chromosome are necessarily linked is a common misconception. Genes on the same chromosome are said to be syntenic but are linked only if they are so close to one another that the frequency of crossovers between them is significantly less than the frequency of recombination for independent assortment. The genetic map distance at which linkage cannot be detected depends on the type of experimental population, the number of individuals in the experimental population, the degree of undetected double crossovers, and several other factors. In most testcross experiments (the most reliable type of linkage experiment), linkage often cannot be distinguished from independent assortment for genes located more than ~60 cM (centiMorgans) apart unless researchers use large numbers of progeny and a mapping function that is appropriate for the species under study. The two genes that Mendel studied on chromosome 1, *i* which governs seed color and *a* which governs flower color, are 204 cM apart, so distant that they assort independently. The same is true for two of the genes that Mendel studied on chromosome 4; *fa* which governs flower position and *le* which governs stem length are 121 cM apart.

Given the traits he studied, Mendel would not have detected linkage if he studied the *p* gene, which governs pod shape, on chromosome 6. He could have observed one case of linkage if he studied the *v* gene, which also governs pod shape. The *v* gene is located 12 cM from the *le* gene on chromosome 4. Novitski and Blixt (1979) compared the arguments favoring the *p* gene with those favoring the *v* gene for Mendel's studies and concluded that either scenario was possible. Because the varieties that Mendel used are not known, the question as to which of these two genes he studied is not likely to be resolved.

However, let's suppose that Mendel did study the *v* gene and thus had the opportunity to observe linkage. As Novitski and Blixt (1979) pointed out, Lamprecht (1968) reported that observed recombination between *le* and *v* may vary from 2.6 to 38.5% and that, according to Lamprecht (1941), the mutation rate for *v* may be as high as 40%. Had either or both of these values been on the higher side in Mendel's experiments, he would not have detected linkage for the *v* and *le* genes.

Also, Mendel conducted his experiments with F_2 progeny, which are not as reliable for detection of linkage as are testcross progeny. Although Mendel did not report data for his experiment with stem length and pod shape, we are fortunate that he described such an experiment in his second letter to Nägeli. The experiment was a tetrahybrid experiment in which one parental variety had green cotyledons, white seed coats, inflated pods, and short stems, and the other had yellow cotyledons, colored seed coats, constricted pods, and long stems. Mendel obtained a true-breeding F_3 plant in 1859, the fourth year of his experiments, that had yellow cotyledons, white seed coats, inflated pods and long stems. Thus, the dominant alleles for pod shape and stem length, if they were linked, were in repulsion conformation in the F_1 generation, and were recombined into coupling conformation in the true-breeding descendent that Mendel described.

Linkage for loci that are 12 cM apart in repulsion conformation may escape detection in an F_2 population because recombination frequencies differ by only 5.89% from those for independent assortment (Fairbanks and Andersen, 1999). If, as Mendel stated, he examined a small number of individuals in this experiment, and he observed at least one recombinant type (the true-breeding descendent mentioned in his letter to Nägeli), he probably did not detect linkage, if indeed the genes he studied were linked.

Di Trocchio (1991, p. 506) raised another question about linkage and from it drew an unusual conclusion to support his case that Mendel's experiments were fictitious: "we must determine why Mendel did not perform other hybridization experiments with an eighth, ninth, or tenth character in order to test the general validity of his law. This is a particularly intriguing question since we know that crosses with a number of characters higher than seven would quite surely have shown linkage." Di Trocchio concluded that "Mendel did find linkage, but he discarded it as senseless in order to concentrate on the only evident regularity—namely, 3:1 ratio. In doing so he thus chose, from among all the characters he experimented on, the famous seven non linked traits" (p. 511).

For us to evaluate such claims, it is useful to determine how Mendel chose the traits he studied. Did he choose them because their inheritance obeyed the laws he wished to illustrate (as Di Trocchio claimed), or did he choose them for other reasons? One way to answer this question is to examine which traits other pea hybridists who had no knowledge of Mendel's work chose to study. Roberts (1929) reviewed the work of plant hybridists who published their results before Mendel and described the traits that they studied. Among the hybridists before Mendel are several who studied pea. Knight (1799, 1823) studied seed coat color and described the phenomenon of dominance. Goss (1824) studied cotyledon color and described dominance, as well as segregation, although in purely qualitative terms. Seton (1824) studied stem length and cotyledon color. Gärtner (1849) reviewed Knight's work with seed coat color, and described his own work in *Pisum* with stem length, flower color, cotyledon color, and seed shape. Mendel studied Gärtner's (1849) book on plant hybridization in detail before and during his experiments, as he indicated in his first letter to Nägeli and as evidenced by his 17 references to it in "Versuche." Thus, he was familiar with both Knight's and Gärtner's pea hybridization experiments. We examined Mendel's copy of Gärtner's book and found numerous marginalia throughout it. On the page facing the back cover are Mendel's handwritten notes about traits in *Pisum*, which, as translated by Olby (1985) read:

Pisum arvense: flowers solitary, wings red.

Pisum arvense et sativum: pods almost cylindrical, in *Pisum umbellatum Mill.* [terminal-flowered varieties] cylindrical and straight; in *saccharatum* Host. [sugar varieties] Straight, ensiform, constricted on both sides. (var. *flexuosum* Willd. sickle-shaped, seeds small, angular); in *Pisum quadratum* Mill. [wrinkled-angular seeded varieties] straight, ensiform, not constricted, seeds pressed tightly together. In *Pisum sativum* and *arvense* the bases of the stipules rounded and denticulate-crenate, stipules cordate. In *saccharatum* and *quadratum*, stipules obliquely incised, pods pressed flat. In *sativium*, *saccharatum* and *umbellatum*, seeds round. (Olby, 1985, pp. 212–213)

Mendel studied all four traits that previous hybridists had studied, seed shape, cotyledon color, stem length, and seed-coat color (also flower color). His handwritten note mentions flower color, pod shape, and seed shape, and flower position is implied in the note by his mentioning *P. umbellatum*.

The following comment from Mendel's paper about the monohybrid experiments reveals information about the order of his monohybrid experiments: "Experiments 1 and 2 [seed shape and cotyledon color] have by now been carried through six generations, 3 and 7 [seed coat color and stem length] through five, and 4, 5, and 6 [pod shape, pod color, and flower position] through four" (Stern and Sherwood, 1966, pp. 15–16). The generation for seed traits can be scored one growing season earlier than that for plant traits, so Mendel must have initiated the monohybrid experiments for seed shape, cotyledon color, seed coat color, and stem length (the same traits that Gärtner studied) in the first year of his hybridization experiments, and the monohybrid experiments for the three remaining plant traits in the following year.

The four traits studied by previous hybridists were the same four traits that Mendel used in his first year of hybridization. The remaining three traits, pod shape, pod color, and flower position, were not among those studied by previous hybridists, and Mendel initiated experiments on them during the following year. This reconstruction argues against Di Trocchio's (1991, p. 508) assertion: "It is likely, in fact, that he [Mendel] planned his hybridization experiments following the checkerboard method. A checkerboard of 22 × 22 squares represents all of the crosses." Several other items in Mendel's paper also argue against Di Trocchio's assertion. According to statements in Mendel's paper, the varieties he used as parents for his monohybrid experiments for seed color did not include varieties with colored seed coats (because the opaque colored seed coats prevent observation of cotyledon color), and those he used for his monohybrid experiments for stem length did not include those with intermediate stem lengths.

The type of analysis that Mendel conducted relies on traits that display discontinuous variation, as does detection of linkage. The traits that Mendel chose to study are among only a few that varied in a discontinuous fashion among commercially available 19th Century varieties. Mendel's varieties certainly differed in more than the seven traits on which he reported data. However, most of the other traits display continuous variation and are governed by multiple genes and environmental influences, and cannot be easily analyzed in a simple Mendelian fashion. Mendel listed a number of these traits in his paper, then stated that he could not clearly analyze such traits: "However, some of the traits listed do not permit a definite and sharp separation, since the difference rests on a 'more or less' which is often difficult to define. Such traits were not usable for individual experiments; these had to be limited to characteristics which stand out clearly and decisively in the plants" (81).

Mendel's choice of traits apparently was based first on those studied by his predecessors, and second on those that had distinct discontinuous phenotypic differences that permitted conclusive analysis. Because of the locations of the genes that governed such traits and the design of his experiments, it is unlikely that he could have detected linkage. There is no botanical or historical evidence to support the claim that Mendel observed and then disregarded linkage.

Did Mendel Support or Oppose Darwin?

We addressed the four previous controversies in botanical contexts. This final controversy is purely historical, but it is widely debated by the same authors who address botanical issues and they have related their conclusions on botanical issues to this issue. Thus, we also are compelled to include it. Mendel and Darwin were contemporaries and both addressed evolutionary questions in their work. Twice in "Versuche" Mendel used the term "*Entwicklungsgeschichte*," which in the English translations of "Versuche" is rendered as "evolution" or "evolutionary history." One of the passages with this term appears near the beginning of "Versuche" where Mendel referred to his experiments as "the one correct way of finally reaching the solution to a question whose significance for the evolutionary history [*Entwicklungs-Geschichte*] of organic forms must not be underestimated" (*000*).

Thus, Mendel was clearly interested in evolution and he considered his experiments as relevant to an understanding of evolution. At the time Mendel wrote his paper, the *Origin* was well known and evolution through natural selection was a popular topic for discussion in scientific societies. Alexander Makowsky, who was a student at the same school as Mendel and was among his closest friends, presented a lecture that favorably treated Darwin's theory of natural selection to the Brünn Natural History Society the month before Mendel presented the first of two installments of his article to the Society (Makowsky, 1866).

Although there is no evidence that Darwin knew of Mendel's work, there is ample evidence that Mendel read some of Darwin's writings and that those writings may have influenced his work. Efforts to elucidate the Mendel-Darwin connection have been underway for nearly a century (for examples, see Bateson, 1913; Iltis, 1924; Fisher, 1936; Olby, 1985; Callender, 1988; Bishop, 1996; Orel, 1996).

In spite of much research, there is no consensus about Mendel's views on Darwinism. The numerous articles and commentaries on this topic begin with a comment in a letter to William Bateson written in 1902 by Mendel's nephew, Ferdinand Schindler, who stated, "He [Mendel] read with great interest Darwin's work in German translation, and admired his genius, though he did not agree with all of the principles of this immortal natural philosopher" (Orel, 1996, p. 188). Bateson (1913, p. 329) wrote, "With the views of Darwin which at that time were coming into prominence Mendel did not find himself in full agreement." Fisher (*120*) believed that Mendel understood his laws to "form a necessary basis for the understanding of the evolutionary process" and that "had he [Mendel] considered that his results were in any degree antagonistic to the theory of selection it would have been easy for him to say this also." Sapp (1990) determined that Fisher's 1936 paper was the turning point for the "modern synthesis" of Mendelism and Darwinism. In the mid-1960s, when a large number of articles were published at the centennial of Mendel's paper, most authors viewed Mendel as a supporter of Darwinism. By contrast, Olby (1979, 1985) studied the historical context of evolutionary thought during Mendel's day and determined that Darwin's "views on the role of hybridization in evolution were very far removed from Mendel's" (Olby, 1979, p. 67). Callender (1988) and Bishop (1996) expressed the most extreme views of Darwin's influence on Mendel. Callender (p. 72) claimed that "Mendelism came into being historically as a sophisticated form of the doctrine of Special Creation" and that it "stood in open conflict with the Darwinian conception of evolution as descent with modification by means of Natural Selection." Bishop (p. 212) proposed that "Mendel's sole objective in writing his *Pisum* paper, published in 1866, was to contribute to the evolution controversy that had been raging since the publication of Darwin's the *Origin of Species* in 1859," and that "Mendel was in favor of the orthodox doctrine of special creation." After a detailed review of the literature on Mendel's perception of evolution, Orel (1996, p. 198) determined that "Mendel came across Darwin's theory as his *Pisum* experiments were drawing to a close. From his notes and from indirect evidence one can suppose that he did not see any conflict between this theory and his own."

The extreme disagreement among scholars about Mendel's view of Darwin's writings is probably because Mendel wrote very little about Darwin, and thus most claims are suppositions about what Mendel must have thought about Darwin. In his surviving writings, Mendel's overtly referred to Darwin only four times, all in 1870, four years after the publication of "Versuche." One reference is in Mendel's (1870) *Hieracium* paper and three are in his eighth and ninth letters to Nägeli (Stern and Sherwood, 1966). All four references are brief and reveal neither strong support of nor opposition to Darwin's theories.

Because Mendel's major contribution to the science of genetics was "Versuche," we will focus on how Darwin may have influenced Mendel before the 1866 publication of "Versuche." Of Darwin's writings, only the *Origin* was available to him before 1866. Several authors have noted that Mendel's personal copy of the *Origin* contains marginalia (Iltis, 1924; Moore, 1963; Voipio, 1987; Hartl and Orel, 1992; Bishop, 1996; Orel, 1996). Mendel purchased a copy of the second German edition, published in 1863, which was translated from the third English edition (Darwin, 1861). This copy contains Mendel's marginalia and is in the collection of the Mendelianum Museum Moraviae in Brno.

When Mendel began his classic experiments with peas in 1856, none of Darwin's works were available for him to read. According to Orel (1971, 1996), Mendel probably first heard of Darwin in September 1861 during a lecture. He might have read the *Origin* during the latter part of 1862 or early 1863 when the Brünn Natural History Society acquired a copy of the German translation of the first English edition (Darwin, 1859). The 1863 publication date of Mendel's personal copy of the *Origin* coincides with the last year of his experiments with peas. Therefore, the *Origin* had no effect on the design or conduct of those experiments, although it may have influenced Mendel's interpretation of those experiments in "Versuche." Orel (1996), de Beer (1964), Fisher (1936), and Bateson (1913) concluded that Darwin's influence on Mendel, primarily from the *Origin*, is evident in "Versuche." Our comparison of Mendel's marginalia in the *Origin* with passages in "Versuche" supports this view.

Mendel's complete marginalia in the *Origin* have not been published, although Orel (1996) discussed a few of them. (The full German and English texts of the complete marginalia along with our commentary, can be accessed electronically at http://ajbsupp.botany.org/v88/fairbanks.html. When quoting from the English version of the *Origin* in this manuscript we use the 3rd edition text [Darwin, 1861] because this is the English edition from which Mendel's German copy was translated.)

The marginalia consist of passages marked in pencil and two very brief notes in script. The marks are either single or double vertical lines in the margins next to passages that Mendel apparently found interesting. Mendel marked passages on only 18 pages. The marked passages are clustered into two groups. Eight of them are in Chapters 1–4 (five in Chapter 2 "Variation Under Nature"), and ten of them are in Chapters 8 and 9 (eight in Chapter 8 "Hybridism").

The marginalia include only two notes in script that can be attributed to Mendel. One of them is the series of numbers "1.6.7.13.16.48.52.57.62. 63.76.78.80" written inside the back cover of the book. These may be page numbers, but our examination of the corresponding pages suggests that the numbers do not refer to pages in the *Origin*. The other note in script is on page 1 and reads "pag 302." The term "pag" probably is an abbreviation of the Latin word *pagina* for page. On page 302 is a passage that Mendel marked with double lines. In the original English it reads: "The slight degree of variability in hybrids from the first cross or in the first generation, in contrast with their extreme variability in the succeeding generations, is a curious fact and deserves attention" (Darwin, 1861, p. 296).

Apparently, Mendel found this to be the most interesting of the passages he marked. It is the only passage he cited by page number and is one of only two passages marked with double lines. Mendel's observations of uniformity in the F_1 generation, and predictable variability in the F_2 generation and his theoretical explanations for these phenomena form one of the key points of his paper. According to Orel (1996, p. 193), "Here Mendel must have felt some gratification in the thought that his theory was soon to explain this curious fact."

Of interest is Darwin's explanation, which immediately follows this passage, and is partially included within Mendel's mark. Darwin's explanation of the uniformity of hybrids in the F_1 generation and the variability of their F_2 offspring differs substantially from Mendel's in that Darwin places the cause on altered reproductive systems rather than constant inherited traits:

> For it bears on and corroborates the view which I have taken on the cause of ordinary variability; namely, that it is due to the reproductive system being eminently sensitive to any change in the conditions of life, being thus often rendered either impotent or at least incapable of its proper function of producing offspring identical with the parent-form. Now hybrids in the first generation are descended from species (excluding those long cultivated) which have not had their reproductive systems in any way affected, and they are not variable; but hybrids themselves have their reproductive systems seriously affected, and their descendants are highly variable. (Darwin, 1861, p. 296)

This is one of many passages in the *Origin* in which Darwin uses the phrase "conditions of life," which in Mendel's German edition of the *Origin* is translated as *Lebens-Bedingungun*. Mendel marked three passages in the *Origin* with this phrase (pages 17, 295, and 302 in his German edition). The most important is the first marked passage in Mendel's copy of the *Origin*:

> It seems pretty clear that organic beings must be exposed during several generations to the new conditions of life to cause any appreciable amount of variation; and that when the organization has once begun to vary, it generally continues to vary for many generations. (Darwin, 1861, p. 7)

This passage appears in the opening remarks of Chapter 1, "Variation Under Domestication," in which Darwin (1861, p. 7) suggested that the higher degree of variations in domesticated species compared to their wild counterparts is due to the "domestic productions having been raised under conditions of life not so uniform as, and somewhat different from, those to which the parent-species have been exposed under nature." Mendel's view of this subject differed from Darwin's. From "Versuche":

> Granted willingly that cultivation favors the formation of new varieties and that by the hand of man many an alteration has been preserved which would have perished in nature, but nothing justifies the assumption that the tendency to form varieties is so extraordinarily increased that species soon lose all stability and their progeny diverge into an infinite number of variable forms. If the change in living conditions [*Lebensbedingungun*] were the sole cause of variability one would expect that those cultivated plants that have been grown through centuries under almost identical conditions should have regained stability. This is known not to be the case, for it is precisely among them that not only the most different but also the most variable forms are found. (106–7)

Commenting on this passage, Fisher (1936, 136) wrote: "The reflection of Darwin's thought is unmistakable, and Mendel's comment is extremely pertinent, though it seems to have been overlooked. He may at this time have read the *Origin*, but the point under discussion may equally have reached his notice at second hand." Indeed, Mendel probably had read the *Origin* at the time he presented "Versuche." This passage from "Versuche" seems to be a direct response to the passage he marked on page 17 of the *Origin*. In it Mendel contradicted Darwin's claim that changing conditions of life were the cause of variation in domesticated species.

In spite of all that has been written about Mendel's views of Darwinism, Mendel's marginalia in the *Origin* and his written comments in "Versuche" are the best indicators of his opinion of Darwin's writings when he wrote "Versuche." Surprisingly, although many authors have addressed Darwin's influence on Mendel, only Orel (1996) used the content of Mendel's marginalia in the *Origin* as a source for his conclusions.

Several writers have claimed that Mendel marked *many* passages in the *Origin* and thus was very interested in and familiar with Darwin's writings (Iltis, 1924; Moore, 1963; Voipo, 1987; Bishop, 1996). In fact, Mendel's marginalia in his copy of the *Origin* are sparse; as mentioned above, the marked passages are found on only 18 pages. Mendel, however, was not one to avoid marking his books. His marginalia are abundant in his copies of Gärtner's (1849) book and Darwin's (1868) the *Variation of Animals and Plants Under Domestication*, which however was published two years after "Versuche" and thus had no influence on it.

Mendel never mentioned Darwin in "Versuche," although he mildly contradicted some of the points that Darwin made in the *Origin* and supported a few others. Mendel also did not mention special creation or deity in "Versuche," even though such a practice was not unusual in his day, especially for a priest. Instead, his paper is a highly focused and objective treatment of his work and its relationship to the work of other plant hybridists. It is devoid of polemics, sweeping conclusions, or speculations about theories that his experiments did not directly address.

Mendel's apparent reserved rather than intense interest in the *Origin* may be due to his well-known concern for detail. Even though Gärtner's descriptions of experiments in plant hybridization are far more detailed than the information Darwin provided in the *Origin*, Mendel lamented in his first letter to Nägeli:

The results which Gärtner obtained in his experiments are known to me; I have repeated his work and have reexamined it carefully to find, if possible, an agreement with those laws of development which I found to be true for my experimental plant. However, try as I would, I was unable to follow his experiments completely, not in a single case! It is very regrettable that this worthy man did not publish a detailed description of his individual experiments, and that he did not diagnose his hybrid types sufficiently, especially those resulting from like fertilizations. (Stern and Sherwood, 1966, p. 57)

Darwin wrote that the *Origin* was merely a "brief sketch" and an "abstract," and has far less detail on plant hybrids than Gärtner's (1849) book. Also, regarding the information he presented on plant hybridization in the *Origin*, Darwin (1861, p. 277) wrote: "The following rules and conclusions are chiefly drawn up from Gärtner's admirable work on the hybridisation of plants." Therefore, most of the information on plant hybridization in the *Origin* was a summary of detailed information that Mendel had already studied.

Perhaps Mendel's apparent lack of engagement with Darwin can shed some light on an old question. Historians have searched in vain for evidence that Darwin knew of Mendel's work, that Mendel contacted Darwin, or that Mendel sent a reprint of his paper to Darwin. According to Iltis (1924), Darwin's son examined his father's belongings and found no copies of Mendel's publications. As Olby (1985) pointed out, Darwin had a copy of Focke's (1881) *Die Pflanzen-Mischlinge* containing references to Mendel including summaries of Mendel's work with *Pisum* and *Phaseolus*, although the pages with the summaries were uncut and therefore unread. Mendel visited London in July and August of 1862, when his *Pisum* experiments were nearly completed, but there is no evidence that he attempted to contact Darwin, who in any case was not in London at the time (Orel, 1996).

Perhaps one of the most obvious reasons that Mendel did not attempt to contact Darwin is the language barrier; Mendel did not speak English, according to Orel (1996). However, apart from the language barrier, the answer may also lie in the nature of Darwin's writings. Although the passages Mendel marked in the *Origin* briefly address phenomena that he observed experimentally, the book had little in the way of detailed results and explanations that would have been useful to him. Mendel may not have contacted Darwin because, under the circumstances, there was little to gain in doing so.

In response to the question that heads this section, we find no evidence that Mendel either strongly supported or opposed Darwin when he wrote "Versuche."

Conclusion

Although Mendel's paper is considered a classic in the history of biology, it generated much controversy throughout the century that elapsed since the rediscovery of Mendelian laws in 1900. Scholars disagree about Mendel's integrity in his presentation, his articulation of the fundamental laws of inheritance, his experimental design, his motives for conducting his experiments, and his conclusions. Our review of Mendel's work in a botanical and historical context leads us to agree with Fisher (1936, *134*) that Mendel's "report is to be taken entirely literally, and that his experiments were carried out in just the way and in much the order that they are recounted." There is no credible evidence to indicate that Mendel was inaccurate or dishonest in his description of his experiments or his presentation of data. The main questions about his results can be resolved by an appeal to botanical principles and historical evidence.

NOTES

This paper was originally published as Fairbanks, Daniel J., and Bryce Rytting. 2001. "Mendelian Controversies: A Botanical and Historical Review." *American Journal of Botany* 88 737–52. Reprinted with permission of *American Journal of Botany*.

The authors thank Anna Matalová, Director of the Mendelianum Museum Moraviae in Brno, Czech Republic, for permitting examination of original copies of books and records, and W. Ralph Andersen, James F. Crow, Blair R. Holmes, Duane E. Jeffery, Alan F. Keele, L. Eugene Robertson, and Marcus A. Vincent for their assistance and suggestions. This research was supported by funds from an Alcuin Fellowship to DJF from Brigham Young University.

LITERATURE CITED

Bateson, W. 1913. *Mendel's principles of heredity.* Cambridge University Press, Cambridge.
Beadle, G. W. 1967. Mendelism, 1965. *In* R. A. Brink and E. D. Styles [eds.]. *Heritage from Mendel*, 335–350. University of Wisconsin Press, Madison.
Bishop, B. E. 1996. Mendel's opposition to evolution and to Darwin. *Journal of Heredity* 87: 205–213.
Blixt, S. 1975. Why didn't Gregor Mendel find linkage? *Nature* 256: 206.
Broad, W., and N. Wade. 1983. *Betrayers of the truth.* Simon and Schuster, New York, New York.
Callender, L. A. 1988. Gregor Mendel: an opponent of descent with modification. *History of Science* 26: 41–57.
Campbell, M. 1976. Explanations of Mendel's results. *Centaurus* 20: 159–174.
———. 1985. *A century since Mendel.* Illert Publications, Adelaide.
Corcos, A., and F. V. Monaghan. 1984. Mendel had no "true" monohybrids. *Journal of Heredity* 75: 499–500.
Correns, C. 1900. G. Mendel's Regel über das Verhalten der Nachkommenshaft der Rassenbastarde. *Berichte der Deutschen Botanischen Gesellschaft* 8: 158–168.
Curtis, N. M. [ed.]. 1889. Seventh annual report of the Board of Control of the New York Agricultural Experiment Station for the year 1888. Troy Press Co., Albany, New York.
Darwin, C. 1859. *On the origin of species by means of natural selection, or the preservation of favoured races in the struggle for life.* John Murray, London.
———. 1861. *On the origin of species by means of natural selection, or the preservation of favoured races in the struggle for life*, 3rd ed. John Murray, London.

———. 1868. *The variation of animals and plants under domestication*. John Murray, London.

De Beer, G. 1964. Mendel, Darwin, and Fisher (1865–1965). *Notes and Records of the Royal Society of London* 19: 192–225.

Di Trocchio, F. 1991. Mendel's experiments: a reinterpretation. *Journal of the History of Biology* 24: 485–519.

Doyle, G. G. 1968. Too many small χ^2's or hanky-panky in the monastery? In J. L. Hodges, D. Krech, and R. S. Crutchfield. *Statlab: an empirical introduction to statistics*, 228–229. McGraw-Hill, New York.

Dunn, L. C. 1967. *A short history of genetics*. McGraw-Hill, New York.

Edwards, A. W. F. 1986. Are Mendel's results really too close? *Biological Review* 61: 295–312.

Fairbanks, D. J., and W. R. Andersen. 1999. *Genetics: The continuity of life*. Brooks/Cole and Wadsworth Publishing Companies, Pacific Grove, California.

Fisher, R. A. 1936. Has Mendel's work been rediscovered? *Annals of Science* 1: 115–137.

Focke, W. O. 1881. *Die Pflanzen-Mischling. Ein Beitrag zur Biologie der Gewächse*. Gebruder Bornträger, Berlin, Germany.

Gardner, M. 1977. Great fakes of science. *Esquire* (October 1977): 88–92.

Gärtner, C.F. (1849) *Versuche und Beobachtungen über die Bastarderzeugung im Pflanzenreiche*. K. F. Herring, Stuttgart, Germany.

Goss, J. 1824. On the variation in the colour of peas, occasioned by cross-impregnation. *Transactions of the Horticultural Society of London* 5: 234.

Hartl, D. L., and V. Orel. 1992. What did Gregor Mendel think he discovered? *Genetics* 131: 245–253.

Hennig, R.M. 2000. *The monk in the garden*. Houghton Mifflin Co., Boston.

Iltis, H. 1924. *Gregor Johann Mendel, Leben, Werk und Wirkung*. Julius Springer, Berlin. (For an English translation, see Iltis, H. 1966. *Life of Mendel*, 2nd ed. E. Paul and C. Paul [trans.] Hafner Publishing Co., New York.)

Knight, T. A. 1799. An account of some experiments of the fecundation of vegetables. *Philosophical Transactions, Royal Society of London*. 1: 195–204.

———. 1823. Some remarks on the supposed influence of the pollen in cross-breeding, upon the colour of the seed-coats of plants, and the qualities of their fruits. *Transactions of the Horticultural Society of London* 5: 377–380.

Lamprecht, H. 1941. Über Genlabilität bei Pisum. *Züchter* 13: 97–105.

———. 1968. *Die Grundlagen der Mendelschen Gesetze*. Paul Parey, Berlin, Germany.

Makowsky, A. 1866. Über Darwins Theorie der organischen Schöpfung. *Verhandlungen des naturforschenden Vereines in Brünn* (Sitzungs-Berichte) 4: 10–18.

Magnello, M. E. 1998. Karl Pearson's mathematization of inheritance: From ancestral heredity to Mendelian genetics (1895–1909). *Annals of Science* 55: 35–94.

Mendel, G. 1866. Versuche über Pflanzen-Hybriden. *Verhandlungen des naturforschenden Vereines in Brünn* (Abhandlungen) 4: 3–47.

———. 1870. Über einige aus künstlicher Befruchtung gewonnenen Hieracium-Bastarde. *Verhandlungen des naturforschenden Vereines in Brünn* 8: 26–31.

Monaghan, F. V., and A. Corcos. 1990. The real objective of Mendel's paper. *Biology and Philosophy* 5: 267–292.

———., and ———. 1993. The real objective of Mendel's paper: A response to Falk and Sarkar's criticism. *Biology and Philosophy* 8: 95–98.

Moore, R. E. 1963. *Man, time, and fossils, the story of evolution*, 2nd ed. Alfred A. Knopf, New York.

Morgan, T. H. 1913. *Heredity and sex.* Columbia University Press, New York.
Nilsson, E. 1951. *Trädsgardsärter.* Svensk Växförädling, Stockholm.
Nissani, M. 1994. Psychological, historical, and ethical reflections on the Mendelian paradox. *Perspectives in Biology and Medicine* 37: 182–196.
Novitski, E., and S. Blixt. 1979. Mendel, linkage, and synteny. *BioScience* 28: 34–35.
Olby, R. C. 1979. Mendel no Mendelian? *History of Science* 17: 53–72.
———. 1985. *Origins of Mendelism.* University of Chicago Press, Chicago.
Orel, V. 1971. Mendel and the evolution idea. *Folia Mendeliana* 6: 161–172.
———. 1996. *Gregor Mendel: The first geneticist.* Oxford University Press, Oxford.
———, and D. L. Hartl. 1994. Controversies in the interpretation of Mendel's discovery. *History and Philosophy of the Life Sciences* 16: 263–267.
Peters, J. A. 1959. *Classic papers in genetics.* Prentice-Hall, Englewood Cliffs, New Jersey.
Piegorsch, W. W. 1983. The questions of fit in the Gregor Mendel controversy. *Communications in Statistics. Theory and Method* 12: 2289–2304.
———. 1986. The Gregor Mendel controversy: Early issues of goodness-of-fit and recent issues of genetic linkage. *History of Science* 24: 173–182.
Roberts, H. F. 1929. *Plant hybridization before Mendel.* Princeton University Press, Princeton.
Sapp, J. 1990. The nine lives of Gregor Mendel. *In* H. E. Le Grand [ed.], *Experimental inquiries: historical, philosophical and social studies of experimentation in science,* 137–166. Kluwer Academic Publishers, Dordrecht.
Seton, A. 1824. On the variation in the colour of peas from cross-impregnation. *Transactions of the Horticultural Society of London* 5: 236.
Stern, C., and E. R. Sherwood. [eds.] 1966. *The origin of genetics: A Mendel source book.* W. H. Freeman and Co., San Francisco.
Sturtevant, A. H. 1965. *A history of genetics.* Harper and Row, New York.
Thoday, J. M. 1966. Mendel's work as an introduction to genetics. *Advancement of Science* 23: 120–124.
Voipio, P. 1987. What did Mendel say about evolution? *Hereditas* 107: 103–105.
Vries, H. de. 1900. Sur la loi de disjonction des hybrides. *Comptes Rendus de l'Academie des Sciences* 130: 845–847.
Weiling, F. 1986. What about R. A. Fisher's statement of the "too good" data of J. G. Mendel's *Pisum* paper? *Journal of Heredity* 77: 281–283.
———. 1989. Which points are incorrect in R. A. Fisher's statistical conclusion: Mendel's data agree too closely with his expectations? *Angewandte Botanik* 63: 129–143.
———. 1991. Historical study: Johann Gregor Mendel 1822–1884. *American Journal of Medical Genetics* 40: 1–25.
———. 1994. Johann Gregor Mendel: Forscher in der Kontroverse. *Medizinische Genetik* 6: 35–50.
Weldon, W. R. F. 1902. Mendel's law of alternative inheritance in peas. *Biometrika* 1: 228–254.
White, O. E. 1917. Studies of inheritance in *Pisum.* II. The present state of knowledge of heredity and variation in peas. *Proceedings of the American Philosophical Society* 56: 487–588.
Wright, S. 1966. Mendel's ratios. *In* C. Stern and E. R. Sherwood [eds.], *The origin of genetics: A Mendel source book,* 173–175. W. H. Freeman, San Francisco.

POSTSCRIPT TO CHAPTER 7

■ Mendelian Controversies
An Update

DANIEL J. FAIRBANKS

The Mendel-Fisher controversy focuses largely on statistical anomalies noted by Fisher in Mendel's data. Although the so-called "too-good-to-be-true" nature of Mendel's data has become the most disputed issue, and is the focus of this book, Fisher dealt with other contentious issues as well. Fairbanks and Rytting (2001; also chapter 7 of this book) address five modern controversies, three of which (1, 2, and 5) were also addressed by Fisher:

1. Are Mendel's data too good to be true?
2. Is Mendel's description of his experiments fictitious?
3. Did Mendel articulate the laws of inheritance attributed to him?
4. Did Mendel detect but not mention linkage?
5. Did Mendel support or oppose Darwin?

I will briefly review all five of these controversies, updating chapter 7 to include information published since 2001, and highlighting Fisher's views where appropriate.

Are Mendel's Data Too Good to Be True?

For this question, I will restrict my comments to botanical issues. As summarized in Allan Franklin's introduction to this volume, and as reviewed by Fairbanks and Rytting (2001), authors who examine Mendel's work repeatedly refer to or dispute botanical issues to explain anomalies in Mendel's data. Two cases, in particular, stand out. The first is the controversy over Mendel's method for genotype testing, in which he cultivated ten F_3 progeny to determine whether or not an F_2 plant with the dominant phenotype was homozygous or heterozygous. The second is repeated reliance in the literature on the tetrad-pollen model—the possibility that during self-pollination the stigma samples male gametes without replacement from a finite number of pollen grains derived from tetrads— to explain the bias toward expectation in Mendel's data.

Let's first examine the controversy over Mendel's F_2 genotype testing with F_3 progeny plants. As noted several times in this book, Fisher showed that the summed data from these experiments (399:201) are exceptionally close to Mendel's expected 2:1 ratio, deviating (significantly according to Fisher's unidirectional calculation) from the predicted 1.7:1 ratio after correction for misclassification. Moreover, Fisher presumed that Mendel used the same progeny-testing method to determine the genotypes of F_2 plants for seed-coat color in his trihybrid experiment. Franklin (chapter 1) rightly concluded, "It was the agreement of Mendel's data with what Fisher regarded as the incorrect 2:1 ratio that was most important for Fisher."

Fisher, as well as others, proposed that Mendel may have planted more than ten F_3 descendents for each F_2 plant in these experiments, which reduces the probability of misclassification. However, some have dismissed this proposal, stating that we should take Mendel at his word and assume that, in fact, he cultivated exactly ten plants grown from ten F_3 seeds in each case ("*von jeder 10 Samen angebaut*"). There is, however, a practical difficulty with this assumption. Had Mendel planted exactly ten seeds, he could not have scored ten plants in each case because of germination losses or failure of plants to reach fruition. Although Mendel did not provide data for losses in his progeny tests, he did provide such data in other experiments, totaling 111 losses out of 1622 seeds planted for an overall loss rate of 6.84%.[1]

Although Fisher's value of 5.63% misclassification is theoretically correct, it cannot be correct for Mendel's actual experiments. Mendel either planted exactly ten F_3 seeds from each F_2 plant, in which case he would have scored fewer than ten plants in some cases, or he planted more than ten F_3 seeds per F_2 plant to ensure that he had at least ten plants in each case. He could then thin the seedlings to ten. If he employed the former method, the misclassification rate exceeds 5.63% and the bias of his data toward the 2:1 ratio is even more serious than that proposed by Fisher. If, instead, he employed the latter method, he could have scored more than ten plants for traits that could be scored as seedlings, rendering the misclassification rate less than 5.63%.

E. Novitksi (2004) recognized this problem and proposed that Mendel may have planted exactly ten seeds from each F_2 plant but that he planted more than 100 sets of ten in each experiment, the excess sets serving as replacements for questionable sets that produced less than ten. Under such a scenario, Mendel could have reliably scored any set with less than ten that had at least one recessive type as a heterozygote but he would have replaced any set of less than ten that had no recessive types with a full set of ten. C. E. Novitski (2004) showed mathematically that such a procedure biases the replacement sets toward heterozygotes, thus compensating for the misclassification of heterozygotes as homozygotes. However, as Hartl and Fairbanks (2007) pointed out, given the loss rate Mendel documented in his experiments, this procedure would, in fact, overcompensate for the misclassification Fisher noted, biasing the data toward a ratio higher than 2:1. It also seems a rather inefficient procedure because it would potentially leave Mendel with excess sets of ten in most, if not all, of the experiments. Mendel would have intentionally excluded some sets, either selectively or arbitrarily, to have exactly 100 sets in each experiment.

Hartl and Fairbanks (2007) argued that "if there is anything to which contemporaries who knew Mendel agreed, it was that he was a superb gardener, and any experienced gardener would know exactly how to do this experiment." Because Mendel had approximately 30 seeds per plant, and therefore approximately 30 seeds available for progeny testing in each experiment, it is highly likely that, instead of planting exactly ten seeds, or planting extra sets of ten, he germinated more than ten (either in the field or the greenhouse) for each set to protect against losses, then selected ten seedlings to grow to maturity in his garden. Two of the plant traits Mendel studied, stem length and axillary pigmentation (which, as Mendel noted, is perfectly correlated with seed-coat color and flower color), can be readily scored in seedlings. Had Mendel planted more than ten seeds, he could have readily scored these traits before discarding the extra seedlings. Utilizing such a procedure, he certainly would have consciously preserved any seedlings with recessive phenotypes in these two experiments to ensure correct classification of heterozygotes.

That Mendel was aware of genetic differences for stem length in seedlings, and that he transplanted those seedlings, is evident in the following passage from his paper: "In this experiment the dwarfed plants were carefully lifted and transferred to a special bed. This precaution was necessary, as otherwise they would have perished through being overgrown by their tall relatives. Even in their quite young state they can be easily picked out by their compact growth and thick dark-green foliage"(87). Not surprisingly, Mendel's monohybrid progeny test for stem length biases the data away from Fisher's expected ratio of 1.7:1 and toward Mendel's expectation of 2:1 more than any other experiment in the series.

In Mendel's trihybrid experiment, his choice of seed shape and seed color as two of the three traits to investigate is obvious—these two traits can be scored in the seeds themselves, obviating the need to plant the seeds. By contrast, his choice of seed-coat color for the third trait seems, at first glance, to be the worst possible choice because Mendel had to remove at least part of the opaque seed-coat to identify the yellow or green seed color in any seed borne on a purple-flowered plant. Offhand, it seems that Mendel should have crossed two white-flowered varieties as parents for his trihybrid experiment (the seed coats of such varieties are transparent allowing the seed color to show through) and selected another plant trait as the third.

E. Novitski (2004) also addressed this issue, presuming from passages in Mendel's paper, and in papers by Correns (1900) and Wright (1966), that phenotypes for seed-coat color in pea display incomplete dominance so that Mendel could distinguish heterozygotes from homozygous-dominant individuals by observing variation for color and spotting patterns on the seed coat. Novitski concluded, "It seems that Mendel rather cleverly used three seed characters together so that the complete determination of the genotype of the parental plant could be made immediately upon inspection of the seeds borne by that plant" (1135).

Unfortunately, Novitski's presumption of incomplete dominance for seed-coat color is incorrect, and is apparently based on a misreading of Mendel and Correns, a mistake that Wright (1966) also made. The relevant passage in Mendel's paper states, "The hybrid seeds in the experiments with seed-coat are often more spotted, and the spots sometimes coalesce into small bluish-violet patches. The spotting also frequently appears even when it is absent as a parental character"(85). Later, Mendel refers to the seed coats of the F_1 seeds in the trihybrid experiment as, "spotted, grey-brown or grey-green" (93). The passage in Correns (1900) reads, "Even in peas, where some traits completely conform to this rule [complete dominance], other trait pairs are also known in which neither trait dominates, as for instance the color of the seed coat being either reddish-orange or greenish-hyaline. In this case the hybrid may show all transitions, (this is especially true for the seed coat of peas)" (Stern and Sherwood 1966, 121). In fact, color and spotting in colored seed coats is a highly variable phenotype. However, the variation for seed-coat color in F_1 hybrids is in part due to a quantitative interaction of alleles of multiple genes. Khvostova (1983) summarized the literature on seed-coat color in pea and concluded that "26 genes have been identified which control the nature and intensity of pigmentation of the seed coat" (65). Moreover, there is considerable non-genetic variation for seed-coat pigmentation. The variation noted by Mendel appeared in seeds borne on F_1 plants, all of which had the same genotype, and is therefore non-genetic. My own observations of variation for seed-coat color in pea experiments confirm the accuracy of Mendel's and Correns's statements on the quantitative, highly non-genetic nature of variation for this trait. The distinction between colored and transparent seed coats in purple and white flowered plants, respectively, is clear and unambiguous but among seeds with colored seed coats variation for color and spotting tends to be highly quantitative. Mendel could readily distinguish colored and transparent seed coats but he could not have distinguished heterozygotes from homozygous dominant types on the basis of variations for seed-coat color.

So why did he choose this trait for the trihybrid experiment? As mentioned by Fairbanks and Rytting (2001), seed-coat color is, in fact, one of the best choices as the third trait. Because of its perfect association with flower color and axillary pigmentation, Mendel could identify which plants would eventually have purple or white flowers, and colored or transparent seed coats, simply by observing the pigmentation, or lack of it, in the leaf axils of seedlings. Had he done so, the delay between his scoring seed traits (round or wrinkled, yellow or green) and axillary pigmentation (pigmented or not pigmented) would have been at most a few weeks, and garden-space limitations would have been minimal because seedlings require little space.

Let's now turn our attention to the pollen-tetrad model. As documented by Franklin (chapter 1), the pollen-tetrad model was first proposed in 1966 and multiple authors have appealed to it as an explanation of Mendel's "too good to be true" data from that time to the present. As discussed by Fairbanks and Rytting (2001), this model can be empirically tested. If the pollen-tetrad model is valid, the distribution of segregating peas within pods should deviate from a binomial distribution with the more probable combinations overrepresented and those less probable underrepresented. In November 2006, I completed data collection for an experiment that tested such a model, tracking F_2 individuals that were segregating for flower color (and therefore axillary pigmentation), partitioned by pods. I chose this trait because phenotypes can be unambiguously scored without any questionable judgment on the part of the researcher. Chi-square analysis of the results showed no significant deviation from a binomial distribution. Furthermore, when expected numbers of pods for every possible combination (based on the binomial model) are plotted on the horizontal axis against observed numbers of pods with those combinations on the vertical axis, a least-squares fit of a line should have a slope of 1 if the binomial model is correct, and a slope greater than 1 if the pollen-tetrad model has an appreciable effect. The slope of such a line in this experiment was 0.97, confirming that the binomial distribution best explains the data with no evidence of a tetrad-pollen effect.[2]

Is Mendel's Description of His Experiments Fictitious?

Mendel clearly stated that in his monohybrid experiments, "plants were used which differed in only one essential character" [*wesentliches Merkmal*] (*90*). Bateson was the first to question this straightforward statement, writing that "it is very unlikely that Mendel could have had seven pairs of varieties such that the members of each pair differed from each other in only one considerable character (*wesentliches Merkmal*) (Bateson 1913, 350)." Fisher (1936) disputed Bateson's claim at some length, concluding with an often-quoted sentence that, "there can, I believe, be no doubt whatever that his [Mendel's] report is to be taken entirely literally, and that his experiments were carried out in just the way and in much the order that they are recounted" (*307*). As highlighted in Franklin's introduction to this volume, several authors have repeated and embellished Bateson's assertion that Mendel's experiments were fictitious. In Fairbanks and Rytting (2001), we provided evidence to support the conclusion: "the nature of variation in pea varieties (both old and modern) facilitates, rather than prevents, the construction of monohybrid experiments" (*282*), and that Fisher's conclusion regarding the literal nature of Mendel's experiments is botanically sound.

Interestingly, had Mendel worked with *Drosophila*, no one would have questioned his ability to design monohybrid experiments. When researchers discover a new mutant phenotype in *Drosophila*, they almost inevitably cross the mutant type with the wild type—a monohybrid cross. Mendel is criticized because he worked with a domesticated plant with multiple variants preserved by artificial selection. Without citing evidence, his critics have assumed that the variations in pea varieties were too great for Mendel to design true monohybrid crosses when, in fact, monohybrid pea experiments can be readily designed.

Did Mendel Articulate the Laws of Inheritance Attributed to Him?

This controversy is a more recent one, most forcefully promulgated by Callender (1988), and Monaghan and Corcos (1990), and more recently by Allchin (2003). All of these authors claimed that Mendel did not articulate the laws of segregation and independent assortment but that they must be attributed to twentieth-century geneticists. As Fairbanks and Rytting (2001), and more recently Westerlund and Fairbanks (2004), responded, Mendel's own words readily dismiss such claims. Mendel quite lucidly articulated the law of segregation: "[I]t is only possible for the differentiating elements to liberate themselves from the enforced union when the fertilising cells are developed. In the formation of these cells all existing elements participate in an entirely free and equal arrangement, by which it is only the differentiating ones which mutually separate themselves" (*111*). Likewise, he clearly stated the law of independent assortment: "[T]he behaviour of each pair of differentiating characters in hybrid union is independent of the other differences between the two original plants" (*111*). Fisher did not address this controversy; indeed decades would elapse from the time he published his 1936 paper until this became a controversy.

Did Mendel Detect but Not Mention Linkage?

Speculation abounds as to why Mendel studied only seven traits when the pea plant has seven chromosomes. The speculation often turns to skepticism of Mendel's integrity with the question, "How could he be so lucky as to find one gene on each chromosome?" Dunn (1965) went so far as to calculate (erroneously) the probability of him doing so as 0.0061. Such skepticism is based on a misunderstanding of linkage, which is defined as a deviation from independent assortment due to the close proximity of two genes on the same chromosome. Those insufficiently familiar with genetics often assume that any two genes on the same chromosome are linked, when in fact genes on the same chromosome are often far enough apart to assort independently due to a high enough rate of crossing over.

In fact, two of the genes that Mendel studied may have been linked, but Fairbanks and Rytting (2001) explained that even in this case, for experimental and statistical reasons, Mendel probably still would not have detected linkage. Moreover, we reviewed compelling evidence of why Mendel chose the traits he did.

The linkage map in figure 1 of Fairbanks and Rytting (2001) (*266*) is an accurate portrayal of the pea genetic map derived from data published by Blixt (1975), Novitski and Blixt (1979), and Khvostova (1983). However, the John Innes Centre posted a brief commentary on its Web site entitled "Some Comments on Fairbanks and Rytting (2001)" (http://www.jic.bbsrc.ac.uk/staff/noel-ellis/fairbanks.htm), showing that the pea chromosome map in Fairbanks and Rytting differs from a more recently published map (Ellis and Poyser 2002). This more recent map suggests that the *r* and *gp* genes may also display linkage, although not consistently in all crosses.

Linkage maps of the pea plant were in their early stages of development in 1936, so it is not surprising that Fisher did not comment on the possible complications linkage may have introduced into Mendel's analysis.

Did Mendel Support or Oppose Darwin?

No question about Mendel engenders more disparate opinions than this one. Some authors claim that Mendel fully agreed with Darwinism, whereas others counter that Mendel was an advocate of special creation and, as such, strongly opposed Darwin. Bateson (1909) suggested that Mendel's disagreement with Darwin's views was the impetus for his classic experiments in peas. Fisher (1936) disputed Bateson's claim with historical evidence, pointing out that Mendel could not have known of Darwin when he commenced his experiments but that he certainly could have read Darwin's writings before he presented his experiments in 1865. Sixty years after Fisher's disproof of Bateson's claim, Bishop (1996) resurrected it, proposing that Mendel initiated his experiments specifically to discredit Darwin's *Origin of Species*, going so far as to claim, with no documentary evidence, that Mendel lied in a letter to Nägeli about the dates of his experiments (1856–1863). In fact, Mendel owned a German translation of *Origin of Species*, published in 1863, and marked passages in it. Thus, it could not have influenced his experiments but there is clear evidence that it influenced his interpretation of them.

Fisher (1936) addressed the Mendel-Darwin connection at some length, commenting on one passage from Mendel's paper, "The reflection of Darwin's thought is unmistakable" (*136*). Indeed, Fairbanks and Rytting (2001) described passages marked by Mendel in his copy of *Origin of Species* that are directly related to his experiments and to the Darwinian passage in Mendel's paper that Fisher highlighted. There is little doubt that Mendel knew nothing of Darwin when he conduced his experiments but had read *Origin of Species* by the time he presented his paper in 1865.

An often-repeated assertion that continues to fuel the controversy over Mendel and Darwin, and one that merits correction here because it is so prevalent in print and on the Internet, is the notion that Darwin had a reprint of Mendel's paper in his library but that it was uncut and, therefore, unread. In fact, there is no evidence that Darwin owned a copy of Mendel's paper. Instead, this assertion confuses Mendel's paper with a book Darwin owned, Focke's (1881) *Die Pflanzen-Mischlinge*, which briefly refers to Mendel's paper but the pages that refer to Mendel in Darwin's copy of Focke's book were uncut.

The mythical "meeting of the minds" between Mendel and Darwin was a one-way street. Mendel clearly was familiar with Darwin but Darwin knew nothing of Mendel, nor did he have a reasonable opportunity to discover Mendel's work. Fairbanks and Rytting (2001) concluded on the basis of Mendel's marginalia in his copy of *Origin of Species*, and his four written references to Darwin (all in 1870), that "we find no evidence that Mendel either strongly supported or opposed Darwin when he wrote [his paper]" (*298*).

Although the statistical anomalies in Mendel's data raised by Fisher constitute the most prominent controversy, other controversies about Mendel's work that Fisher rightly dismissed have also continued to the present. Most of the questions about these controversies can be resolved on the basis of botanical, historical, and statistical evidence often overlooked by Mendel's critics. However, short of a miraculous discovery of Mendel's original notebooks, other questions will forever remain unresolved. The overwhelming weight of evidence indicates, as Franklin puts it in his introduction to this volume, that "the issue of the 'too good to be true' aspect of Mendel's data found by Fisher still stands," albeit, in my opinion, its impact is much reduced by botanical evidence, and that "Mendel was not guilty of deliberate fraud in the presentation of his experimental results." I must agree with Franklin's conclusion: "It is time to end the controversy."

NOTES

1. Mendel obtained 529 F_2 plants from 556 seeds in the dihybrid experiment for seed shape and seed color; 639 F_2 plants from 687 seeds in the trihybrid experiments for seed shape, seed color, and seed-coat color; 90 plants from 98 seeds in Experiment 1, and 87 plants from 94 seeds in Experiment 3 for seed shape and seed color in the section of Mendel's paper entitled "Reproductive Cells of the Hybrids"; and 166 plants from 187 seeds in the dihybrid testcross for flower color and plant height, also in the section entitled "Reproductive Cells of the Hybrids."

2. Detailed results and analysis of this experiment will be published in Fairbanks and Schaalje (2008).

REFERENCES

Allchin, D. 2003. "Scientific Myth-conceptions." *Science Education* 87:329–51.
Bateson, W. 1909. *Mendel's Principles of Heredity*. Cambridge: Cambridge University Press.
———. 1913. *Mendel's Principles of Heredity*. Cambridge: Cambridge University Press.
Bishop, B. E. 1996. "Mendel's Opposition to Evolution and to Darwin. *Journal of Heredity* 87:205–13.
Blixt, S. 1975. "Why Didn't Gregor Mendel Find Linkage?" *Nature* 256:206.
Callender, L. A. 1988. "Gregor Mendel: An Opponent of Descent with Modification." *History of Science* 26:41–57.
Correns, C. 1900. "G. Mendel's Regel über das Verhalten der Nachkommenshaft der Rassenbastarde." *Berichte der Deutschen Botanischen Gesellschaft* 8:158–68.
Dunn, L. C. 1965. *A Short History of Genetics*. New York: McGraw-Hill.
Ellis, T. H. N., and S. J. Poyser. 2002. "An Integrated and Comparative View of Pea Genetic and Cytological Maps." *New Phytologist* 153:17–25.
Fairbanks, D. J., and B. Rytting. 2001. "Mendelian Controversies: A Botanical and Historical Review." *American Journal of Botany* 88:737–52.
Fairbanks, D. J., and B. Schaalje. 2008. "The Tetrad-Pollen Model Fails to Explain the Bias in Mendel's Pea (*Pisum sativum*) Experiments." *Genetics* (forthcoming 2008).
Fisher, R. A. 1936. "Has Mendel's Work Been Rediscovered?" *Annals of Science* 1:115–37.
Focke, W. O. 1881. *Die Pflanzen-mischlinge ein Beitrag zur Biologie der Gewächse*. Berlin.
Hartl, D. L., and D. J. Fairbanks. 2007. "Mud Sticks: On the Alleged Falsification of Mendel's Data." *Genetics* 175:975–79.
Khvostova, V. V. 1983. *Genetics and Breeding of Peas*. New Delhi: Oxonian Press.
Monaghan, F. V., and A. Corcos. 1990. "The Real Objective of Mendel's Paper." *Biology and Philosophy* 5:267–92.
Novitski, C. E. 2004. "Revision of Fisher's Analysis of Mendel's Garden Pea Experiments." *Genetics* 166:1139–40.
Novitski, E. 2004. "On Fisher's Criticism of Mendel's Results with the Garden Pea." *Genetics* 166:1133–36.
Novitski, E., and S. Blixt. 1979. "Mendel, Linkage, Synteny." *BioScience* 28:34–35.
Stern, C., and E. R. Sherwood, eds. 1966. *The Origin of Genetics: A Mendel Source Book*. San Francisco: W. H. Freeman and Co.
Westerlund, J., and D. J. Fairbanks. 2004. "Gregor Mendel and 'Myth-Conceptions.'" *Science Education* 88:754–58.
Wright, S. 1966. "Mendel's Ratios." In *The Origin of Genetics: a Mendel Source Book*, ed. C. Stern and E. R. Sherwood, 173–75. San Francisco: W. H. Freeman.

APPENDIX

Probability, the Binomial Distribution, and Chi-square Analysis

DANIEL J. FAIRBANKS

Mendel discerned patterns in his experiments because he conducted numerous experiments with large numbers of progeny, and his observed data were exceptionally close to theoretical ratios (too close, according to some statisticians). Observed data from genetic experiments often approximate theoretical ratios, but rarely do the data match the theoretical ratios exactly. Geneticists often observe deviations from theoretical ratios because chromosome assortment during meiosis and union of gametes at fertilization are random events.

In most experiments, scientists deal with random events and must use statistical analysis to interpret their data. The science of statistics and probability analysis is important in all experimental sciences, but especially in genetics. In fact, much of the science of statistics was developed by geneticists or scientists who are quite familiar with genetics, R. A. Fisher among them.

In this appendix, we begin with two fundamental rules of probability, the product rule and the sum rule. Then we will explore how the product and sum rules are used to derive the binomial distribution. Finally, we will review how chi-square analysis is applied to data from genetic experiments to test the reliability of interpretations with a focus on chi-square analysis of Mendel's experiments.

The Rules of Probability

Two basic rules of probability explain the patterns of inheritance observed in most genetic experiments. The first rule of probability that we will examine is the product rule, which can be stated as follows:

The Product Rule. When two events are independent of one another, the probability that they will occur together is the product of their individual probabilities.

We can illustrate this rule by flipping two fair coins, such as a penny and a nickel, one time each. The outcome of one flip has no influence on the outcome of the other, meaning that the two coin flips are independent—as events must be for the product rule to be valid. The probability that the penny will turn up heads is 0.5. The probability that the nickel will turn up heads is also 0.5. The probability that both will turn up heads is the product of the individual probabilities, which is $0.5 \times 0.5 = 0.25$.

Let's use the data from Mendel's dihybrid experiment on seed color and seed shape to illustrate application of the product rule. In this experiment, the probability of any seed in the F_2 generation being yellow is 3/4. The probability of any seed in the F_2 generation being round is also 3/4. Because the inheritance of seed color is independent of the inheritance of seed shape, the probability of any seed in the F_2 generation being both yellow and round is $3/4 \times 3/4 = 9/16$. The probability of any seed being green is 1/4, so the probability of any seed being both green and round is $1/4 \times 3/4 = 3/16$. Likewise the probability of any seed being both yellow and wrinkled is $3/4 \times 1/4 = 3/16$. Lastly, the probability of a seed having the two recessive phenotypes, green and wrinkled, is $1/4 \times 1/4 = 1/16$. In this way, the product rule predicts the 9:3:3:1 phenotypic ratio in Mendel's dihybrid experiment.

The other law of probability that often applies in genetics is the sum rule:

The Sum Rule. If two events are mutually exclusive, the probability that one of the two events will occur is the sum of their individual probabilities.

We can also illustrate this rule by flipping a fair coin. When a coin is flipped, it must come up either heads or tails. It cannot come up both heads and tails in the same flip, so heads and tails are mutually exclusive of one another, and we may therefore apply the sum rule. The probability that a coin will come up either heads or tails on a single flip is the sum of the respective probabilities: probability of heads = 0.5, probability of tails = 0.5, and 0.5 + 0.5 = 1.0, so the probability of a single coin flip coming up either heads or tails is 100%, which is obvious because there are no possibilities other than heads and tails. We can also apply the sum rule to rolling a fair die. When a die is rolled, there are six mutually exclusive possibilities: 1, 2, 3, 4, 5, and 6. What is the probability that either a 1 or a 4 will come up in a single roll? Using the sum rule, we determine that the probability of rolling a 1 is 1/6 and the probability of rolling a 4 is also 1/6, so the probability of rolling either a 1 or a 4 is 1/6 + 1/6 = 1/3.

Let's see how the sum rule can be applied in genetics. In the F_2 generation of a monohybrid experiment with alleles A and a, there are three mutually exclusive possibilities for genotypes: AA, Aa, and aa, and the probabilities for these genotypes are 1/4, 1/2, and 1/4, respectively. If A is dominant, the probability that any one F_2 individual will have the dominant phenotype is the sum of the probabilities that the individual will have the genotype AA or Aa, which is 1/4 + 1/2 = 3/4.

The Binomial Distribution

A common illustration of Mendel's experiments depicts a pea pod with four seeds: three yellow and one green. Obviously, we cannot say that on a pea plant that is heterozygous for the alleles that govern seed color every pod with four seeds will have three that are yellow and one that is green. However, we can say that the probability of any one seed being green is 1/4, and the probability of any one seed being yellow is 3/4, using the rules of probability just discussed. For any pod with four seeds, there are five possible combinations: all four yellow, three yellow and one green, two yellow and two green, one yellow and three green, and all four green. All five combinations are possible but not equally probable.

A convenient way for us to determine the probability of any particular combination is to use the binomial distribution. It consists of an equation that applies the product and sum rules to predict the probability of any particular combination in a series of all possible combinations. The binomial distribution is $(p + q)^n$, which is the expansion of $p + q$ raised to the nth power, where p represents the probability of one outcome, q the probability of another mutually exclusive outcome, and n is the total number of events, or individuals, in a finite sample. The distribution is called binomial because it applies to situations in which there are only two mutually exclusive outcomes, or classes, into which an event or individual could fall, such as heads or tails, female or male, yellow or green.

The binomial distribution requires two assumptions, which are the same assumptions required by the product and sum rules. First, the two possible outcomes must be mutually exclusive: an event can have one or the other outcome but not both. Second, each event must be independent of every other event in the series: for example, the fact that one seed is yellow cannot influence the outcome of a second seed. These assumptions are often met in genetic situations.

The probability (P) of any particular outcome in a binomial distribution can be calculated as:

$$P = \frac{n!}{x!y!} p^x q^y$$

where n is the total number of individuals, x is the number of individuals in one class (such as yellow), y is the number in the other class (such as green), p is the probability of falling into the class with x individuals, and q is the probability of falling into the class with y individuals. The symbol "!" stands for "factorial," which means that the number is multiplied by all descending whole numbers down to unity, such as $5! = 5 \times 4 \times 3 \times 2 \times 1 = 120$. Also, $0! = 1$, which can be important in the binomial distribution because x or y may equal zero. The quantity $x + y$ always equals n, and $p + q$ always equals unity in a binomial distribution.

To see how the binomial distribution is derived from the product and sum rules, let's return to the example of a pod with four seeds, three of which are yellow and one of which is green. Starting at the base of the pod and moving toward the tip, the probability that the first seed will be yellow is 3/4. Let's suppose that in fact it is. Likewise the probability that the second seed will be yellow is also 3/4, and let's suppose that it too turns out to be yellow. The same can be said for the third seed, and we'll assume that it too turns out to be yellow. The probability that the fourth seed will be green is 1/4. Because the outcome for each seed is independent of every other seed, we can use the product rule to determine this particular outcome, first three seeds yellow and the fourth green, as $3/4 \times 3/4 \times 3/4 \times 1/4 = 27/256$, which is equivalent to $(3/4)^3 \times (1/4)^1 = 27/256$, and is represented in the binomial distribution equation by the term $p^x q^y$. Let p represent the probability of a yellow seed, q the probability of a green seed, x the number of yellow seeds, and y the number of green seeds in a pod of four seeds, then $p = 3/4$, $q = 1/4$, $x = 3$, and $y = 1$, so $(3/4)^3 \times (1/4)^1 = 27/256$. There are, however, four possible, mutually exclusive, seed orders for a pod of four seeds with three yellow seeds and one green seed: one each with the green seed occupying the first, second, third, or fourth position, and yellow seeds occupying the other three positions. According to the product rule, the probability of each of these orders is $p^x q^y$, or in our example, $(3/4)^3 \times (1/4)^1 = 27/256$. Because each of these orders is mutually exclusive, we can use the sum rule to determine the probability of a pod with four seeds having three yellow seeds and one green seed in any order as $4 \times (3/4)^3 \times (1/4)^1 = 108/256 = 0.421875$. The number 4 in that calculation is derived from the term $\frac{n!}{x!y!}$ in the binomial distribution equation. This term is called the binomial coefficient, and it is a simple mathematical way to calculate all possible orders for a particular combination. In our example, $n = 4$, $x = 3$ and $y = 1$, so $\frac{n!}{x!y!} = \frac{4!}{3!1!} = 4$. We can view the binomial coefficient, $\frac{n!}{x!y!}$, as applying the sum rule and $p^x q^y$, as applying the product rule.

Hypothesis Testing and Chi-square Analysis

The binomial distribution provides an exact probability for any particular outcome when there are only two possible outcomes (such as yellow or green for any one seed). However, that probability is based on a hypothesis. For example, when we calculated the probability of three yellow seeds and one green seed in a pod of four seeds, we hypothesized that for each seed, the probability of it being yellow was 3/4 and the probability of it being green was 1/4. We could just as easily have hypothesized that the probability of a seed being yellow was 1/2 and the probability of it being green was 1/2. However, there is a good reason for us to not use this latter hypothesis. Mendel's principle of segregation predicts a 3:1 ratio for yellow and green seeds in an F_2 generation because of allele segregation, random fertilization during self-pollination, and dominance.

In fact, all of Mendel's monohybrid experiments displayed deviations from exact 3:1 ratios (although the deviations were, for the most part, quite small). When analyzing experimental data, scientists must determine whether there is evidence to suggest that an experimental deviation is just a random fluctuation or whether some non-random cause explains at least some of the deviation from the predicted ratio. We address such questions with statistical hypothesis testing, a procedure that compares experimental results with the results predicted by a hypothesis and determines the likelihood that random fluctuations can explain the observed deviation if the hypothesis is indeed correct.

The deviation from perfect expected ratios due to random variation is often called sampling error. Sampling error is always a possibility with samples of finite size, and its magnitude tends to be greatest in small samples. Statistical tests of experimental results provide researchers with the probability that sampling error can explain any observed differences. If the tests show a high probability of sampling error explaining the differences, the researchers do not reject the hypothesis; in other words, they have no evidence that anything other than sampling error explains the deviation from expectation. On the other hand, if the probability is low, they reject the hypothesis and conclude that something other than sampling error must be responsible for the deviation.

The binomial distribution shows us that any possible combination of genotypes or phenotypes is theoretically possible, but the closer a combination is to expectation, the more likely it is to be observed, if the deviation from expectation is due solely to sampling error. Alternatively, sampling error may not be the sole cause of the observed deviation; instead, the hypothesis used to determine expected values may be incorrect. Presumably, we could use the binomial distribution to test how well sampling error could explain deviations from expected values. However, in practice, the binomial distribution is cumbersome and useful only for relatively small sample sizes because the binomial coefficient becomes excessively large as sample sizes increase. For example, in Mendel's experiment with seed color, he examined 8022 seeds, so in the binomial distribution, the value for n is 8022. To calculate the binomial coefficient, we would need to determine the value of 8022!, which is so large that most modern computers cannot calculate it in the standard way (although they can rely on certain mathematical formulas to estimate the number). Most genetic experiments include sample sizes that are too large for analysis with the binomial distribution. Instead, a statistical test that scientists commonly use to test such hypotheses is called chi-square analysis.

Chi-square analysis, developed in 1900 by Karl Pearson, compares observed values with hypothesized expected values and estimates the probability that the observed deviation from expectation can be explained by sampling error under the assumption that the hypothesis is correct. The equation for chi-square analysis is

$$\chi^2 = \Sigma \frac{(O - E)^2}{E}$$

where χ^2 is chi-square, O is the number of individuals observed in a particular phenotypic (or genotypic) class, and E is the number of individuals expected in that class based on the null hypothesis. Because there must be more than one expected class for variation to be tested (yellow and green seeds for instance), the chi-square values for the classes are summed (the reason for the symbol Σ). After we obtain a chi-square value for an experiment, we can compare that value to theoretical values to see how closely the observed data fit the expectations. The theoretical values can be estimated from a chi-square table, found in most statistics and genetics textbooks, or more often nowadays scientists use computers to identify exact probabilities instead of estimating them with chi-square tables. Many software applications, including spreadsheet and statistical applications, include chi-square analysis and provide exact probability values for any particular chi-square value. In this appendix, we'll rely on the computer-generated probabilities instead of working with tables. A general rule to remember is that there is an inverse (albeit nonlinear) relationship between chi-square values and probability. High deviations from expected values produce high chi-square values and low probabilities; observed results that are very close to expected values produce low chi-square values and high probabilities.

The degrees of freedom for most chi-square calculations are one less than the number of classes being analyzed. For instance, there are two phenotypic classes in the F_2 generation of each of Mendel's monohybrid experiments, so there is one degree of freedom in each experiment. There are four phenotypic classes in the F_2 generation in his dihybrid experiment—yellow, round seeds; yellow, wrinkled seeds; green, round seeds; and green, wrinkled seeds—so there are three degrees of freedom. Think of degrees of freedom in this way: In Mendel's dihybrid experiment, each seed must have one of the four possible phenotypes, so for each seed one phenotype was taken leaving three degrees of freedom.

If the chi-square value calculated from observed data equals or exceeds the chi-square value with a probability of 0.05, then the observed ratio is said to be significantly different from the expected ratio, because the probability that the observation would deviate that much due to sampling error alone is equal to or less than 5%. The probability associated with a particular chi-square value tells us how frequently we should expect to see that deviation due solely to sampling error if the hypothesis is correct. For example, in an experiment with one degree of freedom, the chi-square value associated with a probability of 0.05 probability level is 3.84. In such an experiment, we expect to see an observed chi-square value of 3.84 or greater due to sampling error in only 5% of experiments (on average one out of every 20 experiments). If the chi-square value does not exceed the theoretical value, then there is no statistical evidence for us to reject the hypothesis, because the deviations may be reasonably explained by random sampling error.

Researchers typically use two theoretical values to test significance in statistical analysis: the 0.05 level, which is said to be significant, and the 0.01 level, which is said to be highly significant. At the 0.05 level, we expect to erroneously reject a correct hypothesis in about 5% of experiments. At the 0.01 level, we expect to erroneously reject a correct hypothesis in only about 1% of experiments but run a greater risk of failing to reject an incorrect hypothesis at the 0.05 level. The erroneous rejection of a correct hypothesis is called a type I error. The value chosen for hypothesis testing reflects the level of type I error that the experimenter is willing to tolerate. The 5% level for significance and the 1% level for high significance are generally accepted standards for most, albeit not all, experiments. Computer analyses of genetic data provide an exact probability value associated with the chi-square value, telling the researcher the probability of making a type I error if she or he rejects the hypothesis for any particular experiment, thus allowing the researcher a way to measure the level of confidence for a particular conclusion.

Let's examine Mendel's dihybrid experiment as an example. In this experiment, Mendel hybridized a variety that breeds true for yellow, round seeds with one that breeds true for green, wrinkled seeds. All of the F_1 seeds were yellow and round, as expected. In the F_2 generation, Mendel expected a $9:3:3:1$ ratio. His observed results, and the expected values under the hypothesis of a $9:3:3:1$ ratio are as follows:

Phenotype	Observed	Expected
Yellow, round	315	312.75
Yellow, wrinkled	101	104.25
Green, round	108	104.25
Green, wrinkled	32	34.75

The chi-square calculation is $\chi^2 = [(315 - 312.75)^2 \div 312.75] + [(101 - 104.25)^2 \div 104.25] + [(108 - 104.25)^2 \div 104.25] + [(32 - 34.75)^2 \div 34.75] = 0.47$ with 3 degrees of freedom. This chi-square value has a probability of 0.9254, which substantially exceeds the 0.05 probability level, so there is no evidence for us to reject the hypothesis of independent assortment. In fact, this probability is unusually high, so high that if we repeated Mendel's experiment many times, we should observe a deviation from expectation as great or greater than the one he observed, due purely to sampling error, in 92.54% of the experiments. In other words, the probability of observing results as close as these in any one experiment is less than 8%.

Let's now look at the set of Mendel's experiments that most concerned Fisher. In his monohybrid experiments, Mendel proposed that two-thirds of F_2 plants in each experiment should be heterozygous ("hybrid" in Mendel's words) and one-third should be homozygous ("constant" in Mendel's words). To determine if this was indeed the case, Mendel grew ten F_3 offspring derived from natural self-fertilization of each of 100 F_2 plants in each of the five monohybrid experiments for plant characters, a procedure geneticists call a progeny test. Of the ten F_3 offspring, if one or more had the recessive phenotype, he classified the F_2 parent as heterozygous. If all ten F_3 offspring had the dominant phenotype, he classified the F_2 parent as homozygous.

Fisher noted that this procedure subjects some heterozygous plants to misclassification as homozygous. According to the product rule, the probability that all ten offspring derived from self-fertilization of a heterozygote will have the dominant phenotype is $0.75^{10} = 0.0563$. In other words, using this procedure, Mendel should have misclassified 5.63% of heterozygous F_2 plants as being homozygous because all ten of their F_3 progeny had the dominant phenotype.

Mendel used this procedure to test all five plant traits, expecting a 2:1 ratio for heterozygous:homozygous F_2 plants. In the experiment for pod color, he classified 60 plants as heterozygous and 40 as homozygous. He felt that the deviation in this experiment was too far from expected so he repeated the experiment and the second time classified 65 plants as heterozygous and 35 as homozygous, a result that more closely matched his expectation. Let's see if Mendel was justified in suspecting that a 60:40 observation was too far from expected to be attributed to sampling error. Then, we will test Fisher's hypothesis using expected values that take into account the effect of misclassification.

Mendel expected 2/3 of the 100 F_2 plants to be heterozygous and 1/3 homozygous. To determine the expected values, we multiply the total number of individuals in the experimental sample by the expected fractions. In this case, the expected number of heterozygous plants is $2/3 \times 100 = 66.67$, and the expected number of homozygous plants is $1/3 \times 100 = 33.33$. To calculate the chi-square value, we take the observed value, subtract the expected value, square the result, then divide by the expected value in both cases, then sum the results of both cases: $\chi^2 = [(60 - 66.67)^2 \div 66.67] + [(40 - 33.33)^2 \div 33.33] = 2.00$. Because there are two classes (heterozygous and homozygous) there is one degree of freedom. Computer analysis shows that the probability is 0.1573. This value tells us that if the 2:1 hypothesis is correct, and we were to repeat this experiment many times, we would expect a deviation as great or greater than the one Mendel observed in 15.73% of the experiments due solely to sampling error. Because the probability exceeds 0.05, we have no evidence to reject the hypothesis of a 2:1 ratio. Mendel's suspicion that the results of this experiment are too far from expected is not justified statistically. However, we must remember that such statistical tests were not available in Mendel's day and he had to rely on intuition alone. His suspicion of this experiment, and his repetition of it, is indicative of his cautious nature and willingness to be thorough before drawing a conclusion.

Now let's conduct a chi-square test on the same data but this time with a different hypothesis—the one Fisher proposed with correction for misclassification. In this case, we expect 5.63% of the heterozygotes to be classified as homozygotes, so we must change the expected frequencies accordingly. Under Fisher's assumption, the expected number of plants that Mendel classified as heterozygotes is $[2/3 - (2/3 \times 0.0563)] \times 100 = 62.91$, and as homozygotes is $[1/3 + (2/3 \times 0.0563)] \times 100 = 37.09$. The chi-square analysis is $\chi^2 = [(60 - 62.91)^2 \div 62.91] + [(40 - 37.09)^2 \div 37.09] = 0.36$ with one degree of freedom. Computer analysis shows that the probability associated with this chi-square value is 0.5465. Because this probability also exceeds 0.05, we have no evidence to reject Fisher's hypothesis.

Notice that in neither case did we use chi-square analysis to *accept* a hypothesis. If the probability associated with a chi-square value exceeds 0.05, we conclude that there is no evidence to reject the hypothesis. However, we cannot conclude from the chi-square test alone that the hypothesis is correct; we simply have no evidence to indicate that it is incorrect. As this example shows, we looked at two hypotheses to test the results of the same experiment and did not find evidence to reject either hypothesis. Both hypotheses cannot simultaneously be correct but we have insufficient evidence from this experiment to reject either one.

Let's now look at the entire series of progeny tests in which Mendel expected a 2:1 ratio and Fisher a 1.7:1 ratio. In each of these Mendel tested ten F_3 progeny to determine the genotypes of 100 F_2 plants for the five plant traits in his experiments. He repeated one of them (the pod color experiment we just examined) for a total of 600 F_2 plants tested. The following table summarizes his results:

Experiment	Classified as heterozygous	Classified as homozygous
Seed-coat color	64	36
Pod shape	71	29
Pod color	60	40
Flower position	67	33
Stem length	72	28
Pod color (repeat)	65	35
Totals	399	201

Fisher pointed out that these totals were exceptionally close to Mendel's uncorrected ratio of 2:1. In fact each of the two totals deviated by only one plant from a perfect 2:1 ratio of 400:200. A chi-square test with Mendel's 2:1 ratio as the hypothesis yields a chi-square value of $\chi^2 = [(399 - 400)^2 \div 400] + [(201 - 200)^2 \div 200] = 0.0075$ with one degree of freedom. The probability associated with this chi-square value is 0.9310, which is very high. However, a chi-square analysis of the summed data with Fisher's expectation of 1.7:1 after correction for misclassification is $\chi^2 = [(399 - 377.5)^2 \div 377.5] + [(201 - 222.5)^2 \div 222.5] = 3.3020$ with one degree of freedom and a probability of 0.0692, which is not significant. According to chi-square analysis, there is no evidence for us to reject either Mendel's or Fisher's hypothesis. However, Fisher (1936) stated in reference to this chi-square test, "A deviation as fortunate as Mendel's is to be expected once in twenty-nine trials" (*127*). A probability of 0.05 represents one in twenty trials, so Fisher's statement of one in twenty-nine trials implies statistical significance. How could he make such a claim when the chi-square probability of 0.0692 is not significant? To reconcile this difference, we need to recognize that the probability determined from a chi-square test is based on deviations of equal magnitude in both directions from the expected value. Fisher was concerned that the deviation in Mendel's experiments was biased toward his expectation of 2:1, in other words in one direction rather than both. Given this restriction, Fisher halved the chi-square probability to 0.0346, which equals 1/28.9, corresponding to Fisher's "one in twenty-nine trials."

There is, however, another valid way to analyze Mendel's results with chi-square analysis. Instead of summing the data across all experiments, we can analyze each experiment individually, in which case the chi-square values under both Mendel's 2:1 hypothesis and Fisher's 1.7:1 hypothesis are as follows:

Experiment	Chi-square (2:1)	Chi-square (1.7:1)	Degrees of freedom
Seed-coat color	0.3200	0.0507	1
Pod shape	0.8450	2.8033	1
Pod color	2.0000	0.3635	1
Flower position	0.0050	0.7161	1
Stem length	1.2800	3.5394	1
Pod color (repeat)	0.1250	0.1868	1
Totals	4.5750	7.6598	6

Chi-square values and their degrees of freedom in a series of experiments can be summed and then the probability determined for the summed chi-square value with summed degrees of freedom. In this case, the chi-square value for Mendel's expectation of 2:1 is 4.5750 with six degrees of freedom, which has a probability of 0.5994, which is not significant. The chi-square value for Fisher's expectation of 1.7:1 is 7.6598 with six degrees of freedom, which has a probability of 0.2641, which also is not significant, even when halved to 0.1321. Such an analysis indicates that the deviation in Mendel's data toward the uncorrected expectation is not as serious as implied by Fisher's analysis of the summed data with one degree of freedom.

In a group of experiments, such as this one, random fluctuations due to sampling error should on average produce a chi-square value approximately equal to the degrees of freedom. In this case, there are 6 degrees of freedom, so we expect a chi-square value of 6. Under Mendel's hypothesis of 2:1, the chi-square value is 4.5750, which is a bit less than 6. Under Fisher's hypothesis of 1.7:1, the chi-square value is 7.6598, a little more than 6. However, in neither case is the value substantially greater or less than 6, so there is no reason under either hypothesis, given this analysis, to suspect that the data are questionable.

NOTE

This article is modified with permission from Fairbanks, D. J., and W. R. Andersen. 1999. *Genetics: The Continuity of Life*. Pacific Grove, CA: Brooks/Cole and Wadsworth Publishing Companies.

REFERENCE

Fisher, R. A. 1936. "Has Mendel's Work Been Rediscovered?" *Annals of Science* 1:115–37.

CONTRIBUTORS

A. W. F. EDWARDS was R. A. Fisher's last student. He is professor of biometry (emeritus) at Cambridge University. His work includes *Likelihood* (1972; 1992); *Foundations of Mathematical Genetics* (1977; 2000); *Pascal's Arithmetical Triangle: The Story of a Mathematical Idea* (1987; 2002); and *Cogwheels of the Mind: The Story of Venn Diagrams* (2004). He coedited (with H. A. David) *Annotated Readings in the History of Statistics* (2001), and (with Milo Keynes and Robert Peel) *A Century of Mendelism in Human Genetics* (2004).

DANIEL J. FAIRBANKS is professor of plant and wildlife sciences and dean of undergraduate education at Brigham Young University. He is the coauthor of *Genetics: The Continuity of Life* (1999) and author of *Relics of Eden: The Powerful Evidence of Evolution in Human DNA* (2007). Dr. Fairbanks is U.S. project leader of a collaborative research project on genetic improvement and genetic resource conservation of quinoa, a highly nutritious and important food plant in Andean South America.

ALLAN FRANKLIN is professor in the Department of Physics at University of Colorado. He has done extensive research on the history and philosophy of science and is the author of numerous books, including *Are There Really Neutrinos? An Evidential History* (2000), *Selectivity and Discord: Two Problems of Experiment* (2002), and *No Easy Answers: Science and the Pursuit of Knowledge* (2005).

DANIEL L. HARTL is Higgins Professor of Biology in the Department of Organismic and Evolutionary Biology at Harvard University. He has authored or coauthored 20 books including *Human Genetics, Principles of Population Genetics, Primer of Population Genetics, Genetics: Analysis of Genes and Genomes*, and *Essential Genetics: A Genomics Perspective*.

VÍTĚZSLAV OREL is Emeritus Head of the Mendelianum, Brno. He is the author of the biography *Gregor Mendel—The First Geneticist* (1996) and is recognized as a foremost authority on the members of the Augustinian Monastery in Brno.

BRYCE RYTTING is professor and chair in the Department of Music, Utah Valley State College.

TEDDY SEIDENFELD is H. A. Simon Professor of Philosophy and Statistics in the Department of Philosophy at Carnegie Mellon University. He is the coauthor, with Joseph Kadane and Mark Schervish, of *Rethinking the Foundations of Statistics* and has written extensively on statistics and the philosophy of science.

INDEX

1 : 2 : 1 law for hybrids, 220
2 : 1 ratio, 21, 37, 238–41
3 : 1 ratio, 124, 125, 236

Allchin, D., 308
alternative hypotheses, 164–66, 227
art, nature of, 181
artificial fertilizations, 241
assistants, Mendel's, 36

Babbage, C., 143
Bateson, W.: on Darwin, 118, 190, 293, 295, 309; Fisher on, 19, 119, 120; and Kilby, 148, 236, 238, 245; Mendel's paper, 118, 142, 281, 282, 307; plant characters, 3–4
Beadle, G. W., 36, 144, 276, 277, 279
beans *(Phaseolus)*, 15
Bennett, J. H., 27, 31, 175
Bernard, C., 178
bias, classification of objects, 132–33
bifactorial experiment, 8–10, 128, 132
binomial coefficient, 316–17
binomial distribution, 315–17, 318
Birkhead, T., 65
Bishop, B. E., 283, 293, 294, 309
Blixt, S., 198, 288, 289, 309
Bowler, P. J., 184
Bratranek, T., 180
Broad, W., 54, 280

Callender, L. A., 190, 285, 293–94, 308
Campbell, M., 40, 184, 277
Canguilhelm, G., 194
centenary of Mendel's discovery, 29
characteristics, peas, 3, 81–82, 84, 219, 258, 265–67
cheating, possible model of, 58, 242
chi-squared test: Edwards, 39–40, 47, 150–53; Fairbanks, 318–24; Fisher, 23–24; Pearson, 150; Weldon, 17. *See also* goodness of fit
chromosomal location of genes studied, 265–67
Churchill, F. B., 170
Cock, A., 27
Comenius, J. A., 181
cooking (data), 143
Corcos, A.: Edwards on, 142, 149–50; inheritance laws, 285, 308; and Monaghan, 44, 45, 53; monohybrid experiments, 281–82; Orel and Hartl on, 177, 189, 193
Correlated Pollen model, 57, 232–36, 240–41, 245, 246, 258
Correns, C. E., 1, 167, 192, 215, 264, 285, 305
cotyledon color, 277
Cramér, H,. 225
creativity, of scientists, 181
cultivated plants, 107

Darbishire, A. D., 148, 236–37, 238, 245
Darlington, C. D., 176
Darwin, C.: Bateson on, 118, 190, 293, 295, 309; corn experiments, 30; cultivated plants, 135–36; experimental breeding, 119; heredity, 171, 186; hybridization, 185; influence on Mendel, 283, 292–98, 309–10
de Beer, G., 30, 176, 182, 295
DeGusta, D., 62–63
de Vries, H. *See* Vries, H. de
Dianthus caryophyllus, 107–8
discovery, scientific, 181
Di Trocchio, F., 51–52, 198, 278, 283, 290, 291
Dobzhansky, T., 38, 161, 195
dominance, 4, 84, 85, 87, 88
dominant character, double signification, 5–6
dominant form, characteristics, 3, 219
Doyle, G. G., 280
Dunn, L. C., 31, 288, 308

Edwards, A. W. F.: alternative hypotheses, 164–66; "Are Mendel's Results Really Too Close?" 45–50, 141–63; chi-squared test, 39–40, 47, 150–53; Mendel's data, 47–48, 153–60, 195, 197; on Pilgrim, 43; Seidenfeld on, 217, 242
evolution, 118, 119, 120, 185, 190, 292, 293
experimentalist-statistician paradox, 51

F_1 forms of hybrids, 83–85
F_2 (first generation from hybrids), 3–5, 85–88
F_3 (second generation from hybrids), 5–7, 88–89
factorial system, 135
Fairbanks, D. J., 65, 66, 68, 302–11, 313–24; and Hartl, 64, 66–67, 68, 304
Fairbanks and Rytting, 264–301; Fairbanks on, 306, 307, 308, 309, 310; Franklin on, 8, 52, 60–62
Falk, R., 177
falsification of data (alleged): Campbell, 40; Di Trocchio, 52; Dobzhansky, 38; Fairbanks and Rytting, 281; Fisher, 25, 27, 29, 213–14; Franklin, 2, 67; Orel and Hartl, 55, 197; Piegorsch, 51; Weiling, 36; Weiss, 62; Wright, 34, 194, 209–10. *See also* fictitious data; fraud debate
fertilization, 187, 230, 241, 279. *See also* tetrad-pollen model; urn model
Festetics, E., 173
fictitious data (alleged): Di Trocchio, 52, 290; Fairbanks, 307; Fairbanks and Rytting, 281–84; Fisher, 23, 121, 122, 131, 213, 273; Piegorsch, 198; Pilgrim, 43. *See also* falsification of data; fraud debate
Fisher, R. A.: Darwin, 295, 297, 309; de Beer on, 30; *Design of Experiments, The*, 164; Dunn on, 31; Edwards on, 145, 147, 161, 164; evolution, 190, 293; Fairbanks on, 307; Fairbanks and Rytting on, 270, 272–73, 274, 275, 281; Franklin on, 16, 18–26, 303; Hartl on, 208–9, 211, 212, 213–14; "Has Mendel's Work Been Rediscovered?" 18–26, 117–40; Mendel-Fisher controversy, 1–2, 194; Mendel's experiments, 24–29, 141–42; Orel and Hartl, 175–76, 196, 197, 198, 200; Seidenfeld on, 215–16, 220–22, 222–30, 244–45
Focke, W. O., 25, 137–38, 298, 310
Ford, E. B., 26
Franklin, A., 1–77, 303
fraud debate: DeGusta, 63; Dunn, 31; Edwards, 45, 142, 143; Fairbanks, 310; Fairbanks and Rytting, 265; Fisher, 273; Franklin, 68; O'Kelly, 65; Orel, 280–81; Root-Bernstein, 42. *See also* falsification of data; fictitious data

Galton, F., 143–44, 169
gametic ratio experiments, 36, 44, 55

Gardner, M., 197, 280
Gärtner, C. F.: Darwin, 298; Fairbanks and Rytting, 265; hybrids, 78, 108, 109; Mendel on, 120, 188, 290; pure species, 110; species transformation, 112, 113, 114
genes, 176, 184, 265–67, 288
geneticists, 175
genetics, 125, 176
genetic terms, 285
Geschwind, R., 183
goodness of fit: DeGusta, 63; Edwards, 147, 150–53, 161, 195; Fairbanks and Rytting, 61–62; Fisher, 29, 228; Franklin, 68; Orel and Hartl, 194; Seidenfeld, 217; Wright, 33. *See also* chi-squared test
Goss, J., 290
Grmek, M., 178

Hartl, D. L., 208–14; and Fairbanks, 64, 66–67, 68, 304; and Orel, 52, 53, 54–55, 167–207, 275
hawkweed. See *Hieracium*
Heimans, J., 190
Heinz, D., 260
Hennig, R. M., 276, 286
hereditary elements, 191, 192, 285
heredity, 167, 169–75, 184, 186, 199
Hieracium (hawkweed), 15–16, 38
Hofmeister, W. F. B., 173
Hornschuh, C. F. 183
Huang, E., 260
Huxley, J., 176
hybridization, 137, 168, 177, 183, 185
hybrids: 1 : 2 : 1 law, 220; F_1 forms, 83–85; F_2 (first generation), 3–5, 85–88; F_3 (second generation), 5–7, 88–89; formation and development, 2; offspring with several differentiating characters, 8–11, 90–95; other plant species, 103–8; progeny 187–89; reproductive cells, 11–15, 95–103; subsequent generations, 7–8, 89–90
hypotheses, 164–66, 179, 277
hypothesis testing, 317–24

Iltis, H., 298
independent assortment, law of, 1, 168, 180, 284, 287, 308
inheritance, 284–88, 308; independent assortment, 1, 168, 180, 284, 287, 308; segregation, 1, 168, 180, 284, 285–87, 308

Jindra, J., 192
Judson, H. F., 63

Kalmus, H., 180, 191, 192
Keller, A., 180–81

Khvostova, V. V., 306, 309
Kilby, H., 148, 236, 238, 245
Klácel, M., 174, 180
Knight, T. A., 290
Kölreuter, J. G., 108
Kříženecký, J., 39, 169, 178, 180

Lamprecht, H., 39, 288, 289
Leguminosae, 107
Leonard, T., 41
Leuckart, R., 169, 172
likelihood, 151–52, 164, 165
Lindley, D. V., 159
linkage, 135, 198, 288–90, 292, 308–9
Liu, Y., 65

Makowsky, A., 293
McGee, R. J., 258
Meijer, O., 190, 191, 192
Mendel, G.: assistants, 36; counting seeds, 42; creativity, 181; Darwin's influence, 283, 292–98, 309–10; "Experiments in Plant Hybridisation," 78–116, 264; on Gärtner, 120, 188, 290; heredity, 186; importance of advice to, 174; *Origin of Species* (Darwin), 294, 295; scientific records, 265; statistical nature of data, 14–15. *See also* Mendel's experimental results
Mendel-Fisher controversy, survey of, 29–67
Mendelian genetics, 175
Mendelians, 168, 215
Mendel's experimental results: arrangement of experiments, 81–83; bias, classification, 132–33; chromosomal location of genes, 265–67; conclusions, 108–14; contemporary reaction to, 136–39; discovery's essence, 180; Edwards on, 153–60; evolution, 190; explanations, survey of, 143–46, 276–81; F_1 hybrids, 83–85; F_2 (first generation from hybrids), 85–88; F_3 (second generation from hybrids), 88–89; Fairbanks and Rytting on, 265–70; Fisher on, 119, 122–34, 147; Franklin on, 2–16; heredity, 184, 199; hybridization, 185; hybrid progeny, 187–89; offspring with several characters, 90–95; oral presentation, 168; Orel and Hartl on, 187–89; other plant species, 103–8; plants, 79–81, 133–34, 218; progeny tests 196, 211, 212; research methodology, 178–80, 181–82; segregation, 192; Seidenfeld on, 218–20, 229–32, 238–41; statistical fluctuations 12; statistical nature of data, 14–15; subsequent generations from hybrids, 89–90; subset of experiments, 280; survey of explanations, 143–46, 276–81; symbols, use of, 191–92, 287; variation in cultivated plants, 136. *See also* falsification of data; fictitious data; fraud debate

misclassification: Fairbanks, 303, 321; Fairbanks and Rytting, 61, 272, 274, 277; Fisher, 127; Franklin, 22; Hartl, 211; Novitski, E., 63; Seidenfeld, 56, 217, 220–24, 245. *See also* progeny-testing
Monaghan, F.: and Corcos, 44, 45, 53; Edwards on, 142, 149–50; inheritance laws, 285, 308; monohybrid experiments, 281–82; Orel and Hartl on, 177, 189, 193
Montgomerie, B., 65
Morgan, T. H., 285
Mummy Pea, 283

Nägeli, K. W. von, 25, 137, 171, 186, 280
Napp, F. C., 169
Nestler, J. K., 173
Niessl, G., 185
Nilsson, E., 288
Nissani, M., 53–54, 270
Norton, B., 194
Novitski, C. E., 55, 65, 304
Novitski, E.: linkage, 198, 288, 289, 309; progeny-testing, 63–65, 67, 211–12, 304, 305

O'Kelly, M., 65
Olby, R. C.: Darwin, 293, 298; Fairbanks and Rytting on, 277, 278; Franklin on, 44; heredity, 168, 169, 184, 190–91; inheritance laws, 284–85, 287–88; Orel on, 39; plant hybridization, 176–77, 189; segregation, 192
oral presentation, Mendel, 168
Orel, V.: Darwin, 294, 295, 297, 298; Fairbanks and Rytting on, 277; fraud debate, 280–81; and Hartl, 52, 53, 54–55, 167–207, 275; review of controversy, 29, 39

pangenesis, 172, 186
pea hybridists, 290
Pearl, R., 34, 209
Pearson, E. S., 194
Pearson, K., 150, 152, 318
peas (*Pisum sativum* L.), 2–15; characteristics, 3, 81–82, 84, 219, 258, 265–67; data on, 279; fertilization, 218, 231; as good randomizer, 148; model of self-fertilization (Correlated Pollen model), 232–36; Mummy Pea, 283; plant characters, 3, 219, 267, 272, 277; seed characters, 3, 219, 267, 272, 277; seeds per plant, 278, 279; seeds per pod, 279; selection 219; varieties, 282
Peters, J. A., 185
Phaseolus (beans), 15
Phaseolus multiflorus, 103, 105–6
Phaseolus nanus, 103, 104
Phaseolus vulgaris, 103, 270
Piegorsch, W. W., 41, 50–51, 146, 147, 198, 264

Pilgrim, I., 43, 142, 195
Pillsbury Labs, trial at, 258–63
Pisum, 80, 104, 110, 265
Pisum sativum L. *See* peas
plant characters, 3, 219, 267, 272, 277
plants: Mendel's experiments, 79–81, 133–34, 218; variation in cultivated, 135–36
pollen cells, alternative distribution, 230–31
pollen-tetrad model. *See* tetrad-pollen model
Pontecorvo, G., 200
Popper, K., 181
Pringsheim, N., 187
probability, rules of, 313–15
product rule, 314
progeny-testing: Edwards, 144–46; Fairbanks, 303, 320–21; Fairbanks and Rytting, 267, 269, 273, 276; Hartl, 211–13; Novitski, E., 63–65, 67, 211–12, 304, 305; Orel and Hartl, 196–97; size of tests 134, 216, 274–75. *See also* misclassification
pure species, 110
Purkyně, J. E., 169, 172, 173–74

Rasmussen, J., 278
recessive form, 84, 85, 87, 88
repeatable counts, difficulty of, 34, 41
reproductive cells, hybrids, 11–15, 95–103
Roberts, H. F., 117, 138, 290
Robertson, T., 41
Root-Bernstein, R. S., 41–42, 146, 147
Rytting, B. *See* Fairbanks and Rytting

sampling error, 317
Sandler, I. and L., 174, 178, 182
Sapp, J., 51, 175, 181, 293
Sarkar, S., 177
Schindler, F., 293
science, nature of, 181
Scott, W. F., 159
seed characters, 3, 219, 267, 272, 277
seeds per plant, 278, 279
seeds per pod, 279
segregation, law of, 1, 168, 180, 192, 284, 285–87, 308
Seidenfeld, T., 55–60, 65–66, 68, 215–57, 258–63
self-pollination, 32
Serre, J. L., 194
Seton, A., 290
sheep breeders, 173
Sherwood, E. R., 33, 145, 192–93

species, transformation, 112–14
Spencer, H., 171
statistical and experimental modes of reasoning, 51
Stern, C., 33, 169, 192–93, 194
Sturtevant, A. H., 31–33, 37, 144, 145, 161, 168, 279
subsequent generations from hybrids, 7–8, 89–90
sum rule, 314–15
surprisingly good data, 46, 143, 224–30, 270–81, 302–7. *See also* goodness of fit

test progenies. *See* progeny-testing
tetrad-pollen model: Beadle, 36; Campbell, 40; Edwards, 46; Fairbanks, 65, 66, 303, 306; Franklin, 39; Weiling, 41
Thoday, J. M., 34–36, 279
too good to be true data. *See* goodness of fit; surprisingly good data
traits. *See* characteristics
trifactorial experiment, 10, 21–23, 127, 128–29
trimming (data), 143
Tschermak, E. von, 1, 167, 215, 264
type I error, 319

urn model, pea fertilization, 50, 60, 61, 279

van der Waerden, B. L., 38, 144, 147
von Nägeli, K. W. *See* Nägeli, K. W. von
Vries, H. de, 1, 167, 171–72, 190, 215, 264, 285

Wade, N., 54, 280
Wagner, R., 169, 172, 183
Weiling, F.: Edwards on, 144–45, 147–49; Fairbanks and Rytting on, 273; Franklin on, 36–38, 41, 44–45, 50, 68; Orel and Hartl on, 194–95, 199; pea fertilization, 279; Seidenfeld on, 217–18
Weismann, A., 169–71, 172, 188
Weiss, K., 62
Weldon, W. F. R., 16–18, 142, 270
Westerlund, J., 308
White, O. E., 283
Wichura, M., 78
Wright, S.: Edwards on, 145, 146; Fairbanks and Rytting on, 276; Franklin on, 33–34, 67; Mendel's experiments, 208, 209–10, 213; Orel and Hartl on, 194; progeny-testing, 305

Zirkle, C., 29–30, 176